T5-ACR-654

Electron Theory and Quantum Electrodynamics
100 Years Later

NATO ASI Series

Advanced Science Institutes Series

A series presenting the results of activities sponsored by the NATO Science Committee, which aims at the dissemination of advanced scientific and technological knowledge, with a view to strengthening links between scientific communities.

The series is published by an international board of publishers in conjunction with the NATO Scientific Affairs Division

A	**Life Sciences**	Plenum Publishing Corporation
B	**Physics**	New York and London
C	**Mathematical and Physical Sciences**	Kluwer Academic Publishers Dordrecht, Boston, and London
D	**Behavioral and Social Sciences**	
E	**Applied Sciences**	
F	**Computer and Systems Sciences**	Springer-Verlag
G	**Ecological Sciences**	Berlin, Heidelberg, New York, London,
H	**Cell Biology**	Paris, Tokyo, Hong Kong, and Barcelona
I	**Global Environmental Change**	

PARTNERSHIP SUB-SERIES

1. **Disarmament Technologies**	Kluwer Academic Publishers
2. **Environment**	Springer-Verlag
3. **High Technology**	Kluwer Academic Publishers
4. **Science and Technology Policy**	Kluwer Academic Publishers
5. **Computer Networking**	Kluwer Academic Publishers

The Partnership Sub-Series incorporates activities undertaken in collaboration with NATO's Cooperation Partners, the countries of the CIS and Central and Eastern Europe, in Priority Areas of concern to those countries.

Series B: Physics

Electron Theory and Quantum Electrodynamics

100 Years Later

Edited by

Jonathan P. Dowling

U.S. Army Missile Command
Redstone Arsenal, Alabama

Plenum Press
New York and London
Published in cooperation with NATO Scientific Affairs Division

Proceedings of a NATO Advanced Study Institute on
Electron Theory and Quantum Electrodynamics: 100 Years Later,
held at the International Center for Physics and Applied Mathematics,
September 5 – 16, 1994,
in Edirne, Turkey

NATO-PCO-DATA BASE

The electronic index to the NATO ASI Series provides full bibliographical references (with keywords and/or abstracts) to about 50,000 contributions from international scientists published in all sections of the NATO ASI Series. Access to the NATO-PCO-DATA BASE is possible in two ways:

—via online FILE 128 (NATO-PCO-DATA BASE) hosted by ESRIN, Via Galileo Galilei, I-00044 Frascati, Italy

—via CD-ROM "NATO Science and Technology Disk" with user-friendly retrieval software in English, French, and German (©WTV GmbH and DATAWARE Technologies, Inc. 1989). The CD-ROM contains the AGARD Aerospace Database.

The CD-ROM can be ordered through any member of the Board of Publishers or through NATO-PCO, Overijse, Belgium.

Library of Congress Cataloging-in-Publication Data

Electron theory and quantum electrodynamics : 100 years later / edited
by Jonathan P. Dowling.
      p.   cm. -- (NATO ASI series. Series B, Physics ; v. 358)
   "Published in cooperation with NATO Scientific Affairs Division."
   "Proceedings of a NATO Advanced Study Institute on Electron Theory
and Quantum Electrodynamics: 100 Years Later, held September 5-16,
1994, in Edirne, Turkey"--T.p. verso.
   Includes bibliographical references and index.
   ISBN 0-306-45514-5
   1. Quantum electrodynamics--Congresses. 2. Quantum optics-
-Congresses. 3. Electrons--Congresses. 4. Electromagnetism-
-Congresses.   I. Dowling, Jonathan P. II. North Atlantic Treaty
Organization. Scientific Affairs Division. III. NATO Advanced
Study Institute on Electron Theory and Quantum Electrodynamics: 100
Years Later (1994 : Edirne, Turkey) IV. Series.
QC679.E44 1997
537.6'7--dc21   97-2807
                                                              CIP

ISBN 0-306-45514-5

© 1997 Plenum Press, New York
A Division of Plenum Publishing Corporation
233 Spring Street, New York, N. Y. 10013

http://www.plenum.com

10 9 8 7 6 5 4 3 2 1

All rights reserved

No part of this book may be reproduced, stored in a retrieval system, or transmitted in any form or by any means, electronic, mechanical, photocopying, microfilming, recording, or otherwise, without written permission from the Publisher

Printed in the United States of America

To Asim O. Barut
(1926–1994)

PREFACES

The electron, so named in 1894 by George Stoney, is the first and the most fundamental of all elementary entities. It is absolutely stable. Starting from historical background, this NATO Advanced Study Institute — conducted really as a school — considered the development and the still-open foundational problems of electron theory and electromagnetism, as well as different approaches to treat electromagnetic interactions and their scope. The basic framework is still that of H. A. Lorentz's self-consistent interaction between the field and matter. Lorentz really introduced this approach in his book *The Theory of Electrons* (1905). The EM-field satisfies Maxwell's equations. For matter, we may have either classical particle equations, or Schrödinger or Dirac equations. Hence, we have a coupled Maxwell–Dirac system today in quantum electrodynamics. The problem has been — now for about 100 years — how to solve this system unambiguously, self-consistently, completely, and possibly nonperturbatively. At this NATO ASI School, Radiation Theory was discussed (see Heitler's book) as well as almost standard perturbative QED. Perturbative QED requires renormalization and regularization, but after that it is a very successful theory in purely EM scattering problems such as the calculation of $g-2$, which was thoroughly discussed. Perturbation theory has its limitations. For example, bound-state problems cannot be treated by perturbative QED starting from first principles. Other assumptions must be introduced.

Of course, QED received impetus from experiments that are among the most accurate experiments in all physics, e.g., the measurements of the Lamb-shift, $g-2$, etc. The experimental situation was discussed in detail at this institute, in particular, the latest very accurate experiments in the hydrogen 1S-Lamb shift, as well as experiments in cavity quantum electrodynamics, were presented. Cavity QED has become — both theoretically and experimentally — an important domain to test many of the ideas of field–matter interactions. This is due to the modifications induced in QED processes by the boundary conditions of the cavity. QED processes for high-Z atoms were also discussed theoretically and experimentally.

Needless to say, physicists are trying to overcome the limitations of QED and find nonperturbative techniques that are not divergent. A method to solve coupled Maxwell–Dirac equations, called self–energy electrodynamics, was discussed. This technique has been applied to almost all QED processes, in particular to nonperturbative two-body bound–state problems. Around these main series of lectures, quite a number of special topics were discussed by lecturers and seminar speakers. Some of these topics were: The solutions of classical Lorentz–Dirac electron theory with radiation reaction, quantum electrodynamics and quantum optics, relativistic electron theory, theories of masers and lasers, interference phenomena, electric dipole moment of the electron, radiation pressure, and atom–light interactions.

A. O. Barut[†]
Boulder, Colorado
c. October 1994

It is with a heavy heart that I agreed to edit these proceedings, after the death of Asim O. Barut on December 5th, 1994, in Denver, Colorado, at age 68. Asim was my adviser, colleague, and friend — and we shall all miss him greatly.

I have done the best I could in gathering the contributions for these proceedings, after they became scattered all over the globe, in the confusion following Asim's death. I apologize for the long delay in getting this book out, as well as for any contributions I may have missed. The previous preface was taken from some notes Asim made summarizing the NATO ASI on *Electron Theory and Quantum Electrodynamics — 100 Years Later*, held in Edirne, Turkey, from 5–16 September 1994, for which this book is to be the proceedings. For my part, I have organized the papers into four categories: I. Quantum Electrodynamics, II. Quantum Optics, III. Electron Theory, and IV. New Developments in Electromagnetism. These topics are indicated in the Contents. The section on QED has several papers on cavity QED.[1]

I would like to thank the NATO ASI Programs and its director, Dr. Luis V. da Cunha, for making this meeting and these proceedings possible. I would also like to thank all the contributors and attendees for their patience and their contributions to this work.

Finally, I would like to dedicate this volume to the memory of Asim Orhan Barut, the director of this ASI, without whose efforts these proceedings would not have been possible. The book belongs to us — but Asim now belongs to the ages.

Jonathan P. Dowling
Redstone Arsenal, Alabama

[†] Deceased, October 1994.

[1] Some references to cavity QED that were discussed at the meeting are: V. Sandoghdar *et al.*, Phys. Rev. A **53** (1996), 1919; E. Hinds and V. Sandoghdar, Phys. Rev. A **43** (1991), 398; D. Meschede *et al.*, Phys. Rev. A **41** (1990), 1587; A. Anderson *et al.* Phys. Rev. A **37** (1988), 3594; J. P. Dowling *et al.*, Opt. Commun. **82** (1991), 415; F. B. Seeley *et al.*, Am. J. Phys. **61** (1993), 545; J. P. Dowling, Found. Phys. **23** (1993), 895; J. P. Dowling and C. M. Bowden, Phys. Rev. A **46** (1992), 612; J. M. Bendickson *et al.*, Phys. Rev. E **53** (1996), 4107.

CONTENTS

I. QUANTUM ELECTRODYNAMICS

II. QUANTUM OPTICS

IV. NEW DEVELOPMENTS IN ELECTROMAGNETISM

Theory of the Energy Levels and Precise Two–Photon Spectroscopy of Atomic Hydrogen and Deuterium [1]

K. Pachucki, [2] D. Leibfried, M. Weitz, A. Huber, W. König, and T.W. Hänsch

Max–Planck–Institut für Quantenoptik
Hans-Kopfermann-Straße 1
D-85748 Garching, Germany

Abstract

In last years a significant progress has been achieved both in the experimental technique and the theoretical methods for the determination of the energy levels of simple hydrogenic systems. We review recent two-photon spectroscopic measurements performed in Garching and the relevant theoretical predictions for the hydrogen energy levels. A good agreement is achieved when all theoretical contributions are included, showing the importance of recently calculated higher order corrections.

1 Introduction

The hydrogen atom is one of the basic systems in atomic physics. As the simplest of stable atoms it permits unique confrontation between fundamental experiments and theory. Spectroscopy of hydrogen has played a central role in the development of quantum mechanics and atomic physics. The interpretation of the regular visible Balmer spectrum of hydrogen has inspired several conceptual breakthroughs, from Bohr and the old quantum physics to the theories of Sommerfeld, de Broglie, Schrödinger and Dirac to the discovery of the Lamb shift and the development of modern quantum electrodynamics (QED). Today, the comparison of experimental results with theoretical predictions give a test of quantum electrodynamics, and also allows for a precise determination of physical constants. The recent progress in high resolution two-photon spectroscopy of atomic hydrogen and deuterium has also stimulated further improvements in the theoretical predictions for the Lamb shift.

In this report we describe spectroscopic experiments on atomic hydrogen developed in Garching in the last few years, and present the results of improved calculations of the higher order corrections to the hydrogen Lamb shift. We will review three experiments, a precise frequency comparison of the 1S-2S and 2S-4S transitions [2], the first direct measurement of the 1S-2S isotope shift between hydrogen and deuterium [3] and finally a determination of the absolute 1S-2S transition frequency [1]. The measurement of the frequency difference of 1S-2S and 2S-4S transition:

$$\Delta E = E(4S - 2S) - \frac{1}{4}E(2S - 1S) \tag{1}$$

allows the determination of the 1S Lamb shift and should ultimately become one of the most precise QED tests.

On the theoretical side, one needs an accurate description of the energy levels over the range of 12 orders of magnitude to compare the theoretical results with the experimental ones. There are many

---

[1] reprinted under permission from Journal of Physics B
[2] present address: Institute of Theoretical Physics, Warsaw University, Hoża 69, 00-681 Warsaw, Poland.

Electron Theory and Quantum Electrodynamics: 100 Years Later
Edited by Dowling, Plenum Press, New York, 1997

corrections to the Dirac energies that contribute at this precision level, namely one- and two-loop QED corrections, pure and radiative recoil corrections due to the finite nuclear mass, a correction for the finite nuclear volume and even the nuclear polarizability (for deuterium). The evaluation of these corrections is a highly nontrivial task. It involves the handling of thousands of terms, which can only be managed by computer algebraic programs. Most of them require a careful treatment due to specific cancelations of ultraviolet and infrared divergences. Additionally, different corrections require different treatments. For example in the evaluation of recoil corrections the hydrogen atom has to be considered as a two body system, while for the electron self–energy corrections the proton is treated as a static source of the Coulomb field. Thus for every correction a specific method for its evaluation has to be developed. There are also some contributions that are not purely QED effects, for instance due to the finite nuclear charge radius. The proton charge radius has been measured in electron scattering experiments, however the corresponding measurement uncertainty is now the most limiting factor in QED tests on the hydrogen atom.

In section 2 we briefly describe the laser spectrometers that are used to observe the 1S-2S and the 2S-4S transition. Also the frequency chain that links the 1S-2S transition frequency to the cesium atomic clock is explained. We then review our recent experimental results. Section 3 gives a description of the theoretical methods for the calculation of hydrogen energy levels, together with the most recent results for the different contributions. Section 4 the theoretical values of the quantities measured in our experiments are given.

2 Two–photon spectroscopy on an atomic beam of hydrogen

Doppler-free two-photon spectroscopy allows for the resolution of transitions with small natural decay-width. The atoms absorb two collinear counter propagating photons. In the atomic reference frame the co-propagating photon is shifted to the red, the counter-propagating to the violet. The transition rate depends on the sum of these two frequencies, therefore the linear term in the Doppler shift cancels out and there remains only a small residual purely relativistic Doppler effect of order $(\frac{v}{c})^2$, which leads to inhomogeneous broadening of the resonance. The 1S–2S transition in hydrogen has a natural line-width of only 1.3 Hz and can currently be resolved to 3 parts in 10^{12}. For the 2S-4S transition the natural line-width of 690 kHz is the factor that limits our experimental resolution.

2.1 The 1S-2S spectrometer

To drive the 1S-2S two photon transition [5], we start from a stable dye laser at 486 nm, that is frequency-stabilized to a passive Fabry-Perot resonator (finesse 57000, resonance width 5 kHz) and can be tuned relative to the fixed longitudinal resonator modes with the help of an acousto-optic modulator (AOM). The absolute stability was measured to be 1.5 kHz in 1 s average time. The output of the dye laser is frequency doubled in a BBO crystal within an external enhancement cavity, as shown in Fig. 1. The resulting uv radiation at 243 nm is then coupled into a standing wave resonator inside a vacuum chamber. Atomic hydrogen is produced in a radio-frequency gas discharge and guided to a liquid nitrogen cooled nozzle by a teflon tube. Due to inelastic collisions the atoms thermalize with the nozzle walls. The cold ground state atoms emerge from the nozzle and form an atomic beam. Atoms are excited into the metastable 2S state along a 11.5 cm interaction region by Doppler-free two-photon transitions (see Fig. 1). At the end of the interaction region a small electric field is applied to mix the 2S state with the quickly decaying 2P state inside the Lyman-α detector, and the resulting fluorescence photons are counted by a solar blind photomultiplier. The 1S-2S line shape is determined by the time of flight broadening and the second order Doppler effect, that red-shifts the transition frequency of moving atoms. Due to the interaction with the liquid nitrogen cooled nozzle, the hydrogen atoms reduce their average velocity to 1300 m/s. At T=80 K the 1S-2S two photon resonance can be observed with a line-width of 32 kHz [5].

2.2 The 2S-4S spectrometer

To drive the 2S-4S transition [2] we use radiation at 972 nm, provided by a stable Ti:sapphire laser (absolute frequency stability 10-15 kHz in 1s average time). The hydrogen atoms are dissociated in a microwave gas discharge and excited to the metastable 2S state by electron impact. The metastable atoms fly collinearly to a standing wave of 972 nm light, that is formed by a linear resonator. If they are excited to the 4S state, they quickly decay mainly to the 1S ground state via the 3P and 2P states. The transition is detected in two ways. At the end of the excitation region, the flux of

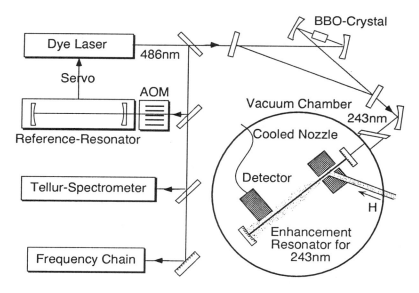

Figure 1. Setup of the hydrogen 1S-2S spectrometer.

metastable atoms is observed. On resonance the flux of 2S atoms decreases. Simultaneously, the 4S-2P fluorescence is detected along the atomic beam path. The observed 2S-4S lineshapes have a width of about 1 MHz, which is slightly above the natural line-width due to power broadening and the spatially inhomogeneous AC-Stark shift [2].

2.3   Absolute frequency measurement of the 1S-2S transition

To perform the frequency measurement [1] of the hydrogen 1S-2S energy separation, we take advantage of the near coincidence between the 28th harmonic of the methane stabilized helium neon laser (HeNe/CH$_4$) at 88 THz and the 1S-2S frequency. Our transportable HeNe/CH$_4$ standard has been built by the group of V.P. Chebotayev at the Institute of Laser Physics in Novosibirsk and was calibrated at the PTB in Braunschweig in a direct comparison to the Cs atomic clock with the help of the PTB frequency chain [6]. The result was (88376181599670±160) Hz with a relative uncertainty of $1.8 \cdot 10^{-12}$. In order to transport the accuracy of the HeNe/CH$_4$ to the 1S-2S frequency measurement, an additional reference frequency at the 8th harmonic (near 424 nm) is synthesized in three steps of second harmonic generation (Fig. 2, left side). To provide enough power for the nonlinear processes that are not very efficient in the optical part of the spectrum, we use additional lasers before each step of second harmonic generation. All these so called "transfer oscillators" are phase locked relative to their predecessors with the help of electronic servo loops. The output power of the HeNe/CH$_4$ standard is only 0.5 mW to avoid power broadening of the CH$_4$ resonance. A second HeNe laser with an output power of 20 mW is phase locked relative to the standard and half of its power is used to generate the second harmonic in a AgGaSe$_2$ crystal. At the second harmonic a NaCl:OH color-center laser provides an output power of 500 mW. The second harmonic of the color-center lasers radiation is generated in a LiIO$_3$ crystal. The laser diode at the 4th harmonic of the standard (848 nm) has an output power of 50 mW. A KNbO$_3$ crystal generates the second harmonic at 424 nm, which is the violet reference for the 1S-2S measurement. To determine the frequency of the dye laser, that is operating close to the 7th harmonic of the HeNe/CH$_4$, we generate the sum frequency of its blue light with the second half of the HeNe lasers radiation (10 mW) in a LiNbO$_3$ crystal. The sum frequency produced near the 8th harmonic of the HeNe/CH$_4$ is then compared to the violet reference by counting their radio-frequency beat note generated on a fast photodetector.

The frequency ratio of 1:28 between the CH$_4$ transition and the 1S-2S frequency does not hold exactly: the HeNe/CH$_4$ is operating at a frequency 300 GHz too high. Therefore a difference frequency $\Delta f=2.1$ THz has to be bridged at the 7 th harmonic of the HeNe/CH$_4$ between 617 THz (486 nm) and 619 THz (484 nm) (Fig. 2, right side). To determine Δf, we use different longitudinal modes of the

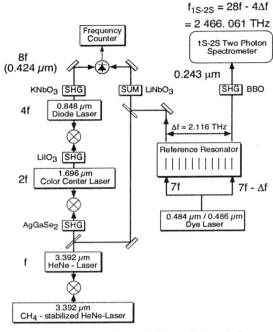

$f_{1S-2S} = 28f - 4\Delta f$

$= 2\,466.\,061$ THz

f = 88 376 181 599. 67 (± 0.16) kHz $(1.8 \cdot 10^{-12})$

Figure 2. The frequency chain.

passive Fabry-Perot resonator, which is used to stabilize the dye laser. With the help of a second dye laser and an electro-optical modulator at 84 GHz we could calibrate the mode spacing of the resonator between the 7 th harmonic of the HeNe/CH$_4$ (619 THz) and 1/4 th of the 1S-2S transition frequency (617 THz) with an accuracy of better than 1 Hz [8].

Our result [1] for the hyperfine corrected frequency

$$f_{1S-2S} = 2466061413.182\,(45)\,\text{MHz}\,,\qquad (2)$$

is 18 times more accurate than the best [9] of the previous measurements. From our result for f$_{1S-2S}$ and the most recent determination of the 1S-Lamb shift (see below and ref.[2]), we deduced a value of the Rydberg constant:

$$R_\infty = 109737,3156849(30)\,\text{cm}^{-1}\,,\qquad (3)$$

which is in good agreement with a recent result of a group at the ENS in Paris: $R_\infty=109737,3156834(24)$ cm^{-1} [10].

With a new fully phase coherent frequency measurement of the 1S-2S transition in atomic hydrogen and deuterium we expect to improve the accuracy of f$_{1S-2S}$ again by more than one order of magnitude in the near future.

2.4   Measurement of the Lamb shift

In a second experiment the 1S-2S frequency is directly compared with that of 2S-4S via the combination:

$$E(4S - 2S) - \frac{1}{4}E(2S - 1S)\,.\qquad (4)$$

For a frequency comparison of the 1S-2S and 2S-4S transitions, we lock the 1S-2S spectrometer to the maximum of the narrow 1S-2S signal. A small part of the infrared light provided by the Ti:Sapphire laser of the 2S-4S spectrometer is frequency doubled in a KNbO$_3$ crystal. The resulting blue light is mixed with the blue light from the dye laser locked to the 1S-2S transition on a fast photodiode,

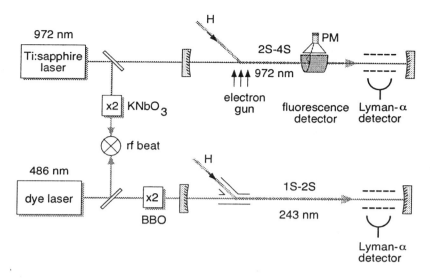

Figure 3. Frequency measurement of $E(4S - 2S) - \frac{1}{4}E(2S - 1S)$.

see Fig. 3. The frequency of the beat signal is measured both by a radio frequency counter and a spectrum analyzer locked to a rubidium standard. The Ti:sapphire laser is scanned over the 2S-4S resonance and the signal of both the metastable and the fluorescence detector is recorded versus the beat frequency. After averaging several such spectra, the obtained 2S-4S resonance lineshape is fitted with a theoretical model to extract the center frequency. In this way we obtained the value [2]:

$$L(4S - 2S) - \frac{1}{4}L(2S - 1S) = 868\,631(10)\,\text{kHz}\,. \tag{5}$$

From this combination the 1S Lamb shift for hydrogen can be derived using the experimental value for the smaller 2S-2P Lamb shift and the theoretical values for the 2P and 4S Lamb shifts shifts which are small and well known. Our result is

$$L_{1S} = 8172.874(60)\,\text{MHz}\,. \tag{6}$$

This value is in good agreement with a more recent measurement of D. Berkeland, E. Hinds and M. Boshier at Yale [11] obtained by a $1S-2S/2S-4P$ comparison $L(1S) = 8172.827(51)$ MHz. A further measurement based on a comparison of $1S-3S/2S-6S$ has recently been communicated by F. Biraben and coworkers.

In a new experiment we hope to increase the accuracy to about 1 kHz by using the optically excited slow metastable beam of the 1S-2S spectrometer for further excitation into 4S state. This would allow one to significantly improve the value for the 1S and 2S Lamb shifts and for the Rydberg constant using the following equations:

$$L(1S) = \frac{64}{7}\left(\left[L(4S - 2S) - \frac{1}{4}L(2S - 1S)\right]_{exp} - \left[L(4S) - \frac{5}{4}L(2S) + \frac{9}{64}L(1S)\right]_{theo}\right), \tag{7}$$

$$L(2S) = \frac{8}{7}\left(\left[L(4S - 2S) - \frac{1}{4}L(2S - 1S)\right]_{exp} - \left[L(4S) - \frac{17}{8}L(2S) + \frac{1}{4}L(1S)\right]_{theo}\right), \tag{8}$$

$$\frac{3}{4}\frac{\mu}{m}R_\infty = E(2S - 1S)_{exp} + 8\left(E(4S - 2S) - \frac{1}{4}E(2S - 1S)\right)_{exp} - \Delta_{theo}\,, \tag{9}$$

$$\Delta_{theo} = \Delta_{rm} + \left[8\,L(4S) - 9\,L(2S) + L(1S)\right]_{theo}\,, \tag{10}$$

where Δ_{rm} is a relativistic and reduced mass (denoted by μ) correction of the energy levels involved in the determination of the Rydberg constant, see Eq.(19). The subscripts exp and $theo$ denote the experimental and theoretical values respectively. The coefficients are chosen in such a way, that all terms that scale with the principal number like $1/n^3$ are automatically cancelled in the energy level combinations of the theoretical part *theo*. It means that the nuclear finite size effect and most of the QED corrections do not contribute to them in the leading orders. Therefore, these combinations can be determined much more reliably and precisely than the Lamb shift for separate states. For example for the noted combination in the Rydberg constant determination we obtain

$$[8\,L(4S) - 9\,L(2S) + L(1S)]_{theo} = -178\,832(5)\,\text{kHz}, \qquad (11)$$

where the error comes from the unknown state dependence of B_{61} coefficient and is estimated as half of the known B_{62} contribution. S. Karshenboim [12] has first pointed out that the determination of the 1S Lamb shift could be significantly improved when the theoretical predictions for the $L(1S) - 8L(2S)$ difference are used, for which the $1/n^3$ terms cancel out. The above formulas will help to fully exploit the future high accuracy of the measurement of $L(4S - 2S) - 1/4\,L(2S - 1S)$. The similar idea has been already applied by M. Eides *et al.* in [27]. Since there is a new improved measurement of the electron-proton mass ratio by Van Dyck [14], the Rydberg constant as determined from (9) will be improved by an order of magnitude, after the new direct frequency measurement of the 1S-2S transition is performed. It will also be very interesting to compare the 2S Lamb shift derived from $L(4S - 2S) - 1/4\,L(2S - 1S)$ with the results obtained from Lundeen-Pipkin [15], Hagley-Pipkin [16] and Palchikov et al. [13], and thus to verify the different experimental methods.

2.5 Measurement of the hydrogen–deuterium isotope shift

In a third experiment we have measured the isotope shift of the 1S-2S transition of hydrogen and deuterium

$$\Delta_{\text{ISO}} = E(2S - 1S)_{\text{D}} - E(2S - 1S)_{\text{H}}. \qquad (12)$$

To overcome the limitations of former indirect measurements relative to an intermediate tellurium standard [9], we directly bridged the large frequency gap between the two resonances with a novel electro-optic phase modulator [8]. The light of a second dye laser (laser 2), stabilized similarly to the laser of the spectrometer, is sent through a nonlinear $LiTaO_3$-crystal inside the focus of an open millimeter-wave Fabry-Perot cavity. The standing millimeter-wave in this cavity modulates the index of refraction for the laser light inside the crystal. The light is guided through the crystal in a zig-zag path by multiple internal total reflections. In this way we achieve phase matching of the optical and millimeter-waves and a modulation index of about 5 % [8]. The millimeter-wave oscillator near 84 GHz is phase-locked to a rubidium frequency standard with an accuracy of better than 10^{-10}. The sideband spacing is twice the modulation frequency Ω,

$$2\Omega = (167347764.14 \pm 0.02)\,\text{kHz}. \qquad (13)$$

To move these sideband frequency markers near the hydrogen and deuterium resonances, the modulated laser carrier is tuned midway in between the resonance frequencies. While the first laser excites the 1S-2S transition of hydrogen or deuterium, the small residual beat frequencies Δn_D and Δn_H with the sidebands of the reference laser are recorded with a fast photodiode, followed by a spectrum analyzer. The spectrometer laser is alternatively stabilized to the 1S-2S resonances of hydrogen (hyperfine level $F = 1$) and deuterium (hyperfine level $F = 3/2$, $m_F = \pm 3/2$), while the beat note between laser 1 and the sidebands of laser 2 is recorded (Fig. 4). The frequency difference between the two-photon resonances at 243 nm is then given by:

$$\Delta_{exp} = 2(2\Omega + \Delta n_D - \Delta n_H), \qquad (14)$$

where the factor of 2 arises from the frequency doubling of the 486 nm light. Because we cannot excite both resonances at the same time, and the reference resonator frequency of laser 2 is slowly drifting (approx. 15 Hz/s), the frequency marks are also drifting. To correct for this effect, we record the drifts of both beat frequencies with time. The drift of $\Delta n_D(t)$ is fitted with a polynomial of fourth order. The difference between measured $\Delta n_H(t)$ and interpolated $\Delta n_D(t)$ at the same times is constant within the measurement accuracy. The mean value has an uncertainty of 5 kHz due to the interpolation procedure described above. We correct for the hyperfine structure to obtain the line centroids. Our final result is

$$\Delta_{\text{ISO}} = 670\,994\,337(22)\,\text{kHz}, \qquad (15)$$

which is in good agreement with the last measurement at Oxford [9], $\Delta_{\text{ISO}} = 670\,994.\,33(64)\,\text{MHz}$, but a factor of 25 more accurate.

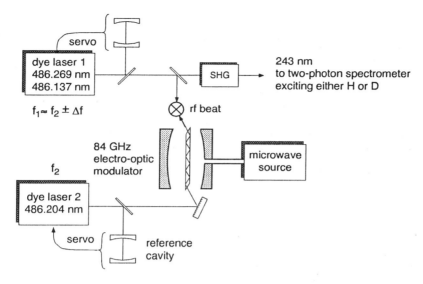

Figure 4. Frequency measurement of the hydrogen-deuterium isotope shift.

3   Theory of the hydrogen energy levels

Our goal is to present here an updated complete collection of known corrections to hydrogen energy levels including that, recently calculated. We concentrate mainly on S-states since these states are involved in our experiments. We base in this paper on the excellent review given by Sapirstein and Yennie in [4].

The hydrogen energy levels as given by Bohr $E = -\frac{m\,\alpha^2}{2\,n^2}$ do not exactly agree with the measured values. The most significant correction of order α^2 is due to relativistic effects described by the Dirac equation. One derives from this equation the following expression for electron energy levels in the Coulomb potential,

$$
\begin{aligned}
E(n,j) &= m\,f(n,j)\,, \\
f(n,j) &= \left(1 + \frac{(Z\,\alpha)^2}{\left[n - j - \frac{1}{2} + \sqrt{(j + \frac{1}{2})^2 - (Z\,\alpha)^2}\right]^2}\right)^{-\frac{1}{2}},
\end{aligned}
\tag{16}
$$

where n is the principal number, m is the mass of the electron, Z is the number of protons in the nucleus, α the fine structure constant, and j is the total angular momentum of the electron. In order to determine the energy levels with an accuracy comparable to Lamb shift experiments, one has to take into account the radiative and recoil corrections. The radiative corrections are due to the electron self-interaction, i.e. photon emission and immediate reabsorption and due to the vacuum polarization The recoil corrections are due to movement of the nucleus. Each photon exchange between the electron and the proton is accociated with a change of the proton kinetic energy. In the nonrelativistic limit, the proton mass dependence is accounted for by the reduced mass of the two-body system. The relativistic treatment is much more complicated and in general it is not even possible to write a two-body relativistic Hamiltonian. As a starting point one considers the approximated effective Hamiltonian, which is

$$
H_{\text{eff}} = \boldsymbol{\alpha} \cdot \mathbf{p} + \beta\,m + \frac{\mathbf{p}^2}{2M} + V_{\text{eff}}\,,
\tag{17}
$$

$$
V_{\text{eff}} = -\frac{\alpha}{r}\left[1 + \frac{1}{M}\left(\delta^{ij} + \frac{r^i\,r^j}{r^2}\right)p^i\,\alpha^j\right]
\tag{18}
$$

where α^i are the Dirac matrices , m and M are the electron and the nuclear mass respectively.

For simplicity we set $Z = 1$ since we are dealing with hydrogen. Apart from the Dirac factor this Hamiltonian contains the Breit interaction V_{eff} and the kinetic energy of the nucleus. The expression for the energy (not taking into account the hyperfine structure), that is valid up to order α^4 but gives the precise mass dependence is [17, 4]:

$$E = m + M + \mu\left[f(n,j) - 1\right] - \frac{\mu^2}{2\,(M+m)}\left[f(n,j) - 1\right]^2 + \frac{\alpha^4\,\mu^3}{2\,n^3\,M^2}\left(\frac{1}{j+\frac{1}{2}} - \frac{1}{l+\frac{1}{2}}\right)(1 - \delta_{l0}),\quad (19)$$

The remaining recoil corrections in higher order require separate treatment and will be described later. The Lamb shift is defined as the energy correction beyond the first four terms in the above formula.

The main contribution to the hydrogen Lamb shift is electron one-loop self-energy described by

$$\Delta E = e^2 \int \frac{d^4k}{(2\pi)^4 i}\,\frac{g_{\mu\nu}}{k^2}\,\langle\overline{\psi}|\gamma^\mu\,\frac{1}{\not{p} - \not{k} + \frac{\alpha}{r}\gamma^0 - m}\gamma^\nu|\psi\rangle - \langle\delta m\rangle,\quad (20)$$

and is represented by the Feynman graph on Fig. 5. Its evaluation has a rich history and in the last

Figure 5. Electron self-energy diagram

years a significant progress has been made [18, 19]. There are mainly two complementary methods: the numerical one and the analytical one. The numerical method [20] is based on the decomposition of the Dirac-Coulomb Green function into partial waves. In the analytical method [21] one expands the expression (20) in α and calculates subsequently all the coefficients up to the order of interest. Both methods have some advantages and disadvantages. The direct numerical calculation for Z=1, (i.e. hydrogen) is extremely difficult, and so far the calculation has been done only for larger Z (Z=5,10, and so on). In the analytical calculation the coefficients in the α expansion are known precisely, but one always faces the estimation of uncalculated higher order terms. In fact these terms could also be obtained from the numerical integration data by extraction of the analytically known lower order terms and extrapolation to Z=1. Since we are here interested in the S-states, we limit the attention to this case. The coefficients in the analytical expansion are defined by:

$$\Delta E = \frac{m\,\alpha\,(Z\,\alpha)^4}{\pi\,n^3}\,F,\quad (21)$$

$$F = A_{40} + A_{41}\,\ln\left[(Z\,\alpha)^{-2}\right] + A_{50}\,Z\,\alpha +$$
$$(Z\,\alpha)^2\left\{A_{60} + A_{61}\,\ln\left[(Z\,\alpha)^{-2}\right] + A_{62}\,\ln^2\left[(Z\,\alpha)^{-2}\right]\right\} + (Z\,\alpha)^3\,G\quad (22)$$

where G incorporates all the higher order terms. The values for the electron self-energy for S-states are given in the following list:

$$A_{40}(n) = \frac{10}{9} - \frac{4}{3}k_0(n)$$

$$\ln k_0(1) = 2.9841285557655,$$

$$\ln k_0(2) = 2.8117698931205,$$

$$\ln k_0(4) = 2.7498118405,$$

$$A_{41} = \frac{4}{3},$$

$$A_{50} = 4\pi\left(\frac{139}{128} - \frac{\ln(2)}{2}\right),$$

$$A_{62} = -1,$$

$$A_{61}(1) = \frac{28}{3}\,\ln(2) - \frac{21}{20},$$

$$A_{61}(2) = \frac{16}{3}\ln(2) + \frac{67}{30},$$

$$A_{61}(4) = \frac{4}{3}\ln(2) + \frac{391}{80},$$

$$A_{60}(1) = -30.92890,$$

$$A_{60}(2) = -31.84047. \tag{23}$$

where $\ln k_0(n)$ denotes the Bethe logarithms [4]. Most of these data are taken from the comprehensive review of Sapirstein and Yennie in [4]. We have added the 4S state and the A_{60} coefficients for 1S and 2S states calculated by (K.P.) in [19]. The value for the A_{61} coefficient for the 4S state is taken from the review by Drake in [22].

$$A_{60}(4) = -31.5(4),$$

$$G(n) = G'(n) + 2\pi\ln(Z\alpha)^{-2},$$

$$G'(1) = 26.9(5),$$

$$G'(2) = 29.8(10),$$

$$G'(4) = 27.2. \tag{24}$$

The term $A_{60}(4)$ and the remaining higher order terms summarized by G are calculated by extrapolation of numerical integration data obtained by P. Mohr [18, 23]. We have found it convenient for the extrapolation to subtract out a large $2\pi\ln(Z\alpha)^{-2}$ term. In this way one significantly decreases the dependence on $(Z\alpha)$. The extrapolation plots for $G'(1)$ and $G'(2)$ are presented on Fig. 6. We do not accociate an error to $G'(4)$, because it is already contained in $A_{60}(4)$.

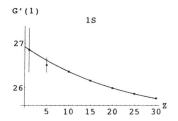
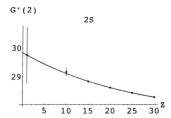

Figure 6. Extrapolation plots for 1S and 2S states. For most of the points the errors bars are inside the drawn size of the dots. The quoted values for Z=1 are the results of these extrapolations.

The next largest contribution to the Lamb shift is the one-loop vacuum polarization. It is given by the formula:

$$\Delta E_{VP} = \int d^3x_1 \int d^3x_2\, \phi^+(\mathbf{x}_1)\,\phi(\mathbf{x}_1)\,\frac{\alpha}{|\mathbf{x}_1 - \mathbf{x}_2|}(-i) \cdot \mathrm{Tr}\left[\gamma_0\, S_F(\mathbf{x}_2, \mathbf{x}_2, t = 0)\right], \tag{25}$$

where S_F is the Dirac-Coulomb propagator. This contribution is represented by the diagram shown in Fig. 7.

The calculation is done by expansion in α. The results are [4]:

$$A_{40} = -\frac{4}{15},$$

$$A_{50} = -\frac{5\pi}{48},$$

$$A_{61} = -\frac{2}{15},$$

9

Figure 7. Vacuum polarization diagram

$$A_{60}(1) = \frac{4}{15}\ln(2) - \frac{1289}{1575} + \left(\frac{19}{45} - \frac{\pi^2}{27}\right),$$

$$A_{60}(2) = -\frac{743}{900} + \left(\frac{19}{45} - \frac{\pi^2}{27}\right),$$

$$A_{60}(4) = -\frac{4}{15}\ln(2) - \frac{5363}{8400} + \left(\frac{19}{45} - \frac{\pi^2}{27}\right). \tag{26}$$

We did not find the coefficient $A_{60}(4)$ for the vacuum polarization in the literature, thus we have calculated it analytically. The unknown higher order terms are expected to contribute few kHz for the 1S state.

Further QED corrections are due to the two-loop diagrams. So far there is no complete direct numerical calculation of this correction. We therefore present the result of a calculation based on expansion in $Z\alpha$,

$$\Delta E = \frac{m\alpha^2(Z\alpha)^4}{\pi^2 n^3}\left\{B_{40} + Z\alpha\, B_{50} + (Z\alpha)^2[B_{63}\ln^3(Z\alpha)^{-2} + B_{62}\ln^2(Z\alpha)^{-2} + \ldots]\right\}. \tag{27}$$

The leading order term B_{40} is determined only by the electron form factors F_1' and F_2 at $q^2 = 0$ and the two-loop vacuum polarization [24],

$$B_{40} = -\frac{2179}{648} - \frac{10\pi^2}{27} + \frac{3\pi^2\log(2)}{2} - \frac{9\zeta(3)}{4} = 0.538941 \tag{28}$$

The evaluation of the next to leading order term B_{50} was completed by one of us (K.P.), with the surprisingly large result

$$B_{50} = -21.4(1). \tag{29}$$

One class of the diagrams, that gives the largest contribution to B_{50}, is shown in Fig. 8. Recently M. Eides and coworkers [26, 27] have also finished the evaluation of this correction, and obtain a more accurate result which converted to our notation is

$$B_{50} = -21.556(3) \tag{30}$$

in agreement with (29). Our value presented previously in the literature in [25] was slightly less accurate. In the meantime (K.P.) has improved and checked this result by an independent numerical integration. The problem in this calculation lies in the fact that every separate diagram is strongly divergent in the infrared, and only the complete sum is finite. Although, in some special gauges these diagrams could be finite, the corresponding expression is then much more complicated. Some details of this calculation and the checking methods are presented in the Appendix. It has been discovered by S. Karshenboim [28] that higher order terms could be also significant, because they are enhanced by powers of $\ln(Z\alpha)^{-2}$. His result for the coefficient B_{63} is $B_{63} = -\frac{8}{27}$ what gives about -3.6 kHz for 2S state. We will not include it in the theoretical value for the Lamb shift until other corrections of this order are known. For the evaluation of the Lamb shift combinations in (7), (8) and (10) the state dependent part of B_{62} could be significant. The expression

$$B_{62} = \frac{16}{9}\left(\psi(n) + \frac{1}{4n^2} - \frac{1}{n} - \ln(n) + C\right), \tag{31}$$

where C is a n independent constant and ψ is a logarithmic derivative of Euler gamma function, was obtained in [12]. It contributes by about 10 kHz to the Lamb shift combination (10).

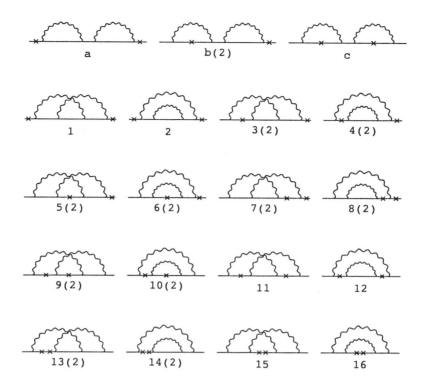

Figure 8. Two-loop diagrams; the cross denotes the insertion of the Coulomb vertex.

At this point we have a complete set of known corrections in the limit of infinite nuclear mass. The additional recoil corrections, due to movement of the nucleus, beyond formula (19) are conventionally divided into pure and radiative recoil corrections. The pure recoil corrections include all diagrams with the exchange of photons between electron and nucleus, neglecting the self-interaction, see Fig. 9:

Figure 9. Some diagrams for pure recoil corrections; waved and dashed lines denote the transverse and Coulomb photon respectively

The exact mass dependence of the pure recoil correction in α^5 order was worked out in [29] using the Bethe–Salpeter formalism.

$$\Delta E = \frac{\mu^3}{mM}\frac{\alpha^5}{\pi n^3} \left\{ \frac{2}{3}\delta_{l0}\ln(\frac{1}{\alpha}) - \frac{8}{3}\ln k_0(n,l) - \frac{1}{9}\delta_{l0} - \frac{7}{3}a_n \right.$$
$$\left. - \frac{2}{M^2 - m^2}\delta_{l0}\left[M^2\ln\left(\frac{m}{\mu}\right) - m^2\ln\left(\frac{M}{\mu}\right) \right] \right\}, \tag{32}$$

where

$$a_n = -2 \left(\ln(\frac{2}{n}) + (1 + \frac{1}{2} + \ldots + \frac{1}{n}) + 1 - \frac{1}{2n} \right) \delta_{l0} + \frac{1 - \delta_{l0}}{l(l+1)(2l+1)}, \tag{33}$$

and μ denotes the reduced mass and $\ln k_0(n)$ the Bethe logarithm.

In order α^6 the m/M correction has been calculated in [30] using the effective potential method. It has been found that this calculation contains some errors. Khriplovich and coworkers [31] calculated the exact mass dependence of the logarithmic term for the case of arbitrary masses and found that the $\log(\alpha)$ term should be absent. A recent calculation of one of us (K.P.) together with H. Grotch [32] gives the complete result for the pure recoil corrections in order $(Z\alpha)^6$,

$$\Delta E(nS) = \frac{m^2}{M} \frac{(Z\alpha)^6}{n^3} \left(4 \ln(2) - \frac{7}{2} \right). \tag{34}$$

The radiative recoil corrections include the diagrams where a photon is emitted and absorbed by the same particle, or the photon propagator is modified by the fermion loop. It has been checked in

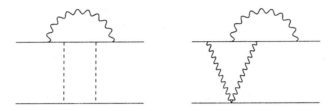

Figure 10. Some diagrams for the radiative recoil corrections.

[33, 4] that for S-states most of these corrections could be incorporated by multiplying the radiative corrections (electron self-energy, vacuum polarization and two-loop corrections) by a factor $(\mu/m)^3$, which comes from the modification of the wave function at the origin, and changing the argument of $\ln(Z\alpha)^{-2}$ to $\ln\left(\frac{m}{\mu}(Z\alpha)^{-2}\right)$ which is twice the logarithm of the ratio of binding energy and electron rest energy. These replacement rules give an exact expression for the recoil correction in order $\alpha(Z\alpha)^4$, but there are some corrections in order $\alpha(Z\alpha)^5$ which are not yet incorporated. Some example diagrams are presented in Fig. 10. This radiative recoil correction has originally been calculated in [33]

$$\Delta E = -\frac{\alpha(Z\alpha)^5 m^2}{M n^3} 1.988(4), \tag{35}$$

but it has been recently recalculated in [34] with the result:

$$\Delta E = -\frac{\alpha(Z\alpha)^5 m^2}{M n^3} 1.36449. \tag{36}$$

Due to different method used in [34] the origin of the discrepancy has not yet been found. A small difference is due to inclusion of vacuum polarization terms, but the greatest difference comes from the so called pole contribution.

There are some conceptional problems with the proton self-energy correction. For a point-like proton it would be

$$\Delta E = \frac{\alpha^5 \mu^3}{\pi n^3 M^2} \left[\left(\frac{10}{9} + \frac{4}{3} \ln \frac{M}{\mu\alpha^2} \right) \delta_{l0} - \frac{4}{3} \ln k_0(n, l) \right] \tag{37}$$

For a finite size proton some corrections in the above are counted twice because they are already incorporated in the proton charge radius. This problem has been considered in [34]. The conclusion is that only the logarithmic term should be excluded from the nuclear finite size contribution. Therefore the complete nuclear size and self-energy correction for S-states and any nuclear charge has the form

$$\Delta E = \frac{2}{3 n^3} (Z\alpha)^4 \mu^3 \langle r^2 \rangle + \frac{4\mu^3}{3\pi n^3 M^2} (Z^2\alpha)(Z\alpha)^4 \left[\ln\left(\frac{M}{\mu(Z\alpha)^2} \right) - \ln k_0(n) \right], \tag{38}$$

12

where $\langle r^2 \rangle$ denotes the nuclear charge radius. The additional correction in the expression above beyond the finite size effect gives 4.6 kHz for the 1S state in hydrogen.

There are some additional corrections that are specific for deuterium. The binding energy of the deuteron 2.2 MeV is comparable to the electron mass, giving rise to a small polarizability correction. We have estimated this correction [35] using the square well potential for the proton-neutron interaction,

$$\Delta E = -m \, \alpha \, \phi(0)^2 \, \alpha_d \left(\frac{19}{6} + 5 \ln(2 \, \bar{E}) \right).$$ (39)

where $\bar{E} = 4.915$ MeV, and $\alpha_d = 0.635 \, \text{fm}^3$ is a deuteron polarizability. This correction gives -22(2) kHz for 1S state in deuterium. A more precise results using different models has later been obtained by J. Martorell *et al.* [36] $\Delta E = \dot{2}2.09(24)$ kHz, and Leidemann *et al.* [37] $\Delta E = $ -21.5(3) kHz, which agree with our estimation.

Since the deuteron has spin 1 the Zitterbewegung term in the $1/c$ expansion is absent. It means that there is an additional recoil correction beyond Eq. (19) for deuterium [38],

$$\Delta E = -\frac{1}{2} \frac{\mu^3}{M^2} \frac{(Z \, \alpha)^4}{n^3} \delta_{l0}$$ (40)

It contributes -13 kHz for the 1S state in deuterium. Both corrections are significant for a determination of the deuteron radius from the hydrogen-deuterium isotope shift [3].

4   Conclusions

The theoretical predictions for the Lamb shifts of hydrogen energy levels, using a proton *rms* radius $r_p = 0.862(12)$ fm [39] are:

$$L(1S) = 8172802(32)(24) \, \text{kHz},$$ (41)
$$L(2S) = 1045003(4)(3) \, \text{kHz},$$ (42)
$$L(4S) = 131675.2(5)(4) \, \text{kHz},$$ (43)
$$L(4S - 2S) - \frac{1}{4} L(2S - 1S) = 868622(4)(3) \, \text{kHz},$$ (44)
$$L(2S - 2P) = 1057839(4)(4) \, \text{kHz},$$ (45)

where the first error comes from proton radius and the second from the estimation of higher order two-loop corrections and three loop contribution. For the evaluation of the 2P Lamb shift we used data given in [4], except for $G_{SE}(2P)$. For this higher order term we perform an extrapolation of improved data obtained by P. Mohr [18] with the result: $A_{60}(2P) = -0.96(7)$, $A_{70}(2P) = 2.6$. The coefficient $A_{71}(2P)$ is exactly zero.

The older proton radius $r_p = 0.805(11)$ [40] will give a result which is 18 kHz smaller for the 2S Lamb shift and 16 kHz smaller for the combined Lamb shift. Comparing the theoretical predictions with the experimental results:

$$L(2S_{1/2} - 2P_{1/2}) = 1\,057\,845(9) \, \text{kHz} \, [15]$$ (46)
$$L(2S_{1/2} - 2P_{1/2}) = 1\,057\,839(12) \, \text{kHz} \, [16],$$ (47)
$$L(4S - 2S) - \frac{1}{4} L(2S - 1S) = 868631(10) \, \text{kHz} \, [2].$$ (48)

a satisfactory agreement is found if the larger value for the proton radius is used.

Although, the recently calculated two-loop correction leads to agreement of the hydrogen experimental results with the theoretical predictions based on the newer proton radius, a strong disagreement appears for the He$^+$ Lamb shift, as first noted by M. Boshier [41]. The theory predicts:

$$L_{He}(2S - 2P) = 14\,041.33(18) \, \text{MHz},$$ (49)

while the experimental result of Wijngaarden et al. [42] is

$$L_{He}(2S - 2P) = 14\,042.52(16) \, \text{MHz}.$$ (50)

These results are presented in Fig. 11. The strong disagreement between He^+ experiment and the-

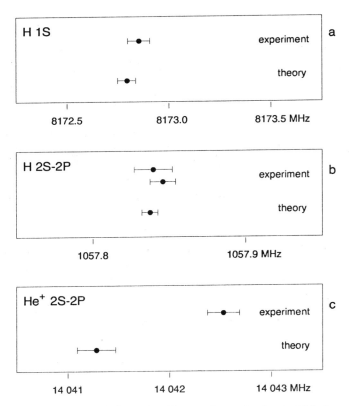

Figure 11. The hydrogen and helium+ Lamb shift, experimental value in 'a' is from [2], in 'b' from [16] and [15], in 'c' from [42]

oretical predictions clearly requires further investigation. In fact the theoretical side has now been confirmed by an independent calculation of M. Eides et. al. [27] and the experimental side will be checked by a measurement of Boshier and Hinds [11], where the hydrogen 1S-2S transition will be compared to the He+ 2S-4S transition.

An even more critical test of QED using L(2S-4S)-1/4 L(1S-2S) is currently limited by the uncertainty in the proton radius which, for example, leads to an error of 4 kHz for the 2S Lamb shift, compared to the predicted accuracy of 1 kHz of the planned measurement of the combined Lamb shift (48). A considerably improved test of QED based on hydrogen requires a more precise determination of the proton charge radius, which should be possible by a Lamb shift measurement in muonic hydrogen. Such an experiment is currently planned at PSI [43].

The most significant uncertainties of the theoretical Lamb shifts and H-D isotope shift are due to

- unknown complete two-loop correction of order $\alpha^2 (Z\alpha)^6$,

- three-loop contribution of order $\alpha^3 (Z\alpha)^4$,

- pure and radiative recoil corrections of order $\frac{m^2}{M} \alpha^7 \ln^2(\alpha)^{-2}$,

which are estimated to be about 3 kHz for 2S state. The evaluation of these corrections becomes now the main problem in the theory of the hydrogen Lamb shift. A precise determination of the state dependent part of B_{61} is important for a possible determination of the Rydberg constant from equation (9) and higher order recoil corrections are significant for the isotope shifts. We have left unevaluated the vacuum polarization correction caused by heavier particles and for example ρ meson contribution to photon formfactor. They could be simply included as a nuclear finite size effect.

From the measurement of the hydrogen-deuterium isotope shift we could determine the value of the deuteron matter radius. Summing up all the reduced mass and Lamb shift contributions, but excluding the nuclear finite size contribution we obtain a theoretical value of

$$[E_D(2S - 1S) - E_H(2S - 1S)]' = 670999572(2)(4)\,\text{kHz}, \tag{51}$$

where the first error comes from the proton-electron mass ratio and the second is the estimate of unknown recoil corrections in order $\frac{m^2}{M}\alpha^7$, which are enhanced by the factor $\ln^2(\alpha)^{-2}$. In this calculation we used the following values for the proton-electron [14] and deuteron-proton [45] mass ratios

$$\frac{m_p}{m_e} = 1836.1526646(58), \tag{52}$$

$$\frac{m_d}{m_p} = 1.9990075013(14). \tag{53}$$

The difference

$$\Delta E = 5235(22)\,\text{kHz} \tag{54}$$

between the theoretical value (51) and experimental value (15) is caused by the nuclear finite size effect

$$\Delta E = \frac{2}{3\,n^3}(Z\,\alpha)^4\,\mu^3\langle r^2\rangle. \tag{55}$$

Using this equation we obtain for the difference of the deuteron-proton charge square radii,

$$r_d^2 - r_p^2 = 3.822(16)\,\text{fm}^2, \tag{56}$$

where r_d and r_p respectively denote the deuteron and proton rms nuclear charge radii. The error of $0.016\,\text{fm}^2$ in (56) is dominated by the 22 kHz error in the measurement of the hydrogen-deuterium isotope shift. The deuteron matter radius r_m is defined by the formula [46]

$$r_d^2 = r_p^2 + r_m^2 + r_n^2 + \frac{3}{4}\frac{1}{m_p^2}, \tag{57}$$

where $r_n^2 = -0.1192\,\text{fm}^2$ is the neutron charge radius. From the experimental isotope shift of 1S-2S, we obtain a deuteron matter radius r_m

$$r_m = 1.977(4)\,\text{fm}. \tag{58}$$

This value disagrees with the matter radius obtained from electron scattering experiments (see [46] and references therein)

$$r_m = 1.953(3)\,\text{fm}, \tag{59}$$

but is in better agreement with that calculated using the effective nucleon-nucleon interaction potential [47],

$$r_m = 1.968(1)\,\text{fm}. \tag{60}$$

These results suggest that the interpretation of low energy electron–deuteron scattering data may require an additional analysis to obtain the correct deuteron charge radius.

The present theory of hydrogen-like systems shows that one perfectly understands all the QED effects, which in principle could be calculated up to any order (except for the nuclear structure corrections). The difficulties are technical, due to complicated expressions exploding in the higher order calculations. Also, there is still no unified treatment of two-body systems in the sense that higher orders terms can not be calculated automatically. Usually one has to perform some approximation, i.e. expansions in α, before the expression can be calculated. This is the trickiest part of all these calculations. There exist several methods for different types of corrections, see [44] and [19] for example, which exploit the fact that different photon gauges are appropriate for the low and high energy regions. It allows for the elimination of spurious singularities at the earlier stage of calculations. But still there is no method for the bound state problems where all corrections are treated in an unified way. One example for these difficulties is the orthopositronium decay rate, where the calculation of the correction in order α^2 is still not completed.

Appendix

In this appendix we briefly describe the calculation of the most difficult class of two-loop binding contributions that was performed by one of us (K.P.) [25], and consists of the diagrams shown in Fig. 8. We also explain how this method has been tested.

There are three momenta involved in the expression generated from the Feynman diagrams: two photon four-momenta k_1 and k_2 and the three-dimensional momentum p transfered through the Coulomb vertices. The correction to the energy described by the coefficient $F = B_{50}$ can be written as

$$\Delta E = \left(\frac{\alpha}{\pi}\right)^2 \frac{(Z\alpha)^5}{n^3} F,$$

$$F = \int \frac{d^3p}{\pi^2} \int \frac{d^4k_2}{\pi^2} \int \frac{d^4k_1}{\pi^2} \frac{1}{4p^4} \left(f(p,k_1,k_2) - f(0,k_1,k_2)\right). \qquad (61)$$

The problem in the calculation lies in the fact that the integration in p is divergent for separate diagrams. The evaluation proceeds as follows. Assuming that the k_1 and k_2 integrations are done, F could be rewritten to the form

$$F = \frac{1}{4} \int \frac{d^3p}{\pi^2} \frac{1}{p^4} \left(f(p^2) - f(0)\right), \qquad (62)$$

where f is an analytical function with a branch cut along the negative axis. From the Cauchy theorem we have

$$f(p^2) = -\int_0^\infty d(q^2) \frac{1}{p^2 + q^2} f^A(q^2),$$

$$f(p^2) - f(0) = p^2 \int_0^\infty d(q^2) \frac{1}{q^2(p^2 + q^2)} f^A(q^2), \qquad (63)$$

$$F = \int_0^\infty dq \frac{f^A(q^2)}{q^2}. \qquad (64)$$

In this way the integral is converted to a form suitable for the further analytical treatment. It is calculated analytically the expansion of $f^A(q^2)$ up to q^{10}. The first terms in this expansion are

$$f^A(q^2) = -\frac{32}{27}q^2 + \frac{104}{9}q^2 \ln(q^2) + \dots \qquad (65)$$

The integral (64) with respect to q from 0 to $q^2 = \frac{1}{2}$ is performed using this analytical expansion, while for $q^2 > \frac{1}{2}$ the integral is evaluated numerically using the Feynman parameter approach. For the final value we obtained

$$B_{50} = -21.4(1). \qquad (66)$$

To check this result several tests were performed. The first logarithmic term in the q^2 expansion of f^A gives the most significant contribution to the B_{50}. It was independly calculated by S. Karshenboim in Yennie gauge, and by us in the Feynman and Coulomb gauge. The next test was the comparison of the numerical part with the analytic expansion at $q^2 = \frac{1}{2}$. The further and the most significant test was done using the fact that for a similar integral

$$f(0) = -\int_0^\infty d(q^2) \frac{1}{q^2} f^A(q^2), \qquad (67)$$

the analytical result is known. It is given by the on shell vertex functions, similar to the leading two-loop contribution (diagrams without closed fermion loop).

$$f(0) = 32 F_1'(0) + 8 F_2(0). \qquad (68)$$

Using the values for the appropriate diagrams [48] the analytical value is

$$f(0) = -\frac{163}{9} - \frac{85\pi^2}{27} + 12\pi^2 \ln(2) - 18\zeta(3) = 11.27. \qquad (69)$$

The combined analytical-numerical code gives

$$f(0) = 11.14(12), \qquad (70)$$

at photon mass $\mu^2 = 0.4 \cdot 10^{-4}$.

The improved value for B_{50} is presented. It is in agreement with the previous one [25], and was obtained by recalculation of the numerical part.

16

Acknowledgments

We acknowledge the valuable contributions of V.P. Chebotayev, F. Schmidt-Kaler, R. Wynands, T. Andreae, C. Zimmermann and M. Prevedelli. The experimental part of this work has been supported in part by the Deutsche Forschungsgemeinschaft and within the frame of an EU science program cooperation, contract No. SCI*-CT92-0816.

References

[1] T. Andreae, W. König, R. Wynands, D. Leibfried, F. Schmidt-Kaler, C. Zimmermann, D. Meschede, and T. W. Hänsch, Phys. Rev. Lett., **69**, 1923 (1992).

[2] M. Weitz, A. Huber, F. Schmidt-Kaler, D. Leibfried, and T.W. Hänsch, Phys. Rev. Lett. **72**, 328 (1994); M. Weitz, A. Huber, F. Schmidt-Kaler, D. Leibfried, W. Vassen, C. Zimmermann, K. Pachucki, T.W. Hänsch, L. Julien, and F. Biraben, Phys. Rev. A, *in press*

[3] F. Schmidt-Kaler, D. Leibfried, M. Weitz, and T. W. Hänsch, Phys. Rev. Lett. **70**, 2261 (1993).

[4] J. R. Sapirstein and D. R. Yennie, in *Quantum Electrodynamics*, Editor T. Kinoshita, World Scientific (1990).

[5] F.Schmidt-Kaler, D. Leibfried, S. Seel, C. Zimmermann, W. König, M. Weitz, and T.W. Hänsch, Phys. Rev. A **51**, 2789 (1995).

[6] C.O. Weiss, G. Kramer, B. Lipphardt, E. Garcia, IEEE J. Quantum Electron. **QE-24**, 1970 (1988).

[7] B. Dahmani, L. Hollberg, and R. Drullinger, Opt. Lett. **12**, 876 (1987).

[8] D. Leibfried, F. Schmidt-Kaler, M. Weitz, and T.W. Hänsch, Appl. Phys. B **56**. 65 (1993).

[9] M.G. Boshier, P.G.E. Baird, C.J. Foot, E.A. Hinds, M.D. Plimmer, D.N. Stacey, J.B. Swan, D.A. Tate, D.M. Warrington, G.K. Woodgate, Phys. Rev. A **40**, 6169 (1989).

[10] F. Nez, M.D. Plimmer, S. Bourzeix, L. Julien, F. Biraben, R. Felder, Y. Milleroux, and P. de Natale, Europhys. Lett **24**, 635 (1993).

[11] D.J. Berkeland, E.A. Hinds and M. Boshier, Phy. Rev. Lett. **75** 2470 (1995).

[12] S.G. Karshenboim, JETP, *in press*.

[13] V. G. Pal'chikov, Yu. L. Sokolov, and V. P. Yakovlev, JETP Lett. **38**, 418 (1983).

[14] R.S. van Dyck, Jr., *et al.*, *private communication*.

[15] S. R. Lundeen, F. M. Pipkin, Phys. Rev. Lett. **46**, 232 (1981).

[16] E.W. Hagley and F.M. Pipkin, Phys.Rev.Lett. **72**, 1172, (1994).

[17] W.A. Barker and F.N. Glover, Phys. Rev. **99**, 317 (1955).

[18] P. J. Mohr, Phys. Rev. A **46**,4421, (1992).

[19] K. Pachucki, Phys. Rev. A, **46** 648, (1992), Ann. Phys. (N.Y.), **226** 1, (1993).

[20] P. J. Mohr, Ann. Phys. (N.Y) **88**, 26, 52 (1974).

[21] G. W. Erickson and D. R. Yennie, Ann. Phys. (N.Y) **35**, 271, 447 (1965).

[22] G.W.F. Drake, Adv. At. Mol. Opt. Phys. **31**, 1 (1993).

[23] P. J. Mohr and Y.-K. Kim, Phys. Rev. A **45**, 2727 (1992).

[24] R. Barbieri, J. A. Mignaco, and E. Remiddi, Nuovo Cim. Lett. **3**, 588 (1970) B. E. Lautrup, A. Peterman, and E. de Rafael, Phys. Lett. B **31** , 577 (1970).

[25] K. Pachucki, Phys. Rev. A, **48**, 2609 (1993); Phys. Rev. Lett. **72**, 3154 (1994).

[26] M.I. Eides and H. Grotch, Phys. Lett. B **301**, 127 (1993), Phy. Lett. B **308**, 389 (1993) M. Eides, H. Grotch, and P. Pebler, Phys. Rev. A **50**, 144 (1994).

[27] M. Eides and V. Shelyuto, Phys. Rev. A **52**, 954 (1995), and references therein.

[28] S. G. Karshenboim, JETP **76**, 541 (1993).

[29] E. E. Salpeter, Phys. Rev. **87**, 328 (1952), G.W. Erickson, J. Phys. Chem. Ref. Data **6**, 833 (1977).

[30] M. Doncheski, H. Grotch and G.W. Erickson, Phys. Rev. A **43**, 2125 (1991).

[31] I. B. Khriplovich, A. I. Milstein and A. S. Yelkhovsky, Phys. Scrip. **T46**, 252 (1993), R. N. Fell, I. B. Khriplovich, A. I. Milstein and A. S. Yelkhovsky, Phys. Lett. A **181**, 172 (1993).

[32] K. Pachucki and H. Grotch, Phys. Rev. A **51**, 1854 (1995).

[33] G. Bhatt and H. Grotch, Ann. Phys. (NY) **178**,1 (1987).

[34] K. Pachucki, Phys. Rev. A **52**, 1079 (1995).

[35] K. Pachucki, D. Leibfried, and T.W. Hänsch, Phys. Rev. A **48**, R1 (1993). K. Pachucki, M. Weitz, and T.W. Hänsch, Phys. Rev. A. **49**, 2255 (1994).

[36] J. Martorell, D.W. Sprung, and D.C. Zheng, Phys. Rev. C **51**, 1127 (1995).

[37] W. Leidemann and R. Rosenfelder, Phys. Rev. C **51**, 427 (1995).

[38] K. Pachucki and S. Karshenboim, J. Phys. B **28**, L221 (1995).

[39] G. G. Simon, C. Schmidt, F. Borkowski, and V. H. Walther, Nucl. Phys. A **333**, 381 (1980).

[40] L. N. Hand, D. J. Miller, and R. Wilson, Rev. Mod. Phys. **35**, 335 (1963).

[41] M. Boshier, *private communication.*

[42] A. van Wijngaarden, J. Kwela, and G.W.F. Drake, Phys. Rev. A **43**, 3325 (1991).

[43] D. Taqqu, Paul Scherrer Institut, Villingen Suisse, *private communication.*

[44] T. Kinoshita and P. Lepage, in Quantum Electrodynamics, edited by T. Kinoshita, (World Scientific, Singapore, 1990), p. 560.

[45] G. Audi and A.H. Waspra, Nucl. Phys. A **565**, 1 (1993).

[46] S. Klarsfeld *et al*, Nucl. Phys. **A456**, 373 (1986).

[47] J. L. Friar, G.L. Payne, V.G.J. Stoks, and J.J. de Swart, Phys. Lett. **B311**, 4 (1993).

[48] R. Barbieri, J.A. Mignaco, and E. Remiddi, Nuovo Cim. **11A**, 824 (1972).

18

QUANTUM ELECTRODYNAMICAL CORRECTIONS IN HIGHLY CHARGED IONS

G. Soff, C. R. Hofmann, and G. Plunien

Institut für Theoretische Physik, Technische Universität,
D-01062 Dresden, Germany

S. M. Schneider

Institut für Theoretische Physik, J. W. Goethe-Universität,
D-60054 Frankfurt am Main, Germany

INTRODUCTION

In this report we will present the results of our calculations concerning the quantum electrodynamical (QED) corrections for highly charged hydrogen-like atoms. Knowledge of all these contributions will allow for a precise test of quantum electrodynamics in the presence of a strong Coulomb field.

For a hydrogen atom the ground state hyperfine structure is one of the best measured quantities in physics. Its experimental value amounts to:

$$\Delta\nu_{\text{HFS}} = 1\,420\,405\,751.7667 \pm .0009 \text{ Hz} \quad . \tag{1}$$

But theoretically only a relative accuracy of 10^{-6} has been achieved. The remaining uncertainty is assumed to arise mainly from the unknown internal structure of the proton, as there are the finite size of its charge distribution, its extended magnetization distribution, and excitation effects, i.e., the polarizability of the proton.

Here we want to stress another topic of QED, namely the contributions to the binding energy of an electron moving in a strong Coulomb field as provided by a nucleus of high charge number Z. In this case another coupling constant to the external field of strength $Z\alpha$ appears in the Dirac equation. For highly charged ions as for instance uranium this coupling constant is of the order of one. Perturbation theory with respect to the coupling to the external field would lead to inaccurate calculations of the QED contributions. We rather have to take into account the influence of the external potential by employing the solutions of the Dirac equation for an electron moving in the Coulomb potential of the nucleus. This will provide us

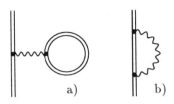

Figure 1. Feynman diagrams in the lowest order in α for a) the vacuum polarization and b) the self energy

with the wave functions we need to calculate the QED effects on the binding energy of this electron.

Evidently, nuclear effects will now contribute on a higher level than in the hydrogen case, thus limiting QED tests in strong fields by the exactness of experimentally determined nuclear parameters and the validity of the nuclear models which are utilized to evaluate these effects.

In the following we discuss the single contributions to the Lamb shift, which will mainly arise from the self energy and the vacuum polarization displayed in Figure 1. Both contributions will be treated in the next two sections. Then we will have to say a few words about the higher-order contributions in the coupling constant α, e.g., the Källén-Sabry energy shift. Finally we will examine the nuclear effects into the computation of atomic binding energies, consequently discussing the constraints of QED tests in the heavy–ion scenario. At the end we will come to the role of the QED corrections in critical fields. We will especially discuss, how these contributions to the binding energy can influence the diving of the $1s$ bound state into the negative energy continuum.

VACUUM POLARIZATION IN EXTERNAL FIELDS

First we have to speak about the electron wave functions that will be employed in the following. For a spherical symmetrical potential these wave functions follow from the solutions of the radial Dirac equation [1]:

$$
\begin{aligned}
\frac{\mathrm{d}g}{\mathrm{d}r} &= -\frac{\kappa}{r}g + (E + 1 - V)f \quad, \\
\frac{\mathrm{d}f}{\mathrm{d}r} &= -(E - 1 - V)G + \frac{\kappa}{r}f \quad,
\end{aligned}
\tag{2}
$$

where κ is the Dirac angular momentum quantum number ($\hbar = c = m_e = 1$).

We consider three simple models to describe the nuclear charge distribution. As a most simple ansatz we start from the charge density of a spherical shell

$$
\rho(r) = \frac{Z}{4\pi R^2}\,\delta(r - R)
\tag{3}
$$

with the resulting potential

$$
V(r) = \begin{cases} -Z\alpha/r & \text{for } r > R \quad, \\ -Z\alpha/R & \text{for } r \leq R \quad. \end{cases}
\tag{4}
$$

The homogeneously charged sphere is described by

$$\rho(r) = \frac{3Z}{4\pi R^3} \Theta(R - r) \quad . \tag{5}$$

The corresponding potential reads

$$V(r) = \begin{cases} -Z\alpha/r & \text{for } r > R \quad , \\ -Z\alpha/(2R) \cdot (3 - r^2/R^2) & \text{for } r \le R \quad . \end{cases} \tag{6}$$

A more realistic charge distribution for the nucleus is the two-parameter Fermi distribution

$$\rho(r) = \frac{N}{1 + \exp\left(4\ln 3 \, \frac{r - c}{t}\right)} \tag{7}$$

where c denotes the half-density radius and t indicates the radial distance over which the charge density declines from 90% to 10% of its value at the origin [2]. The normalization factor N is chosen such that

$$4\pi \int_0^\infty \rho(r) \, r^2 \, dr = Z \quad . \tag{8}$$

The r. m. s. radius R is fitted to the experimentally measured nuclear radii according to the formula:

$$R = 0.83585 \, A^{1/3} + 0.56952 \tag{9}$$

which is valid for atomic mass numbers $A > 9$.

The Feynman diagram for the lowest-order vacuum polarization is displayed in Fig. 1a). The double lines indicate the exact propagators and wave functions in the Coulomb field of an extended nucleus.

Fig. 2 shows an $Z\alpha$-expansion of the vacuum polarization graph. The dominant effect is provided by the Uehling contribution being visualized by the first diagram on the right hand side. It is linear in the external field and thus of order $\alpha(Z\alpha)$.

This is the ultraviolet divergent part of the vacuum-polarization. Fortunately, the Uehling part can be evaluated separately. What remains to be calculated are the diagrams with $n = 3, 5, \ldots$ external photon fields. This is done by computing the difference of the exact vacuum polarization graph minus the Uehling part. The occuring divergencies cancel then term by term for each value of the angular momentum κ.

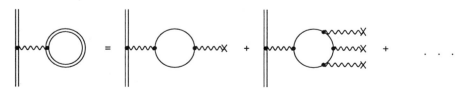

Figure 2. $Z\alpha$-expansion of the vacuum polarization diagram

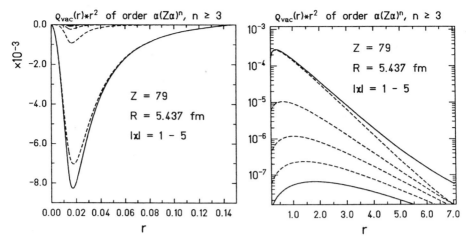

Figure 3. Radial vacuum polarization charge density $\rho_{|\kappa|} \cdot r^2$ of order $\alpha(Z\alpha)^n$ with $n \geq 3$ for the system $Z = 79$ with a nuclear radius $R = 5.437$ fm versus the radial coordinate r in natural units. The various contributions for $|\kappa| = 1, 2, 3, 4$ and 5 are shown separately by the dashed lines. $\rho_{|\kappa|} \cdot r^2$ is given in units of the elementary charge e. The solid line indicates the sum $\rho \cdot r^2$ of the various angular momentum components. a) Linear scale for the range in which the charge density is negative. b) Logarithmic scale to demonstrate the large distance behaviour of $\rho_{|\kappa|} \cdot r^2$. Here the vacuum polarization charge density is positive.

The result is the finite contribution to the binding energy of the order $\alpha(Z\alpha)^n$ with $n \geq 3$. This program is sketched in the following.

From bound state QED [3] the energy shift corresponding to the total vacuum polarization is given by

$$\Delta E = 4\pi i \alpha \int d(t_2 - t_1) \int d\vec{x}_2 \int d\vec{x}_1 \, \bar{\phi}_n(x_2) \gamma^\mu \phi_n(x_2) D_F(x_2 - x_1) \, \mathrm{Tr}[\gamma_\mu S_F(x_1, x_1)] \tag{10}$$

where the photon propagator reads

$$D_F(x_2 - x_1) = \frac{-i}{(2\pi)^4} \int d^4k \, \frac{e^{-ik(x_2 - x_1)}}{k^2 + i\epsilon} \tag{11}$$

and the Feynman propagator for the electron can be represented by

$$S_F(x_2, x_1) = \frac{1}{2\pi i} \int_{C_F} dz \sum_n \frac{\phi_n(\vec{x}_2)\bar{\phi}_n(\vec{x}_1)}{E_n - z} e^{-iz(t_2 - t_1)}$$

$$= \frac{1}{2\pi i} \int_{C_F} dz \, \mathcal{G}(\vec{x}_2, \vec{x}_1, z)\gamma^0 \, e^{-iz(t_2 - t_1)} \quad . \tag{12}$$

It obeys the equation

$$[\gamma^\mu(i\partial_\mu - eA_\mu(x_1)) - m] \, S_F(x_1, x_2) = \delta^4(x_1 - x_2) \tag{13}$$

which implies that external field effects are included to all orders. ϕ_n denotes the electron wave function.

The level shift can be expressed as an expectation value of an effective potential U with

$$
\begin{aligned}
U(\vec{x}_2) &= 4\pi i \alpha \int d(t_2 - t_1) \int d\vec{x}_1 \, D_F(x_2 - x_1) \, \mathrm{Tr}[\gamma_0 \, S_F(x_1, x_1)] \\
&= \frac{i\alpha}{2\pi} \int d\vec{x}_1 \, \frac{1}{|\vec{x}_2 - \vec{x}_1|} \int_{C_F} dz \, \mathrm{Tr}\, \mathcal{G}(\vec{x}_1, \vec{x}_1, z) \quad .
\end{aligned}
$$

$$(14)$$

With the vacuum polarization charge density ρ

$$
\rho(\vec{x}) = \frac{e}{2\pi i} \int_{C_F} dz \, \mathrm{Tr}\, \mathcal{G}(\vec{x}, \vec{x}, z)
$$

$$(15)$$

it simply follows

$$
U(\vec{x}_2) = -e \int d\vec{x}_1 \, \frac{\rho(\vec{x}_1)}{|\vec{x}_2 - \vec{x}_1|} \quad .
$$

$$(16)$$

The formal expression for ρ still contains the infinite unrenormalized charge. As already mentioned, a regularization procedure [4, 5] for the total vacuum polarization charge density is to subtract the Uehling contribution which can then be renormalized separately. Expansion of the Green function in eigenfunctions of angular momentum yields for the vacuum polarization charge density of order $\alpha(Z\alpha)^n$ with $n \geq 3$

$$
\begin{aligned}
\rho(x) - \rho^{(1)}(x) &= \frac{e}{2\pi^2} \int_0^\infty du \left(\sum_{\kappa=\pm 1}^{\pm\infty} |\kappa| \, \mathrm{Re}\left\{ \sum_{i=1}^2 \mathcal{G}_\kappa^{ii}(x, x, iu) \right. \right. \\
&\quad \left. \left. + \int_0^\infty dy \, y^2 \, V(y) \sum_{i,j=1}^2 [\mathcal{F}_\kappa^{ij}(x, y, iu)]^2 \right\} \right) \\
&\quad + \frac{e}{2\pi} \sum_{\substack{\kappa=\pm 1 \\ -m < E < 0}}^{\pm\infty} |\kappa| \left\{ f_1^2(x) + f_2^2(x) \right\} \quad .
\end{aligned}
$$

$$(17)$$

This equation includes terms from any bound-state pole on the negative real z axis, which are picked up as residues in the rotation of the contour of integration. Such terms only appear for superheavy systems where the binding energy of the electron exceeds the electron rest mass. $f_1(x)$ and $f_2(x)$ denote components of the radial Dirac wave function, normalized according to

$$
\int_0^\infty dx \, x^2 \, [f_1^2(x) + f_2^2(x)] = 1.
$$

$$(18)$$

$\mathcal{F}_\kappa^{ij}$ are components of the free Dirac Green functions [6, 7]. According to Wichmann and Kroll [8] the radial Coulomb Green function components $\mathcal{G}_\kappa^{ij}$ may be represented by solutions of the radial Dirac equation. For $y > x$ these Green functions read

$$
\begin{aligned}
\mathcal{G}_\kappa^{11}(x, y, z) &= g_0(x)\, g_i(y)/W \quad , \\
\mathcal{G}_\kappa^{12}(x, y, z) &= g_0(x)\, f_i(y)/W \quad , \\
\mathcal{G}_\kappa^{21}(x, y, z) &= f_0(x)\, g_i(y)/W \quad , \\
\mathcal{G}_\kappa^{22}(x, y, z) &= f_0(x)\, f_i(y)/W \quad ,
\end{aligned}
$$

$$(19)$$

23

while the free Green functions can be written in terms of Bessel functions and Hankel functions of first kind and half-integer order:

$$
\begin{aligned}
\mathcal{F}_\kappa^{11}(x,y,iu) &= -(iu+1)c\,j_{|\kappa+1/2|-1/2}(icx)\,h^{(1)}_{|\kappa+1/2|-1/2}(icy) \ , \\
\mathcal{F}_\kappa^{12}(x,y,iu) &= -ic^2\,\frac{\kappa}{|\kappa|}\,j_{|\kappa+1/2|-1/2}(icx)\,h^{(1)}_{|\kappa+1/2|-1/2}(icy) \ , \\
\mathcal{F}_\kappa^{21}(x,y,iu) &= -ic^2\,\frac{\kappa}{|\kappa|}\,j_{|\kappa-1/2|-1/2}(icx)\,h^{(1)}_{|\kappa-1/2|-1/2}(icy) \ , \\
\mathcal{F}_\kappa^{22}(x,y,iu) &= -(iu-1)c\,j_{|\kappa-1/2|-1/2}(icx)\,h^{(1)}_{|\kappa-1/2|-1/2}(icy) \ .
\end{aligned}
\tag{20}
$$

Here we abbreviated $c=(1+u^2)^{1/2}$. For $y<x$ we have to replace

$$
\begin{aligned}
\mathcal{G}_\kappa^{nm}(x,y,z) &= \mathcal{G}_\kappa^{mn}(y,x,z) \ , \\
\mathcal{F}_\kappa^{nm}(x,y,z) &= \mathcal{F}_\kappa^{mn}(y,x,z) \ .
\end{aligned}
\tag{21}
$$

The Green function in the Coulomb case can be tested by making use of the relation

$$
(E_n-z)\int_0^\infty dx_2 \int_0^\infty dx_1\, x_2^2\, x_1^2 \sum_{i,j=1}^2 \phi_i(x_2)\,\mathcal{G}_{\kappa n}^{ij}(x_2,x_1,z)\,\phi_j(x_1) = 1 \ .
\tag{22}
$$

Equation (17) has been solved numerically. The result is depicted in Fig. 3 for a nuclear charge $Z=79$ (Au). The nuclear radius R is indicated. The various contributions for the angular momentum components $|\kappa|=1\text{--}5$ are shown separately. The radial distance r is given in units of the electron Compton wavelength. $\rho_{|\kappa|}\cdot r^2$ is measured in units of the elementary charge e and the inverse Compton wavelength. Part a) shows on a linear scale $\rho_{|\kappa|}\cdot r^2$ in the range where the charge density is negative. The large distance behaviour of $\rho_{|\kappa|}\cdot r^2$ can be taken from Figs. 3b. Here the radial charge density is positive and displays almost an exponential decline in the depicted range. Obviously the ($\kappa=\pm1$)–contribution to the vacuum polarization charge density dominates by about an order of magnitude. From the order of magnitude of the κ terms one gets an impression of the rapid convergence of the κ-summation. For large distances ($2\le r\le 7$) $\rho_{|\kappa|}\cdot r^2$ decreases rapidly with different decline constants for the various κ-components.

SELF ENERGY IN EXTERNAL FIELDS

We turn now to the calculation of the electron self-energy contribution, which is represented by the diagram b) of Fig. 1. Electronic self–energy corrections for high–Z systems were first studied in the pioneering work by Brown and co–workers [9, 10, 11]. Their method was further refined and successfully applied in computations of electron energy shifts in high–Z elements by Desiderio and Johnson [12] as well as by Cheng and Johnson [13]. Recently, nuclear size corrections were taken into account by Mohr and Soff [14]. In our calculations for superheavy systems we employed these methods, which may be slightly simplified by restriction to K–shell electrons. The energy shift of a $1s_{1/2}$–electron due to the quantum–electrodynamical self energy formally is given by

$$
\Delta E = 4\pi i\alpha \int d(t_2-t_1) \int d\vec{x}_2 \int d\vec{x}_1\, \overline{\phi}_n(x_2)\gamma^\mu S_F(x_1,x_2)\gamma^\nu \phi_n(x_2) D^F_{\mu\nu}(x_2-x_1)
\tag{23}
$$

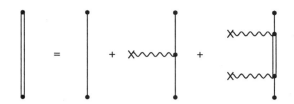

Figure 4. Graphical expansion of the exact electron propagator in an external field

Propagators and wave functions are transformed into momentum space. This admits a decomposition of the self–energy diagram, so enabling infinite mass terms to be identified and removed, leaving the finite observable part of the self energy. We introduce the following Fourier transforms

$$\phi(p) = \int d^4x\, \phi(x)\, e^{ipx} \quad , \tag{24}$$

$$A_\mu^{ex}(p) = \int d^4x\, A_\mu^{ex}(x)\, e^{ipx} \quad , \tag{25}$$

$$S_F(p_2, p_1) = \int d^4x_2 \int d^4x_1\, S_F(x_2, x_1)\, e^{-i(p_1 x_1 - p_2 x_2)} \quad . \tag{26}$$

The full Feynman propagator in momentum space obeys an integral equation which is visualized in Fig. 4. A double line denotes the exact propagator S_F and a single line the free propagator S_0. The This decomposition may be inserted into Eq. (23) to obtain the expansion of the self–energy graph (Fig. 5).

The energy shift of a $1s_{1/2}$ electron due to the quantum–electrodynamical self–energy correction finally can be expressed in a form amenable to direct numerical evaluation

$$\Delta E = \Delta E_0'(Z' = Z) - \Delta E_0'(Z' = 0) + i\pi R_0 + \Delta E^{(2)} + \Delta E_c \quad . \tag{27}$$

Here the residuum term is given by

$$i\pi R_0 = -\alpha \int_0^\infty dr \int_0^r dx \left\{ \frac{2}{3} \frac{x}{r^2} Q(r)Q(x) - \frac{1}{r} P(r)P(x) \right\} \tag{28}$$

with

$$Q(x) = 2 G_{1s}(x)\, F_{1s}(x) \,, \quad P(x) = G_{1s}^2(x) + F_{1s}^2(x) \quad . \tag{29}$$

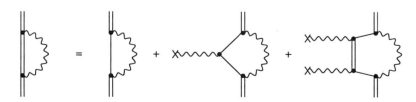

Figure 5. Expansion of the self energy by inserting the expansion of the electron propagator (Fig. 4)

$\Delta E^{(2)}$ is

$$\Delta E^{(2)} = -\frac{5\alpha}{4\pi}\langle V(x)\rangle_{1s} - \frac{2\alpha}{\pi^2}\int_0^\infty Q^-(p)\,\frac{\xi\ln\xi}{\xi-1}\,p^2\,dp$$

$$+\frac{\alpha}{\pi^2}\int_0^\infty G_{1s}(p)\,F_{1s}(p)\,Z(\xi)\,p^2\,dp + \frac{\alpha E_{1s}}{2\pi^2}\int_0^\infty Q^+(p)\,Z(\xi)\,p^2\,dp \qquad (30)$$

with $\xi = p^2 - E_{1s}^2 + 1$ and

$$Z(\xi) = \frac{\xi}{\xi-1}\left(1 + \frac{\xi-2}{\xi-1}\ln\xi\right) \quad . \qquad (31)$$

$G_{1s}(p)$ and $F_{1s}(p)$ denote Bessel transforms of the radial component of the Dirac wave function,

$$G_{1s}(p) = \int_0^\infty G_{1s}(x)\,j_0(px)\,x\,dx \ , \quad F_{1s}(p) = \int_0^\infty F_{1s}(x)\,j_1(px)\,x\,dx \ , \qquad (32)$$

and

$$Q^\pm(p) = G_{1s}^2(p) \pm F_{1s}^2(p) \quad . \qquad (33)$$

$\langle V(x)\rangle_{1s}$ is the expectation value of the potential energy and E_{1s} the energy eigenvalue of the K–shell electron. The counter term ΔE_c is determined by

$$\Delta E_c = -\frac{\alpha}{2\pi}\langle V(x)\rangle_{1s}\int_0^\infty \frac{d\omega}{(\omega^2+1)^{1/2}} \quad . \qquad (34)$$

The contribution of the main term $\Delta E_0'(Z')$ for a given nuclear charge Z' can be written as

$$\Delta E_0' = -\frac{2\alpha}{\pi}\int_0^\infty \omega\,d\omega \int_0^\infty dr \int_0^r dx \sum_{\kappa=\pm 1}^{\pm\infty}$$

$$\mathrm{Re}\left\{\frac{|\kappa|}{\Delta_\kappa(E_{1s}-i\omega)}\left[\frac{(\kappa-1)^2}{\bar{l}(\bar{l}+1)}\,B_{\bar{l}}Q^{\infty+}(r)Q^{0+}(x)\right.\right.$$

$$+\frac{l}{2l+1}\,B_{l-1}\left(Q^{\infty-}(r) + \frac{\kappa+1}{l}\,Q^{\infty+}(r)\right)\left(Q^{0-}(x) + \frac{\kappa+1}{l}\,Q^{0+}(x)\right)$$

$$+\frac{l+1}{2l+1}\,B_{l+1}\left(Q^{\infty-}(r) - \frac{\kappa+1}{l+1}\,Q^{\infty+}(r)\right)\left(Q^{0-}(x) - \frac{\kappa+1}{l+1}\,Q^{0+}(x)\right)$$

$$\left.\left.-B_l\,P^\infty(r)\,P^0(x)\right]\right\} \quad , \qquad (35)$$

with l being the orbital angular momentum related to κ and $\bar{l}$ related to $-\kappa$, respectively. Here we have used the abbreviations

$$Q^{\infty,0\pm}(x) = G_{1s}(x,E_{1s})\,F_\kappa^{\infty,0}(x,E_{1s}-i\omega) \pm F_{1s}(x,E_{1s})\,G_\kappa^{\infty,0}(x,E_{1s}-i\omega) \ ,$$

$$P^{\infty,0}(x) = G_{1s}(x,E_{1s})\,G_\kappa^{\infty,0}(x,E_{1s}-i\omega) + F_{1s}(x,E_{1s})\,F_\kappa^{\infty,0}(x,E_{1s}-i\omega) \ ,$$

$$B_l = h_l^{(1)}(i\omega r)\,j_l(i\omega x) \quad . \qquad (36)$$

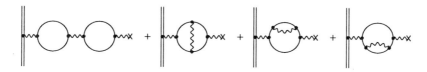

Figure 6. Diagrams contributing in the order $\alpha^2(Z\alpha)$ to the vacuum polarization

$\Delta_\kappa(E)$ denotes the Wronskian for a given complex energy E and angular momentum quantum number κ:

$$\Delta_\kappa(E) = F_\kappa^\infty(x, E)\, G_\kappa^0(x, E) - F_\kappa^0(x, E)\, G_\kappa^\infty(x, E) \quad . \tag{37}$$

$F_\kappa^{0/\infty}$ and $G_\kappa^{0/\infty}$ are solutions of the radial Dirac equation for complex energies which are regular either at the origin ($x = 0$) or at infinity ($x = \infty$). In (36) j_l and $h_l^{(1)}$ are the spherical Bessel function and Hankel function of first kind for purely imaginary arguments.

HIGHER-ORDER CONTRIBUTIONS AND HYPERFINE–SPLITTING

Theoretical values for the Lamb–shift of hydrogen–like high-Z atoms do not only require a precise determination of the vacuum polarization and the self–energy but additionally higher order effects have to be taken into account. Källén and Sabry [15] first investigated the correction of order $\alpha^2(Z\alpha)$ to the vacuum polarization. The corresponding Feynman diagrams are the two–loop diagram and three diagrams with an additional photon line within a single electron-positron loop (Fig. 6). This part of the Lamb–shift is generated by the vacuum polarization potential of order $\alpha^2(Z\alpha)$ [16, 17, 18]. All quantum electrodynamical (QED) corrections to energy eigenvalues in atoms are evidently affected by the nuclear charge distribution.

Although the Källén-Sabry contribution to the Lamb–shift is rather small, it is of more than academic interest to consider the finite size effect to this energy correction. These calculations are motivated by the possible improvement of experimental precision up to about 0.1 eV in future atomic energy measurements. Furthermore, hydrogen–like high-Z atoms are now subject of extended spectroscopic studies, e.g. at GSI in Darmstadt.

In a similar manner to the derivation of the Uehling–potential from the vacuum polarization amplitude, the higher order vacuum polarization potential can be explicitly calculated [17]. For a spherical symmetric extended charge distribution $\rho(r)$ the expression reads

$$V_{21}(r) = -(Z\alpha)\frac{\alpha^2}{\pi r}\int\limits_0^\infty dr'\, r'\rho(r')\left[L_0(2|r - r'|) - L_0(2|r + r'|)\right] , \tag{38}$$

with

$$L_0(x) = \int\limits^{-x} dx\, L_1(x) \tag{39}$$

and

$$L_1(x) = \int\limits_1^\infty dt\, e^{-xt} \left\{ \left(-\frac{13}{54t^2} - \frac{7}{108t^4} - \frac{2}{9t^6} \right) \sqrt{t^2 - 1} \right.$$
$$+ \left(\frac{44}{9t} - \frac{2}{3t^3} - \frac{5}{4t^5} - \frac{2}{9t^7} \right) \ln\left[t + \sqrt{t^2 - 1} \right]$$
$$+ \left(-\frac{4}{3t^2} - \frac{2}{3t^4} \right) \sqrt{t^2 - 1} \ln\left[8t(t^2 - 1) \right] + \left(\frac{8}{3t} - \frac{2}{3t^5} \right)$$
$$\left. \times \int\limits_t^\infty dy \left(\frac{3y^2 - 1}{y(y^2 - 1)} \ln\left[y + \sqrt{y^2 - 1} \right] - \frac{1}{\sqrt{y^2 - 1}} \ln\left[8y(y^2 - 1) \right] \right) \right\}.$$

$$(40)$$

For large distances $(r \to \infty)$ or for a point nucleus V_{21} reduces to

$$V_{21}^{\text{pt}}(r) = -\frac{(Z\alpha)\alpha^2}{\pi^2 r} L_1(2r). \tag{41}$$

Thus, the potential is represented by a three-dimensional integral which has to be computed numerically. Figure 7 depicts $g_2(r) = (Z\alpha)^{-1} V_{21}(r)$ for a point nucleus (full line) and a finite-size nucleus (dashed line) for small values of the radial coordinate r. In contrast to the divergent point nucleus potential at the origin, the finite size potential reduces to a constant value at radial values r corresponding to the radius of the uniform sphere.

The energy shift is then calculated in first-order perturbation theory assuming the nucleus as a homogeneously charged sphere. The difference of the point nucleus energy shift $E(\text{p.n.})$ and the energy shift of an extended nucleus $E(\text{f.s.})$ is plotted in Fig. 8 for nuclear charge numbers ranging from 10 to 100 and several electron states. The result is normalized to the point nucleus energy shift. The finite nuclear size effect on the Källén-Sabry energy shift increases with the nuclear charge number. But it is only of definite interest for high-Z atoms, where this contribution reaches the order of 0.1 eV [23].

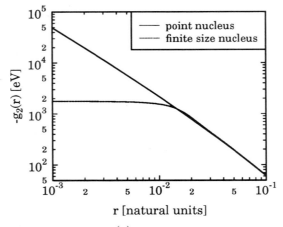

Figure 7. The potential $g_2(r)$ displayed versus the radial distance r

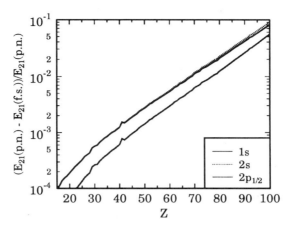

Figure 8. The difference of the energy shift for a point nucleus and a finite size nucleus, respectively, caused by the Källén–Sabry potential, normalized to the point nucleus energy shift. $(E(p.n.) - E(f.s.))/E(p.n.)$ is shown as function of the nuclear charge number Z. The full, dotted and dashed line refer to the $1s$, $2s$ and $2p_{1/2}$ electron state, respectively. The apparent peak–like structures can be traced back of the chosen radii of the nuclei and do not occur if the radii are assumed to be $\langle r^2 \rangle^{1/2} = 1.2 \cdot A^{1/3}$ fm

LIMITATIONS OF QED TESTS

Not only the quantum electrodynamical corrections contribute to the Lamb shift. It is also effected by the unknown structure of the nucleus. The finite size correction to the Coulomb energy is a large contribution to this shift, but formally to be distinguished from the radiative corrections. A more limiting uncertainty for a precision test of QED in a strong magnetic field is assigned to the extended nuclear magnetization distribution (Bohr–Weisskopf effect) [19]. This means that a real nucleus could not be approximated by a magnetic dipole since finite nuclear magnetization effects enter in the first–order calculation as well as modifications due to the extended nuclear charge distributions.

To compute the energy splitting of the electron levels in $^{209}_{83}\text{Bi}^{82+}$ caused by the Bohr–Weisskopf effect we utilized the more realistic probability density of the odd proton of the Bi–nucleus as magnetization distribution [20] $w(R) = |\Psi(R)|^2$. This is supposed to provide a good approximation, since the odd proton is separated from the Pb core of the Bi nucleus by an energy gap. Therefore virtuell exitations might not play an important role for the distribution function. The probability density is calculated within a mean–field theory model [21, 22]. We evaluate the effect of a nuclear magnetization distribution according to the prescription of Bohr and Weisskopf [19]. The decrease of the energy splitting is determined by a parameter ε

$$\Delta E_{\text{finite}} = \Delta E_{\text{point}}(1 - \varepsilon) \quad , \tag{42}$$

where ΔE_{point} denotes the energy splitting due to a magnetic point dipole and ΔE_{finite} signifies the energy splitting for the extended magnetization distribution. The parameter ε can be separated into a spin part ε_S and into an orbital angular momentum part ε_L

$$\varepsilon = \varepsilon_S + \varepsilon_L \quad . \tag{43}$$

29

The total magnetization correction amounts then to about [23] $\Delta\lambda^{BW} = 3.5\,\mathrm{nm}$. Taking into account configuration mixing contributions, this approximation should be accompanied with an error of about 30%. More suptle calculations yield a Bohr–Weisskopf correction of about [24] $\Delta\lambda^{BW} \simeq 4.3\,\mathrm{nm}$. Including all finite size effects, the final first order perturbation theory result reads $\Delta\lambda^{1.O.} \simeq 243.1\,\mathrm{nm}$.

The first order radiative corrections for a bound electron are given by the self energy and the vacuum polarization. Apparently, these QED corrections are also affected by the extended nuclear charge, but it was reported that finite size modifications to the first order radiative corrections [25, 14] could be handled with extreme accuracy. Uncalculated higher order QED effects are estimated to sum up to a total contribution in the order of about 1 eV.

Nuclear polarization effects should be of the same magnitude. Hence, the limitation of the testability of QED effects is given by the accuracy one is able to reach in the determination of nuclear polarization.

Recently, calculations of the nuclear polarization effects were presented by Plunien et al. [26, 27, 28], utilizing the concept of effective photon propagators with nuclear polarization insertions. Taking into account virtual excitations of collective rotational and vibrational modes as well as the dominant giant dipole resonance, energy shifts are computed for $1s_{1/2}$, $2s_{1/2}$ and $2p_{1/2}$ states for various even–A nuclei. Corresponding values are presented in Table 1. The absolute numbers, however, should be envisaged as an estimate, mainly for two reasons: firstly, only a finite number of nuclear excitations is taken into account utilizing experimental parameters (excitation energies and $B(EL)$–values), which are affected with inherent uncertainties. Secondly, only the contribution due to the giant dipole resonance is estimated. Higher order multipole resonances should be included. Accordingly, a typical error of about 25% is assigned.

In the case of the $1s_{1/2}$–state the obtained shifts are considerable corrections, while for the $2s - 2p$ splitting the nuclear polarization modifications are just at the limit of present high–precision experiments. Nevertheless, these values represent the status of precision one may reach to test strong field QED effects within the framework of Lamb–shift measurements. In other words, a test of QED effects below the 0.1 eV level is unaccessible with our todays knowledge about the nuclear structure.

Table 1. Total energy shifts (meV) of the $1s_{1/2}-$, $2s_{1/2}-$ and the $2p_{1/2}$–state and the contribution to the $2s_{1/2} - 2p_{1/2}$ Lamb shift due to the implemented nuclear excitations are shown.

isotope	$|\Delta E_{1s_{1/2}}|$ (meV)	$|\Delta E_{2s_{1/2}}|$ (meV)	$|\Delta E_{2p_{1/2}}|$ (meV)	$|\Delta E_{2s-2p}|$ (meV)
$^{230}_{90}$Th	738.7	135.1	14.6	120.1
$^{232}_{90}$Th	790.7	145.9	15.8	130.1
$^{234}_{92}$U	1043.0	195.9	22.6	173.3
$^{236}_{92}$U	1087.9	204.3	23.5	180.8
$^{238}_{92}$U	1128.4	212.0	24.5	187.5
$^{238}_{94}$Pu	1360.1	260.3	32.1	228.2
$^{240}_{94}$Pu	1397.3	267.4	32.9	234.5
$^{242}_{94}$Pu	1418.9	271.5	33.8	237.7
$^{244}_{94}$Pu	1459.9	279.6	34.2	245.4
$^{246}_{96}$Cm	1808.1	352.8	46.4	306.4
$^{248}_{96}$Cm	1836.2	350.3	46.0	304.3
$^{250}_{98}$Cf	3634.8	824.3	201.5	622.8
$^{252}_{98}$Cf	2187.0	435.1	61.0	374.1

RADIATIVE CORRECTIONS IN CRITICAL FIELDS

We also investigated field–theoretical corrections, such as vacuum polarization and self energy to study their influence on strongly bound electrons in superheavy atoms. In critical fields ($Z \approx 170$) for spontaneous $e^+ e^-$ pair creation the coupling constant of the external field $Z\alpha$ exceeds 1 thereby preventing the ordinary perturbative approach of quantum electrodynamical corrections which employs an expansion in $Z\alpha$. For heavy and superheavy elements radiative corrections have to be treated to all orders in $Z\alpha$.

The K–electron binding energy E_{1s}^b increases strongly as a function of the nuclear charge Z. For $Z = 150$, E_{1s}^b amounts to about the electron reast mass and hence one enters the truly relativistic domain. For $Z > 170$ the binding energy exceeds twice the electron rest mass and the K–shell electron gets imbedded as a resonance in the negative energy continuum, which opens the possibility of spontaneous positron production.

Our investigations have been motivated by the question whether these QED corrections could strongly modify or even prevent the diving of the electron $1s$–state into the negative energy contiuum of the Dirac equation and thus spontaneous positron emission.

The influence of the attractive Uehling potential on electronic binding energies for $Z > 100$ has been calculated already by Werner and Wheeler [29]. For the critical nuclear charge number Z_{cr} at which $E_{1s} = -mc^2$ the Uehling potential leads to an $1s$– energy shift of $\Delta E_{vp}^{(n=1)} = -11.8$ keV [30], which decreases Z_{cr} by one third of a unit. First evaluations of the Wichmann–Kroll contribution were presented by Gyulassy [31, 32, 4] and by Rinker and Wilets [33, 34, 35].

We have examined the vacuum polarization in the field of a high–Z finite size nucleus. The polarization charge density in coordinate space of order $\alpha(Z\alpha)^n$ with $n \geq 3$ is calculated [5, 36] for $Z = 170$. The computed vacuum polarization charge density for the almost critical system $Z = 170$ is depicted in Fig. 9. The nuclear radius R is indicated. The various contributions for the angular momentum components $|\kappa| = 1 - 5$ are shown again separately.

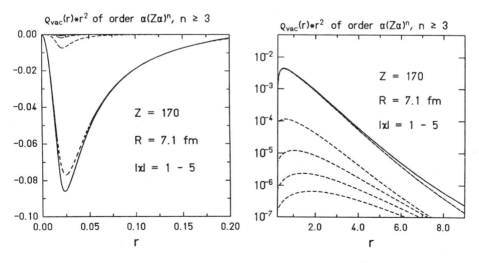

Figure 9. Same as in Figure 3 assuming a nuclear charge $Z = 170$

Again we also observe a rapid convergence in the κ–summation. For $Z = 170$ the binding energy of the strongest bound electron state amounts to $E_{1s} = -1020.895$ keV. The effect of the higher–order vacuum polarization on a K–shell electron in the superheavy system $Z = 170$ results in $\Delta E_{1s} \approx 1.46$ keV [5, 36], which is completely negligible.

Furthermore we estimated the influence of the vacuum polarization potential of order $\alpha^2(Z\alpha)$ [18], i. e., the Källén–Sabry contribution of the strongest bound electron states in superheavy atoms. To simulate nuclear–size corrections we computed the vacuum polarization potential at about the nuclear radius R and employed this constant value also inside the nucleus. For the nuclear charge distribution we assumed a spherical shell. For the almost critical system $Z = 170$ with $R = 7.1$ fm, in which the $1s_{1/2}$–state almost reaches the negative energy continuum, we calculated $\Delta E(1s_{1/2}) = -88.9$ eV. This small value can be completely omitted compared with the huge binding energy of $E_{1s}^b = -1020.895$ keV.

Our most important result was the self–energy shift for $1s$–electrons in the superheavy atom with the critical nuclear charge number $Z = 170$. Here the nuclear radius was adjusted so that the K–electron energy eigenvalue differed only by 10^{-3} eV from

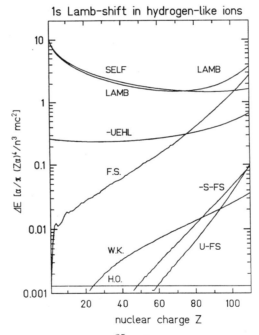

Figure 10. Contributions to the Lamb shift [25] of $1s_{1/2}$ electrons in hydrogen–like atoms versus the nuclear charge number Z. The energy shift ΔE is presented in units of $(\alpha/\pi)(Z\alpha)^4/n^3 mc^2$. LAMB indicates the sum of all contributions. The dominant term (SELF) is provided by the point–nucleus self–energy shift. UEHL denotes the level shift caused by the Uehling potential for point–like nuclei. The energy correction F.S. results from the finite size of the nucleus. The slight irregularities reflect the noncontinuous dependence of the nuclear radius R on the charge number Z. The finite nuclear size correction to the self energy and to the Uehling potential lead to energy shifts S–FS and U–FS, respectively. W.K. denotes the Wichmann–Kroll term and H.O. signifies higher–order corrections incorporating the exchange of two photons. Most of the contributions as well as the total Lamb shift are repulsive. Attractive contributions are indicated by a minus sign

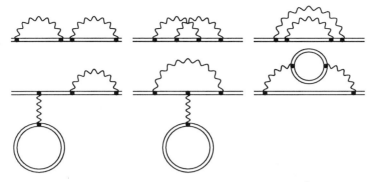

Figure 11. The self-energy diagrams of order α^2

the borderline of the negative energy continuum. Our numerical calculations [37] for $Z = 170$ yielded $\Delta E_{\rm se} = 10.989$ keV, which still represents only a 1% correction to the total K–electron binding energy. The sum of all radiative corrections thus almost cancels completely at the continuum boundaries.

We conclude that radiative corrections such as vacuum polarization or self energy may not prevent the K–shell binding energy from exceeding $2mc^2$ in superheavy systems with $Z > Z_{\rm cr} \approx 170$.

SUMMARY

In these lectures we wanted to point out that a critical test of quantum electrodynamics in strong fields can be performed by precise measurements of the Lamb shift in highly charged hydrogen-like atoms. We introduced various contributions that have been calculated up to now.

Fig. 10 displays a comparison of the contributions to the $1s_{1/2}$ Lamb shift [25]. This figure illustrates the well-known fact that the point-nucleus self energy and the Uehling potential yield the dominant contributions to the Lamb shift for low and intermediate values of Z. The finite size corrections to the Uehling energy shift (U-FS) and to the self energy (S-FS) become as important as the self energy for high nuclear charges.

A possible improvement of the calculations can be obtained by taking into account the higher-order self-energy corrections. The contributing diagrams up to order α^2 are depicted in Fig. 11.

As discussed in the last two sections, the precision tests of quantum electrodynamics suffer from the uncertainties that arise from the nuclear degrees of freedom, such as nuclear radii and nuclear excitations.

REFERENCES

1. M. E. Rose, *Relativistic Electron Theory*, (Wiley, New York, 1961)

2. Y. N. Kim, *Mesic Atoms and Nuclear Structure*, (North-Holland, Amsterdam, 1971)

3. S. S. Schweber, *An Introduction to Relativistic Quantum Field Theory*, (Harper & Row, New York, 1961)

4. M. Gyulassy, Nucl. Phys. A244, 497 (1975)

5. G. Soff, P. Mohr, Phys. Rev. A 38, 5066 (1988)

6. P. J. Mohr, Ann. Phys. (NY) 88, 26 (1974)

7. P. J. Mohr, Ann. Phys. (NY) 88, 52 (1974)

8. E. H. Wichmann, N. M. Kroll, Phys. Rev. 101, 843 (1956)

9. G. E. Brown, G. W. Schaefer, Proc. Roy. Soc. London Ser. A233, 527 (1956)

10. G. E. Brown, J. S. Langer, G. W. Schaefer, Proc. Roy. Soc. London Ser. A251, 92 (1959)

11. G. E. Brown, D. F. Mayers, Proc. Roy. Soc. London Ser. A251, 105 (1959)

12. A. M. Desiderio, W. R. Johnson, Phys. Rev. A3, 1267 (1971)

13. K. T. Cheng, W. R. Johnson, Phys. Rev. A14, 1943 (1976)

14. P. J. Mohr, G. Soff, Phys. Rev. Lett. 70, 158 (1993)

15. G. Källén, A. Sabry, Mat.-Fys. Meddr. Dansk. Vidensk. Selsk. 29, 17 (1955)

16. J. Blomqvist, Nucl. Phys. B 48, 95 (1972)

17. L. Wayne Fullerton, G. A. Rinker, Phys. Rev. A 13, 1283 (1976)

18. T. Beier, G. Soff, Z. Physik D8, 129 (1988)

19. A. Bohr, V. F. Weisskopf, Phys. Rev. 77, 94 (1977)

20. S. M. Schneider, G. Soff, W. Greiner, Phys. Rev. A 50, 118 (1994)

21. P.-G. Reinhard, M. Rufa, J. Maruhn, W. Greiner, J. Friedrich, Z. Phys. A 323, 13 (1986)

22. J. Schaffner, C. Greiner, H. Stöcker, Phys. Rev. C 46, 322 (1992)

23. S. M. Schneider, W. Greiner, G. Soff, J. Phys. B 26, L529 (1993)

24. M. Tomaselli, S. M. Schneider, E. Kankeleit, T. Kühl, to be published

25. W. R. Johnson, G. Soff, Atomic Data and Nuclear Data Tables 29, 453 (1985)

26. G. Plunien, B. Müller, W. Greiner, G. Soff, Phys. Rev. A 39, 5428 (1989)

27. G. Plunien, B. Müller, W. Greiner, G. Soff, Phys. Rev. A 43, 5853 (1991)

28. G. Plunien, G. Soff, to be published

29. F. G. Werner, J. A. Wheeler, Phys. Rev. 109, 126 (1958)

30. G. Soff, B. Müller, J. Rafelski, Z. Naturforsch. 29a, 1267 (1974)

31. M. Gyulassy, Phys. Rev. Lett. 33, 921 (1974)

32. M. Gyulassy, Phys. Rev. Lett. 32, 1393 (1974)

33. G. A. Rinker, L. Wilets, Phys. Rev. Lett. 31, 1559 (1973)

34. L. Wilets, G. A. Rinker, Phys. Rev. Lett. 34, 339 (1975)

35. E. Borie, G. A. Rinker, Rev. Mod. Phys. 54, 67 (1982)

36. G. Soff, Z. Physik D11, 29 (1989)

37. G. Soff, P. Schlüter, B. Müller, W. Greiner, Phys. Rev. Lett. 48, 1465 (1982)

GAUGE INVARIANCE ON BOUND STATE ENERGY LEVELS

Antonio Vairo

I.N.F.N. - *Sez. di Bologna and Dip. di Fisica,*
Università di Bologna, Via Irnerio 46, I-40126 Bologna, Italy

1   Introduction

In this paper I review some basics concerning the gauge invariance on bound state [1].

In the first section some simple examples about gauge invariance on mass-shell are given. In the second section there are some general considerations about gauge invariance in a off shell problem and up to order α^4 the calculation of some contributions to the energy levels of positronium in Feynman gauge. The full cancellation of the spurious $\alpha^3 \log \alpha$, α^3 terms which arise typically in this gauge, is performed. The calculations are done in the Barbieri-Remiddi bound state formalism [2], [3]. K_c, ψ_c are the Barbieri-Remiddi zeroth-order kernel and the corresponding wave function.

2   Gauge invariance on mass-shell

It is generally easy to verify the gauge invariance in a scattering process. The incoming and outcoming particles are on mass-shell and the related wave functions are not α-depending. This involves that each Feynman graph contributes to the scattering amplitude only at the order in the fine structure constant determined by his number of vertices. In the following I will verify the gauge invariance at the leading order in α in two very simply cases: the Compton and the $e^+ e^-$ scattering.

At the order α only the two graphs of Fig. 1 contribute to the Compton scattering amplitude.

The contribution to the scattering amplitude coming from the graphs of Fig. 1 is:

$$A_{Com}(p_1, k_1; p_2, k_2) = -(2\pi)^4 \delta^{(4)}(p_1 + k_1 - p_2 - k_2)\, \alpha \qquad (2.1)$$

$$\cdot \left(\bar{u}(p_2)\not{\epsilon}(k_2) S_F(p_1 + k_1)\not{\epsilon}(k_1) u(p_1) \right.$$

$$+ \left. \bar{u}(p_2)\not{\epsilon}(k_1) S_F(p_1 - k_2)\not{\epsilon}(k_2) u(p_1) \right) ,$$

Electron Theory and Quantum Electrodynamics: 100 Years Later
Edited by Dowling, Plenum Press, New York, 1997

Figure 1. Compton scattering at the tree level.

where k_1, p_1 and k_2, p_2 are the incoming and outcoming momenta (on mass shell: $p_j^2 = m^2$ for the electron momenta and $k_j^2 = 0$ for the photon momenta), ϵ^μ is the photon polarization, S_F the fermion free propagator:

$$S_F(p) = \frac{i}{\not{p} - m + i\epsilon} \; , \tag{2.2}$$

and u, $\bar{u}$, v, $\bar{v}$ are the Dirac spinors $((\not{p} - m)u(p) = 0$, $(\not{p} + m)v(p) = 0)$. The gauge variation (e.g. on the second external photon) of A_{Com} i.e. $A_{Com}^{Gauge\#1} - A_{Com}^{Gauge\#2}$ is:

$$
\begin{aligned}
\delta A_{Com}(p_1, k_1; p_2, k_2) &= -(2\pi)^4 \delta^{(4)}(p_1 + k_1 - p_2 - k_2)\, \alpha \\
&\quad \cdot\; \bar{u}(p_2)\big\{ \not{k}_2 S_F(p_2 + k_2)\not{\epsilon}(k_1) \\
&\quad + \not{\epsilon}(k_1) S_F(p_1 - k_2)\not{k}_2 \big\} u(p_1) \\
&= -(2\pi)^4 \delta^{(4)}(p_1 + k_1 - p_2 - k_2)\, \alpha \\
&\quad \cdot\; \bar{u}(p_2)\big\{ (i - (\not{p}_2 - m)S_F(p_2 + k_2))\,\not{\epsilon}(k_1) \\
&\quad + \not{\epsilon}(k_1)(-i - S_F(p_1 - k_2)(\not{p}_1 - m))\big\} u(p_1) \\
&= -(2\pi)^4 \delta^{(4)}(p_1 + k_1 - p_2 - k_2)\, \alpha \\
&\quad \cdot\; \bar{u}(p_2)\big\{ i\not{\epsilon}(k_1) - i\not{\epsilon}(k_1) \big\} u(p_1) \;=\; 0 \; ,
\end{aligned}
\tag{2.3}
$$

which verifies the gauge invariance.

The second example is the $e^+ e^-$ scattering amplitude. At the order α contributions to the scattering amplitude arise only from the graphs of Fig. 2. The photon propagator $D_{\mu\nu}$ is gauge dependent; in Feynman and Coulomb gauge one can write it as:

$$D_{\mu\nu}^{Fey}(p) = -i\frac{g_{\mu\nu}}{p^2 + i\epsilon} \; ; \tag{2.4}$$

$$D_{\mu\nu}^{Cou}(p) = -i\frac{1}{p^2 + i\epsilon}\left\{ g_{\mu\nu} - \frac{\eta \cdot p(p_\mu \eta_\nu + p_\nu \eta_\mu) - p_\mu p_\nu}{\vec{p}^{\,2}} \right\}$$

$$\eta^\mu = (1, \vec{0}\,) \; . \tag{2.5}$$

The difference between the propagators in the above gauges can be expressed as:

$$D_{\mu\nu}^{Cou}(p) - D_{\mu\nu}^{Fey}(p) = b_\mu(p)p_\nu + b_\nu(p)p_\mu \; ,$$

$$b_\mu(p) \equiv \frac{i}{2(p^2 + i\epsilon)}\frac{1}{\vec{p}^{\,2}}(-p_\mu + 2\delta_{0\mu}p_0) \; . \tag{2.6}$$

Not only the sum of the graphs in Fig. 2 is gauge invariant but each graph also. To verify the gauge invariance of the annihilation graph one needs only equation (2.6) and

Figure 2. $e^+ e^-$ scattering at the tree level.

Figure 3. One-loop vertex corrections to the annihilation graph.

the Dirac equation for the spinors u and v:

$$
\begin{aligned}
\delta A_{ann}(p_1, k_1; p_2, k_2) &= -(2\pi)^4 \delta^{(4)}(p_1 + k_1 - p_2 - k_2)\,\alpha \qquad\qquad (2.7)\\
&\quad \cdot\ \bar{v}(k_1)\gamma^\mu u(p_1)\big\{ D^{Cou}_{\mu\nu}(p_1 + k_1)\\
&\quad -\ D^{Fey}_{\mu\nu}(p_1 + k_1)\big\}\bar{u}(p_2)\gamma^\nu v(k_2)\\
&= -(2\pi)^4 \delta^{(4)}(p_1 + k_1 - p_2 - k_2)\,\alpha\\
&\quad \cdot\ \bar{v}(k_1)\gamma^\mu u(p_1)\big\{ b_\mu(p_1 + k_1)\,(p_{1\nu} + k_{1\nu})\\
&\quad +\ b_\nu(p_1 + k_1)\,(p_{1\mu} + k_{1\mu})\big\}\bar{u}(p_2)\gamma^\nu v(k_2)\\
&= -(2\pi)^4 \delta^{(4)}(p_1 + k_1 - p_2 - k_2)\,\alpha\\
&\quad \cdot\ \big(\bar{v}(k_1)\slashed{b}(p_1 + k_1)\,u(p_1)\bar{u}(p_2)(\slashed{p}_2 + \slashed{k}_2)v(k_2)\\
&\quad +\ \bar{v}(k_1)(\slashed{p}_1 + \slashed{k}_1)u(p_1)\bar{u}(p_2)\slashed{b}(p_1 + k_1)\,v(k_2)\big)\\
&= 0\ .
\end{aligned}
$$

In the same way it is possible to prove the gauge invariance of the second graph of Fig. 2.

3 Gauge invariance on bound state

The bound state wave function is α-dependent. This is well-known for the Schrödinger-Coulomb wave function, but is also true for the Barbieri-Remiddi solution which reproduces it in the non relativistic limit. As a consequence of the non trivial α-dependence each Feynman graph contributes to the energy levels perturbative expansion with a series in α. Also the leading order of this series is not deductable in a trivial manner from a vertices counting. As a consequence in a relativistic bound state problem it is not possible to verify in a simply way, order by order in α, the gauge invariance.

Although the difficulties to reconstruct sets of gauge invariant contributions, since some graphs give series which converge faster in α in a particular gauge than in an other, different gauges are used normally in the energy levels calculation. Typically binding photons (photons connecting fermion lines each other) are calculated in Coulomb gauge while annihilation and radiative ones in covariant gauge as the Feynman gauge. This not only in different graphs but also, where considered possible, in the same Feynman graph.

Such an approach is not free from ambiguities.

In Fig.3 there are drawn the vertex corrections at the one-photon annihilation graph. After regularization in order to obtain the UV-divergences cancellation it is necessary to consider together these three graphs, this means to use the same gauge for all the photons (see [4]). If we use a covariant gauge for the radiative photons in the first two graphs the third one has to be also calculated in a covariant gauge. It follows that there are in any case some binding photons (as the no-annihilation photon of the third graph) which will be calculated in a covariant gauge.

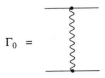

$$\Gamma_0 =$$

Figure 4. One-photon exchange graph.

In order to avoid these ambiguities and taking in account that the apparent simplicity of the Coulomb gauge seems to disappear in higher order calculations, a reference calculation of the energy levels has to be done in the Feynman gauge (or in an other covariant gauge). In the following I will show how the individuation of gauge invariant set of contributions can be helpful to cancel the low-order spurious terms arising in such an *ab initio* reference calculation.

The first contribution to the energy levels shift is coming from the one-photon exchange graph of Fig.4. On the positronium $1S$ state this graph contributes, in the Feynman gauge, as $(m = 1)$:

$$\langle \Gamma_0 \rangle_{1S}^{Fey} = -\frac{1}{4}\alpha^2 - \frac{1}{2\pi}\alpha^3 \log \alpha + \frac{1}{4\pi}\alpha^3 - \frac{3}{16}\alpha^4 , \qquad (3.1)$$

while in Coulomb gauge:

$$\langle \Gamma_0 \rangle_{1S}^{Cou} = -\frac{1}{4}\alpha^2 - \frac{3}{16}\alpha^4 . \qquad (3.2)$$

As expected, and unlike the $e^+ e^-$ scattering, $\langle \Gamma_0 \rangle$ is not gauge invariant: in Feynman gauge there are some contributions (the $\alpha^3 \log \alpha$ and α^3 terms) which are not in (3.2) and in the energy levels itself (it is well-known that the first correction to the Balmer's levels is of order α^4). To restore the gauge invariance there must exist some other contributions in the energy levels expansion which cancel in Feynman gauge these spurious terms. My principal purpose is from now to determine these other contributions. I preliminarily give the following general result.

If the zeroth-order kernel K_c is local,

$$K_c(\vec{p}, \vec{q}\,) = K_c(\vec{p} - \vec{q}\,) , \qquad (3.3)$$

then:

$$\delta \langle \Gamma_0 + 2 \sum_{n=1}^{\infty} I_n \rangle = 0 . \qquad (3.4)$$

In other words the sum of the contributions to the energy levels coming from Γ_0 and from the graphs of Fig.5 ($\times$ 2) is gauge invariant.

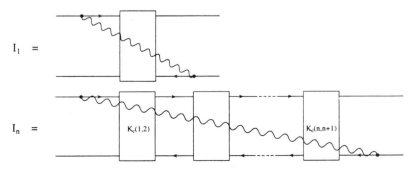

$$I_1 =$$

$$I_n = \qquad K_c(1,2) \qquad \qquad K_c(n,n+1)$$

Figure 5. n kernel K_c crossed by one photon.

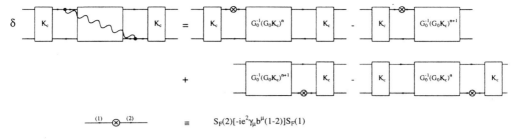

$$\frac{\quad}{\overset{(1)}{} \otimes \overset{(2)}{}} \quad \equiv \quad S_F(2)[-ie^2\gamma_\mu b^\mu(1\text{-}2)]S_F(1)$$

Figure 6. $\delta(\,K_c G_0 I_n G_0 K_c) = A_n G_0 K_c - A_{n+1} + B_{n+1} - K_c G_0 B_n\;$.

In order to prove (3.4) first I give the gauge variation of $\langle I_n\rangle$:

$$
\begin{aligned}
\delta\langle I_n\rangle &= \delta\langle K_c G_0 I_n G_0 K_c\rangle \\
&= \langle A_n\rangle - \langle A_{n+1}\rangle + \langle B_{n+1}\rangle - \langle B_n\rangle\;, \quad (3.5)
\end{aligned}
$$

where the first identity is a consequence of the Bethe-Salpeter equation for the zeroth-order wave function ($\psi^c = G_0 K_c \psi^c$, G_0 is the two-fermion free propagator), and the second one is graphically represented in Fig.6. Fig.6 also gives the definitions of A_n and B_n.
From (3.5) it follows:

$$\delta\sum_{n=1}^{\infty}\langle I_n\rangle = \langle A_1\rangle - \langle B_1\rangle\;, \quad (3.6)$$

and:

$$\delta\langle\Gamma_0\rangle = \langle A_0\rangle - \langle A_1\rangle + \langle B_1\rangle - \langle B_0\rangle\;. \quad (3.7)$$

It is easy to verify that:

$$\langle A_0\rangle = \langle B_1\rangle\;, \quad (3.8)$$

$$\langle B_0\rangle = \langle A_1\rangle\;, \quad (3.9)$$

then from (3.6), (3.7) one obtains (3.4).
    Similar to the artificial graphs of Fig.5 are the graphs of Fig.7 which effectively contribute to the kernel K of the Bethe-Salpeter equation.
Since,

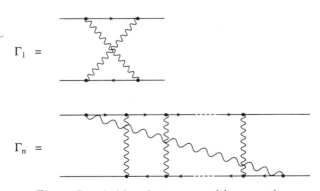

Figure 7. n ladder photon crossed by one other.

41

$$\Gamma_0 G_0 \Gamma_0 =$$

Figure 8. $\Gamma_0 G_0 \Gamma_0$ graph.

$$\langle \Gamma_1 \rangle^{Fey} = 2 \langle I_1 \rangle^{Fey} + \langle T \rangle^{Fey} + O(\alpha^4) , \qquad (3.10)$$

$$\langle \Gamma_n \rangle^{Fey} = \langle I_n \rangle^{Fey} + O(\alpha^4) \qquad n > 1 , \qquad (3.11)$$

and up to order α^4 the Barbieri-Remiddi kernel is local, using the result (3.4), one expects that the spurious terms in (3.1) are cancelled by the ones coming from (3.10) and (3.11). In fact,

$$2 \langle I_1 \rangle_{1S}^{Fey} = \frac{1}{2\pi} \alpha^3 \log \alpha - \frac{5}{2\pi} \alpha^3 + \frac{4}{\pi} \log 2 \, \alpha^3 + O(\alpha^4) , \qquad (3.12)$$

$$\sum_{n=2}^{\infty} \langle I_n \rangle_{1S}^{Fey} = \frac{9}{8\pi} \alpha^3 - \frac{2}{\pi} \log 2 \, \alpha^3 + O(\alpha^4) . \qquad (3.13)$$

The sum of (3.12) and (3.13) ($\times 2$ taking in account the symmetric graphs) cancel completely the α^3 terms in (3.1).

Up to order α^3 the term $\langle T \rangle^{Fey}$, which comes from Γ_1 subtracting from each photon the zeroth-order kernel K_c, contributes also to $\langle \Gamma_1 \rangle^{Fey}$:

$$\langle T \rangle_{1S}^{Fey} = \frac{1}{2\pi} \log 2 \, \alpha^3 + O(\alpha^4) . \qquad (3.14)$$

To obtain the cancellation of (3.14), the graph of Fig.8, arising from the second order perturbations at the energy levels, must be considered; in fact,

$$\langle (\Gamma_0 - K_c) G_0 (\Gamma_0 - K_c) \rangle_{1S}^{Fey} = -\frac{1}{2\pi} \log 2 \, \alpha^3 + O(\alpha^4) . \qquad (3.15)$$

References

[1] S. Love, Ann. Phys. **113** (1978) 153; G. Feldman, T. Fulton and D. L. Heckathorn, Nucl. Phys. **B 167** (1980) 364 and **B 174** (1980) 89;

[2] R. Barbieri and E. Remiddi, Nucl. Phys. **B 141** (1978) 413;

[3] E. Remiddi in these Proceedings;

[4] W.Buchmüller and E.Remiddi, Nucl. Phys. **B 162** (1980) 250.

RENORMALIZATION OF THE SELF FIELD QED

Nuri Ünal

Akdeniz University,
Physics Department
P.K. 510, 07200
Antalya, Turkey

ABSTRACT

A new form of the regularization of self energy term is derived in the QED based on the self field formalism. In this new form of regularization final result is finite and renormalized.

INTRODUCTION

The electrodynamics is the interaction of charged particles with the radiation fields. In the classical approach to the problem the interaction of the charged particles with the radiation fields is considered nonperturbatively. In standard formulation of the quantumelectrodynamics there are seperately quantized electron and photon fields and their interaction can be added as a second step and perturbatively. In order to understand the theory better other approachs are proposed. The self field approach is one of them.[1]

The self field approach is similiar to the classical electrodynamics. In this approach the electron is interacting with the external field and its self field non-perturbatively. In order to formulate the problem the interacting electron is quantized by the first quantization and the photon field is quantized by its source (electron).

In the self field electrodynamics, we consider the interaction of the electron with external field (classical or quantum) and its self radiation field and formulate the radiation reaction. Then, there appears an important question. Does the free particle have radiation reaction? The physical answer have to be no. But we know from the classical electrodynamics that Lorentz-Dirac equation does not satisfy this condition and this is one of the reason for the existence of runaway solutions. In a physical theory the radiation reaction must go to zero when external field goes to zero.

In standart quantumelectrodynamics all the radiative processes are formulated in terms of free quantized electron and photon fields or Green's functions. In this formulation it is not easy to answer the question mentioned above. In the self field quantumelectrodynamics we can choose our physical quantities such that they go to zero when the external field becomes zero.

This approach is very similiar to the scattering theory. In quantum mechanics we have the scattering solutions. These solutions include the infinite plane wave solution. In order to obtain the physical quantities such as the scattering amplitude , we substract the plane wave or free particle solution from the scattering solution.

The main radiative processes are the self energy of the electron, anomalous magnetic moment and spontaneous decay in the free space or in the cavity for different external fields.[2,3,4,5] The contribution of the vacuum polarization term to the Lamb shift of the bound state electron have been investigated.[6,7] This is the most divergent term in the standart formulation of the QED. This new calculation gives the standart result by using a new regularization mechanism. In the first order iteration self-field QED gives exactly the same result as the standart QED calculation.[8] In the formulation of the self energy problem we have also formally divergent integrals. The source of these terms is the sum all over the intermediate quantum states and it includes intermediate bound states and continuous scattering states. The contributions of the bound states to this sum goes to zero when external field vanishes. But the contributions of the scattering states do not satisfy this criteria, because they include infinite plane wave or free particle solutions. In the next section we develope a new method how to regularize these integrals and obtain a finite result for the self energy of the electron.

SELF ENERGY TERM

Self energy is apart of the general energy shift ΔE_n of a quantum level n of system due to the radiative self energy effects. It is given by

$$\Delta E_n^{S.E.} = \frac{e^2}{2} \int\int d^3x \overline{\psi}_n(\mathbf{x}) \, \gamma_\mu \psi_s(\mathbf{x}) \int d^3y \overline{\psi}_s(\mathbf{y}) \, \gamma^\mu \psi_n(\mathbf{y}) \times$$

$$\int \frac{d^3k}{(2\pi)^3} \frac{e^{i\mathbf{k}\cdot(\mathbf{x}-\mathbf{y})}}{2k} P\left(\frac{2k}{(E_s - E_n)^2 - k^2}\right) \tag{1}$$

Where ψ_n is a fixed level and we sum on the over all levels ψ_s, discrete and continuous.

We summarize first the spin algebra and the angular integrations. The relativistic Coulomb functions are written as

$$\psi_n(\mathbf{r}) = \begin{pmatrix} f_n \Omega_n \\ i g_n \Omega_{n'} \end{pmatrix}, \quad \begin{array}{l} n = (j_n, l_n, m_n) \\ n' = (j_n, l_n', m_n); \quad l_n' = l_n + 1 \end{array} \tag{2}$$

where f_n and g_n are the "large" the "small" components respectively. The product of two currents is

$$\overline{\psi}_n \gamma_\mu \psi_s \overline{\psi}_s \gamma^\mu \psi_n = \psi^+_n(\mathbf{r}) \psi_s(\mathbf{r}) \psi^+_s(\mathbf{r}') \psi_n(\mathbf{r}') - $$

$$\psi^+_n(\mathbf{r}) \alpha \psi_s(\mathbf{r}) \cdot \psi^+_s(\mathbf{r}') \alpha \psi_n(\mathbf{r}') \tag{3}$$

After the angular integrations we obtain

$$\Delta E_n^{S.E.} = \frac{e^2}{2} \sum_s \int k dk \int dr \int dr' \, j_i(kr) j_i(kr') \times P\left[\frac{2k}{(E_s - E_n)^2 - k^2 + i\varepsilon}\right]$$

$$\{[w_{1n}^*(r)W_{ns}^{lm}w_{1s}(r)+w_{2n}^*(r)W_{n's'}^{lm}w_{2s}(r)]\cdot[w_{1s}^*(r')(W_{ns}^{lm})^*w_{1n}(r')+w_{2s}^*(r')(W_{n's'}^{lm})w_{2n}(r')]-$$
$$[w_{1n}^*(r)K_{ns'}^{lm}w_{2s}(r)+w_{2n}^*(r)K_{n'}^{lm}w_{1s}(r)]\cdot[w_{2s}^*(r')(K_{ns'}^{lm})^*w_{1n}(r')+w_{1s}^*(r')(K_{n's'}^{lm})w_{2n}(r')]\}$$

$$(4)$$

where

$$
W_{ns}^{lm} = \int d\hat{r}\left[\frac{k_n k_s}{|k_n\|k_s|}\sqrt{\frac{k_n+m_n\mp\frac12}{2k_n\mp1}\frac{k_s+m_s\mp\frac12}{2k_s\mp1}}\,Y_{|k_n\mp\frac12|-\frac12}^{*\,m_n-\frac12}Y_l^m Y_{|k_s\mp\frac12|-\frac12}^{m_s-\frac12}\right.
$$
$$
\left.+\frac{k_n k_s}{|k_n\|k_s|}\sqrt{\frac{k_n-m_n\mp\frac12}{2k_n\mp1}\frac{k_s-m_s\mp\frac12}{2k_s\mp1}}\,Y_{|k_n\mp\frac12|-\frac12}^{*\,m_n+\frac12}Y_l^m Y_{|k_s\mp\frac12|-\frac12}^{m_s+\frac12}\right]
$$

$$
\mathbf{K}_{ns'}^{lm} = \int d\hat{r}\left(\frac{ik_n}{|k_n|}\sqrt{\frac{k_n+m_n-\frac12}{2k_n-1}}\,Y_{|k_n-\frac12|-\frac12}^{*\,m_n-\frac12}\quad,\quad i\sqrt{\frac{k_n-m_n-\frac12}{2k_n-1}}\,Y_{|k_n-\frac12|-\frac12}^{*\,m_n+\frac12}\right)\cdot
$$
$$
\sigma\cdot Y_l^m\cdot\left(\begin{array}{c}-\dfrac{k_s}{|k|}\sqrt{\dfrac{k_n+m_n-\frac12}{2k_n-1}}\,Y_{|k_n-\frac12|-\frac12}^{m_n-\frac12}\\[2mm]\sqrt{\dfrac{k_n+m_n-\frac12}{2k_n-1}}\,Y_{|k_n-\frac12|-\frac12}^{m_n-\frac12}\end{array}\right)
$$

$$(5)$$

We can extend the sum over the intermediate ψ_s states also to the negative energy solutions in order to introduce the energy dependent radial Green's functions $G(r,r';z)$ of the relativistic Coulomb problem, because the negative-energy solutions are equivalent to positive-energy solutions with $-e$. Then $\Delta E_n^{S.E}$ becomes

$$
\Delta E_n^{S.E.} = -4\alpha\sum_s\int\frac{dz}{2\pi i}\int dr\int dr'\int k\,dk\,j_l(kr)\,j_l(kr')P\left[\frac{2k}{(z-E_n)^2-k^2+i\varepsilon}\right]\times R
$$

where R is

$$
R=\left[w_{1n}^*(r)G_{11}(r,r';z)w_{1n}(r')\left|W_{ns}^{lm}\right|^2+w_{2n}^*(r)G_{22}(r,r';z)w_{2n}(r')\left|W_{n's'}^{lm}\right|^2+\right.
$$
$$
w_{1n}^*(r)G_{12}(r,r';z)w_{2n}(r')W_{ns}^{lm}W_{n's'}^{*lm}+w_{2n}^*(r)G_{21}(r,r';z)w_{1n}(r')W_{n's'}^{lm}W_{ns}^{lm}-
$$
$$
w_{1n}^*(r)G_{22}(r,r';z)w_{1n}(r')\mathbf{K}_{ns'}^{lm}\cdot\mathbf{K}_{ns'}^{*lm}-w_{1n}^*(r)G_{21}(r,r';z)w_{2n}(r')\mathbf{K}_{ns'}^{lm}\cdot\mathbf{K}_{n's'}^{*lm}-
$$
$$
w_{2n}^*(r)G_{12}(r,r';z)w_{1n}(r')\mathbf{K}_{n's}^{lm}\mathbf{K}_{ns'}^{*lm}-w_{2n}^*(r)G_{11}(r,r';z)w_{2n}(r')\mathbf{K}_{n's}^{lm}\cdot\mathbf{K}_{n's}^{*lm}\right]
$$

$$(6)$$

Where $G(r,r';z)$ are the matrix elements of the Green's function of the relativistic Coulomb problem and the contour of z-integration is shown in Figure 1. The Green's function $G(r,r';z)$ has the poles corresponding to the bound states, plus the cuts beginning at $\pm m$ corresponding to positive and negative continuous spectra. The other cuts come from the photon Green's function.

45

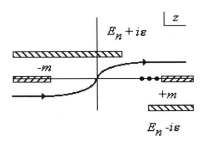

Figure 1. Contour of the z- integration.

The Green's function $G(r,r';z)$ can be constructed in terms of the solutions of the radial problem. It is in the following form:

$$G(r,r';z) = \frac{1}{K(z)} \left(\frac{w_1^{(2)}(r_>;z)}{w_2^{(2)}(r_>;z)} \right) \left(w_1^{(1)}(r_<;z), w_2^{(1)}(r_<;z) \right) \tag{7}$$

Where $w^{(1)}(r_<;z)$ and $w^{(2)}(r_>;z)$ are the regular solutions of the radial problem at the origin and at the infinity respectively.[6] They are given by Wichmann and Kroll in terms of confluent hypergeometric functions as

$$w_{\frac{1}{2}}^{(1)}(r_<;z) = \left[2r_<(z^2-1)^{\frac{1}{2}} \right]^{\gamma} \left[\frac{i\sqrt{z+1}}{\sqrt{z-1}} \right] \left[\left(\kappa - iZ\alpha / (z^2-1)^{\frac{1}{2}} \right) \phi \left(\gamma - iv, 2\gamma + 1, -2i(z^2-1)^{\frac{1}{2}} r_< \right) \right.$$
$$\left. \pm (\gamma - iv)\phi \left(\gamma - iv, 2\gamma + 1, -2i(z^2-1)^{\frac{1}{2}} r_< \right) \right]$$

$$w_{\frac{1}{2}}^{(2)}(r_>;z) = \left[2r_>(z^2-1)^{\frac{1}{2}} \right]^{\gamma} \left[\frac{i\sqrt{z+1}}{\sqrt{z-1}} \right] \left[\left(\kappa - iZ\alpha / (z^2-1)^{\frac{1}{2}} \right) \chi(\gamma - iv, 2\gamma + 1, -2i(z^2-1)^{\frac{1}{2}} r_>) \right.$$
$$\left. \pm (\gamma - iv)\chi(\gamma - iv, 2\gamma + 1, -2i(z^2-1)^{\frac{1}{2}} r_>) \right] \tag{8}$$

where $K(z)$ is

$$K(z) = -2(z^2-1)^{\frac{1}{2}} \left[\kappa - iZ\alpha / (z^2-1)^{\frac{1}{2}} \right] \frac{\Gamma(-\gamma - iv)\Gamma(2\gamma+1)}{\Gamma(\gamma - iv)\Gamma(-2\gamma)} \exp\left[\frac{i\pi}{2}(2\gamma+1) \right] \tag{9}$$

and the regular solutions of the confluent hypergeometric equation at the origin and the infinity are given by

$$\phi(\alpha,\gamma;z) = \frac{\Gamma(\gamma)}{\Gamma(\alpha)\Gamma(\gamma-\alpha)} \int_0^1 dt \, e^{zt} t^{\alpha-1}(1-t)^{\gamma-\alpha-1} \tag{10}$$

and

$$\chi(\alpha,\gamma;z) = \frac{\Gamma(\alpha+1-\gamma)}{\Gamma(\alpha)\Gamma(1-\gamma)} \int_0^\infty dt\, e^{-zt} t^{\alpha-1}(1+t)^{\gamma-\alpha-1}$$ (11)

respectively. After the k-integration Eq (6) becomes

$$\Delta E_n^{S.E.} = -4\alpha \sum_s \int \frac{dz}{2\pi i} \int dr \int dr' \left[-\frac{i\pi}{2} \omega j_l(\omega r_\langle) h_l^{(2)}(\omega r_\rangle) \right] R$$ (12)

Then by deforming the contour of energy integration we can separate $\Delta E_n^{S.E.}$ into low energy and high energy parts:

$$\Delta E_n^{S.E.} \text{(low energy)} = -4\alpha \sum_s \int_0^{E_n} \frac{dz}{2\pi i} \int dr \int dr' \left[-\frac{i\pi}{2} \omega j_l(\omega r_\langle) j_l(\omega r_\rangle) \right] R$$ (13)

$$\Delta E_n^{S.E.} \text{(high energy)} = 4\alpha \sum_s \int_{-i\infty}^{+i\infty} \frac{dz}{2\pi i} \int dr \int dr' \left[\frac{i\pi}{2} \omega j_l(\omega r_\langle) h_l^{(2)}(\omega r_\rangle) \right] R$$ (14)

In order to do r and r' integrations we represent $G(r,r';z)$ as a double Mellin-Barnes type complex integral. Mellin- Barnes type integral representation of the regular solutions are

$$\phi(\alpha,\gamma,z) = \int_{C_t-i\infty}^{C_t+i\infty} \frac{dt}{2\pi i} \Gamma(-t) \frac{\Gamma(\alpha+t)\Gamma(\gamma)}{\Gamma(\alpha)\Gamma(\gamma+t)}(-z)^t$$ (15)

and

$$\chi(\alpha,\gamma;z) = \int_{C_s-i\infty}^{C_s+i\infty} \frac{ds}{2\pi i} \Gamma(-s) \frac{\Gamma(\alpha+s)\Gamma(1-\gamma-s)}{\Gamma(\alpha)\Gamma(1-\gamma)} z^s$$ (16)

where C_t is choosen such that the poles of $\Gamma(-t)$ and the poles of $\Gamma(\alpha+t)$ and $\Gamma(\gamma+t)$ are seperated. Similiarly C_s is choosen such that the poles of $\Gamma(\alpha+s)$ and poles of $\Gamma(-s)$ and $\Gamma(1-\gamma-s)$ are seperated. For the Coulomb problem these conditions are satisfied except the free particle limit. When $Z\alpha$ goes to zero these two sets of poles are not seperated. We discuss this limit in the Appendix A.

Radial Green's function of the photon is

$$j_l(\omega r_\langle) h_l^{(2)}(\omega r_\rangle)$$ (17)

Mellin-Barnes type integral representations of j_l is

$$j_l(\omega r_\langle) = -\frac{1}{2}\sqrt{\frac{\pi}{2\omega r_\langle}} \int_{C_s-i\infty}^{C_s+i\infty} \frac{ds}{2\pi i} \frac{\Gamma\left(\frac{l+\frac{1}{2}+s}{2}\right)}{\Gamma\left(1+\frac{l+\frac{1}{2}-s}{2}\right)}(\omega r_\langle)^{-s}$$ (18)

This representation gives the Taylor expansion of j_l . Mellin-Barnes type integral representation of $h_l^{(2)}$ is

$$h_l^{(2)}(\omega r_\rangle) = \frac{i}{\pi} e^{i(l+\frac{1}{2})\frac{\pi}{2}} \sqrt{\frac{\pi}{2\omega r_\rangle}} \int_{C_s-i\infty}^{C_s+i\infty} \frac{ds}{2\pi i} \Gamma(-s)\Gamma(-l-\frac{1}{2}-s)(i\omega r_\rangle)^{l+\frac{1}{2}+2s} \qquad (19)$$

This representation gives the asymptotic expansion of $h_l^{(2)}$.
Finally, we represent the bound state solutions as

$$\begin{pmatrix} f_n(r) \\ g_n(r) \end{pmatrix} = U_n(r) \begin{pmatrix} \sqrt{1+\varepsilon_n} \sum_{n_1}[A_{n_1}+B_{n_1}] \\ \sqrt{1-\varepsilon_n} \sum_{n_1}[A_{n_1}-B_{n_1}] \end{pmatrix} (2p_N r)^{n_1} \qquad (20)$$

where

$$\begin{pmatrix} A_{n_1} \\ B_{n_1} \end{pmatrix} = \frac{1}{(2\gamma_n+1)_{n_1} n_1!} \begin{pmatrix} n_r(1-n_r)_{n_1} \\ (N_n-\kappa_n)(-n_r)_{n_1} \end{pmatrix} \qquad (21)$$

and $_n(r)$ is given by

$$_n(r) = \left[\frac{\Gamma(2\gamma_n+n_r+1)}{4N_n(N_n-\kappa_n)n_r!} \right]^{\frac{1}{2}} \frac{(2p_N)^{\frac{3}{2}}(2p_N r)^{\gamma_n-1}}{\Gamma(2\gamma_n+1)} \qquad (22)$$

with

$$p_N = \frac{Z\alpha}{N_n}, \quad \varepsilon_n = \frac{E_n}{m}, \quad N = [n^2-2n_r(|\kappa_n|-\gamma_n)]^{\frac{1}{2}},$$

$$n_r = n_r - |\kappa_n|, \quad \kappa_n = \pm(j_n+\tfrac{1}{2}), \quad \gamma_n = [\kappa_n^2-(Z\alpha)^2]^{\frac{1}{2}} \qquad (23)$$

Then we can represent the $\Delta E_n^{S.E}$ as a four dimensional complex integral and energy integral:

$$\Delta E_n^{S.E.} \text{ (high energy)} = 4\alpha \int \frac{dz}{2\pi i} \sum_s \{$$

$$T_{\alpha\alpha'}^{11} R_{\alpha\alpha'}[A_{n_1}A_{n_2}|W_{ns}^{lm}|^2(1+\varepsilon_n)-B_{n_1}B_{n_2}\mathbf{K}_{n's}^{lm}\cdot\mathbf{K}_{n's}^{*lm}(1-\varepsilon_n)]$$

$$+T_{\alpha\alpha'}^{22} R_{\alpha\alpha'}[B_{n_1}B_{n_2}|W_{n's'}^{lm}|^2(1-\varepsilon_n)-A_{n_1}A_{n_2}\mathbf{K}_{ns'}^{lm}\cdot\mathbf{K}_{ns'}^{*lm}(1+\varepsilon_n)] \qquad (24)$$

$$+\left[A_{n_1}B_{n_2}W_{ns}^{lm}W_{n's'}^{*lm}(1-\varepsilon_n^2)^{\frac{1}{2}}-B_{n_1}A_{n_2}\mathbf{K}_{n's}^{lm}\cdot\mathbf{K}_{ns'}^{*lm}(1-\varepsilon_n^2)^{\frac{1}{2}}\right]T_{\alpha\alpha'}^{12}R_{\alpha\alpha'}$$

$$+\left[B_{n_1}A_{n_2}W_{n's'}^{lm}W_{ns}^{*lm}(1-\varepsilon_n^2)^{\frac{1}{2}}-A_{n_1}B_{n_2}\mathbf{K}_{ns'}^{lm}\cdot\mathbf{K}_{n's}^{*lm}(1-\varepsilon_n^2)^{\frac{1}{2}}\right]T_{\alpha\alpha'}^{21}R_{\alpha\alpha'}\}$$

where $R_{\alpha\alpha'}$, $T^{mn}_{\alpha\alpha'}$, b and c are defined by

$$R_{\alpha\alpha'} = -\sum_{n_1 n_2} a_{n_1 n_2} \frac{\pi}{2^3} e^{i\frac{\pi}{2}(l+\frac{1}{2})} \times$$

$$\frac{\Gamma(l+\frac{1}{2}+v_1)}{\Gamma\left(1+\frac{l+\frac{1}{2}-v_1}{2}\right)}\Gamma(-v_2)\Gamma(-l-\frac{1}{2}-v_2)\frac{\Gamma(-s_1)\Gamma(\alpha+s_1)\Gamma(2\gamma+1_1)}{\Gamma(\alpha_1)\Gamma(2\gamma+1+s_1)} \tag{25}$$

$$\times\frac{\Gamma(-s_2)\Gamma(\alpha'+s_2)\Gamma(-2\gamma-s_2)}{\Gamma(\alpha')\Gamma(-2\gamma)}k(z)\left(\frac{2p_N}{2ip}\right)^{2\gamma_n+n_1+n_2+1}\frac{\Gamma(b)\,{}_2F_1\left(1,b,c+1;\frac{1}{2}\right)}{c\left(-1+\frac{p_N}{ip}\right)^b}$$

$$T^{mn}{}_{\alpha\alpha'}=\left[\left(\kappa-\frac{iZ\alpha}{p}\right)\delta_{\alpha,\gamma-iv}-\theta\left(2m-\frac{3}{2}\right)\delta_{\alpha,\gamma+1-iv}\right]$$

$$\times\left[\left(\kappa-\frac{iZ\alpha}{p}\right)\delta_{\alpha',\gamma-iv}-\theta\left(2n-\frac{3}{2}\right)\delta_{\alpha',\gamma+1-iv}\right] \tag{26}$$

$$b=2\gamma_n+n_1+n_2+1+2\gamma+s_1+s_2-s_3+2s_4+l+\frac{1}{2} \tag{27}$$

$$c=\gamma_n+n_1+s_2+2s_4+l \tag{28}$$

Regularization of $R_{\alpha\alpha'}$:

In the electron Green's function when $Z\alpha$ goes to zero $G(r,r';z)$ becomes the free particle Green's function. But when we used $G(r,r';z)$ in the calculation of self energy we must be carefull. Because the self energy of the free particle is included *in the definition of the particle itself.* In order to get rid of this problem, we must substract the free particle contribution from G. This is a new kind of *renormalization.*

We also have this ambiguity when we examined the Mellin-Barnes type integral representation of the Green's function. In $R_{\alpha\alpha'}$ we have s_1,s_2,v_1 and v_2 integrals. In v_1 and v_2 integrals, the contours are well defined. As it has been pointed out in the above in s_1 and s_2 integrals, the contours are not well defined. They are well defined when $Z\alpha\neq0$. But they are not well defined when $Z\alpha=0$, because two sets of the poles are not seperated and in this limit they coincide .That means the free particle limit of the transition amplitudes or the matrix elements of Green's function of the relativistic Coulomb problem are not well defined. If we use the direct product of the contours of s_1 and s_2 integrals we get formally divergent series.

Generally, when there is a double complex integral we cannot define the integration contour as a direct product of two separate contours of the one dimensional complex integrals.[9] In order to understand the physical meaning of this formal divergence and in order to regularize these integrals we examine the poles of the in the complex s_1 and s_2 planes.

In the Appendix A we discuss the regularization of the scattering solutions. We know from the scattering theory that the scattering solutions of the Coulomb problem always include plane wave or free particle solutions. Scattering probabilities or cross sections are physically measurable quantities. In order to calculate the physically measurable quantities we change the boundary conditions of the scattering solutions. The scattering amplitutes or cross sections are defined by substracting the plane waves from the scattering solutions. Then the final results are finite.

Here we also have the same problem. We are using the transition amplitudes and they also include the plane wave solutions. In order to use the transition amplitudes in the calculation of physical measurable self-energy we regularize them in the same way. That is equivalent to find an integration contour such that it seperates the two sets of poles or zeros in the s_1 and s_2 planes.

The poles of the $R_{\alpha\alpha'}(s_1, s_2)$ in the complex s_1 and s_2 planes are shown in Fig. 2 and 3. When $Z\alpha = 0$ the poles of $\Gamma(\gamma - iv + s_2)$ and $\Gamma(-2\gamma - s_2)$ are coincide. In order to separate the poles we regularize the integrals as follows:

$$\frac{\Gamma(\gamma - iv)\Gamma(\alpha + s_1)}{\Gamma(\alpha)} =$$

$$\lim_{\varepsilon \to 0}\left\{\frac{\Gamma(\gamma - iv)\Gamma(\alpha + s_1)}{\Gamma(\alpha)} - \frac{\Gamma(\gamma - i\varepsilon)\Gamma(\alpha + s_1 - i\varepsilon + iv)}{\Gamma(\alpha + iv - i\varepsilon)} + \frac{\Gamma(\gamma + i\varepsilon)\Gamma(\alpha + s_1 + i\varepsilon + iv)}{\Gamma(\alpha + iv + i\varepsilon)}\right\}$$

$$(29)$$

and

$$\frac{\Gamma(\gamma - iv)\Gamma(\alpha' + s_2)}{\Gamma(\alpha')\Gamma(-\gamma - iv)} =$$

$$\lim_{\varepsilon \to 0}\left\{\frac{\Gamma(\gamma - iv)\Gamma(\alpha' + s_2)}{\Gamma(\alpha')\Gamma(-\gamma - iv)} - \frac{\Gamma(\gamma - i\varepsilon)\Gamma(\alpha' + s_2 - i\varepsilon + iv)}{\Gamma(\alpha' + iv - i\varepsilon)\Gamma(-\gamma - i\varepsilon)} + \frac{\Gamma(\gamma + i\varepsilon)\Gamma(\alpha' + s_2 + i\varepsilon + iv)}{\Gamma(\alpha' + iv + i\varepsilon)\Gamma(-\gamma - i\varepsilon)}\right\}$$

$$(30)$$

Then we choose the contours such that the poles of $\Gamma(-s_1)$ and the poles of $\Gamma(\alpha + s_1)$ and the zeros of $(\Gamma(2\gamma + 1 + s_1))^{-1}$ are seperated. In the same way the poles of $\Gamma(-2\gamma - s_2)$ and $\Gamma(-s_2)$ and the poles of $\Gamma(\gamma - iv + s_2)$ are seperated.

We substitute R_{reg} into $\Delta E_n^{S.E}$ and calculate the complex integrals. We calculate the integral of $R_{\alpha\alpha'}$ as a sum of the residues at the poles $s_1 = -\alpha - p_1$ and $s_2 = -\alpha' - p_2$ where p_1 and p_2 range over $0,1,2,...$. In the similiar way we calculate v_1 and v_2 integrals also as residue integrals. They can be written as the sum over the residues at $v_1 = -l - \frac{1}{2} - q_1$ and $v_2 = q_2$ and $v_2 = -l - \frac{1}{2} + q_2$ where q_1 and q_2 range over $0,1,2,....$ By using these expressions we do z-integration. Thus the self energy contribution to the Lamb shift becomes a finite expression. All of the series in this expression are convergent. In order to compare this result with the experiment we need numerical sum of these series.

APPENDIX A

We discuss the relation between scattering solutions and scattering amplitudes of the Dirac Coulomb problem. In order to obtain a relation between them we examine the regular solutions of the Dirac-Coulomb problem. For the Dirac-Coulomb problem we have a regular solution around the origin and the asymptotic forms of this solution are function of sine or cosine. Here we use the regular solutions at the infinity. The asymptotic form of this solutions give exponantial waves.

Regular solutions at the infinity were developed by Wichmann and Kroll. They are

$$
\binom{rf}{rg} = N\left(\frac{i\sqrt{z-1}}{\sqrt{z+1}}\right)e^{ipr}
$$

$$
\times\left[\left(\kappa - \frac{iZ\alpha}{p}\right)\chi(\gamma - iv, 2\gamma + 1; -2ipr) \mp (\gamma - iv)\chi(\gamma + 1 - iv, 2\gamma + 1; -2ipr)\right]
$$

(A-1)

where $\chi(a,c;x)$ is the regular solution of the confluent hypergeometric differential equation at the infinity, and it is the linear combination of the regular and irregular solutions of the confluent hypergeometric differential equation at the origin: Asymptotic form of the regular solution at the infinity gives spherical waves which includes the free particle solutions or unscattered plane waves. The difference between the scattering solutions and the free particle solution is the scattering amplitude and it goes to zero when $Z\alpha$ goes to zero.

$\chi(a,c;x)$ function can be represented as Mellin-Barnes type integral:

$$
\chi(\gamma - iv, 2\gamma + 1; -2ipr) = \int_{C_s - i\infty}^{C_s + i\infty} \frac{ds}{2\pi i}\Gamma(-s)\frac{\Gamma(\gamma - iv + s)\Gamma(-2\gamma - s)}{\Gamma(\gamma - iv)\Gamma(-2\gamma)}z^s
$$

$$
\equiv \int_{C_s - i\infty}^{C_s + i\infty} \frac{ds}{2\pi i}\Gamma(-s)M(\gamma, v; s)z^s
$$

(A-2)

where C_s is the integration contour which can be choosen such that the set of the poles $\Gamma(-2\gamma - s)$ and $\Gamma(-s)$ and the poles of $\Gamma(\gamma - iv + s)$ are seperated. However, Mellin-Barnes type integral representation of $\chi(\gamma - iv, 2\gamma + 1; -2ipr)$ is not well defined when iv goes to zero. In this limit some of the poles of $\Gamma(\gamma - iv + s)$ and $\Gamma(-2\gamma - s)$ coincide and the seperation of two sets of poles is not clear. In order to solve this ambiguity we regularize the $M(\gamma, v; s)$ in the following way:

$$
M(\gamma, v; s) = \lim_{\varepsilon \to 0}\left[M(\gamma, v; s) - M(\gamma, \varepsilon; s) + M(\gamma, -\varepsilon; s)\right]
$$

(A-3)

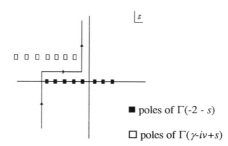

Figure 2. Poles of $M(\gamma, v, s)$

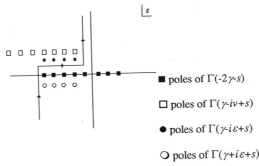

poles of $\Gamma(-2\gamma\text{-}s)$

poles of $\Gamma(\gamma\text{-}iv+s)$

poles of $\Gamma(\gamma\text{-}i\varepsilon+s)$

poles of $\Gamma(\gamma+i\varepsilon+s)$

Figure 3. Poles of $M_{reg}(\gamma,\nu,s)$.

The new set of poles are in Fig.3. Where we have choosen the integration contour such that one of the regularizing terms gives contribution to the integral. We close the integration contour from left hand side. Then the asymptotic form of this solution becomes

$$\chi_{reg}(\gamma-i\nu,2\gamma+1;-2ipr) \cong (-2ipr)^{-\gamma+i\nu}\frac{\Gamma(-\gamma-i\nu)}{\Gamma(-2\gamma)} - (-2ipr)^{-\gamma+i\varepsilon}\frac{\Gamma(-\gamma-i\varepsilon)}{\Gamma(-2\gamma)}$$

(A-4)

The asymptotic solution of the Dirac Coulomb problem is

$$\begin{pmatrix} rf \\ rg \end{pmatrix} = N\begin{pmatrix} i\sqrt{z-1} \\ \sqrt{z+1} \end{pmatrix}\frac{e^{ipr}}{\Gamma(-2\gamma)}\left[e^{\frac{\pi\nu}{2}+i\arg\Gamma(-\gamma-i\nu)+i\delta_\nu}\left|\Gamma(-\gamma-i\nu)\right| - \left|\Gamma(-\gamma)\right|\right]$$

(A-5)

where

$$\delta_\nu = \nu\ln 2pr$$

(A-6)

This solution represents a spherical wave and and when ν goes to zero it becomes zero. The second term corresponds to the spherical wave expansion of $e^{ik\cdot r}$. By this regularization we obtain a transformation from the scattering solution to the scattering amplitudes.

AKNOWLEDGMENTS

This work was partially supported by TUBITAK, Scientific and Technical Research Council of Turkey.

REFERENCES

1. A.O.Barut and J. Kraus, *Phys.Rev.* D **16,** 161 (1988); *Found. Phys.* **13,** 189 (1983).

2. A.O.Barut and J.F. van Heule, *Phys. Rev.* A **32,** 3187 (1985).

3. See the review in *The Electron, New Theory and Experiment*, edited by D. Hestenes and A Weingartshofer (Kluwer, Dordrecht ,1991), pp. 105-148.

4. A.O.Barut and J. Dowling, *Z. Naturforsch. Teil A* **44,** 1051 (1989).

5. A.O.Barut and Y. Salamin, *Phys.Rev.* *A* **37**, 2284 (1988).

6. E. Wichmann and N. Kroll, *Phys.Rev.* **101,** 83 (1956).

7. A.O.Barut and N. Ünal, *Phys.Rev.* *D* **41,** 3822 (1990).

8. A. Açıkgöz, A.O.Barut and N. Ünal, to be published.

9. R.J.Sasiela and J.D. Shelton, *J. Math. Phys.* **34**, 2572 (1993).

QUANTUM FLUCTUATIONS AND INERTIA

Marc-Thierry Jaekel[1] and Serge Reynaud[2]

[1] Laboratoire de Physique Théorique
de l'Ecole Normale Supérieure[1](CNRS),
24 rue Lhomond, F-75231 Paris Cedex 05, France
[2] Laboratoire Kastler Brossel[2](UPMC-ENS-CNRS), case 74,
4 place Jussieu, F-75252 Paris Cedex 05, France

INTRODUCTION

Fundamental problems are raised by the mechanical effects associated with radiation pressure fluctuations in vacuum. The instability of motions when radiation reaction is taken into account, and the existence of "runaway solutions" [1], can be avoided for mirrors by recalling that they are actually transparent to high frequencies of the field [2]. However, partially transmitting mirrors, and cavities, introduce scattering time delays which result in a temporary storage of part of the scattered vacuum fluctuations [3]. In particular, the energy related to Casimir forces [4] identifies with the energy of field fluctuations stored in the cavity [3]. This revives the questions of the contribution of vacuum fluctuations to inertia and gravitation [5], and of its consistency with the general principles of equivalence and of inertia of energy.

Vacuum fluctuations of quantum fields have been known for long to correspond to an infinite energy density [6], or at least to a problematically high energy density, if only frequencies below Planck frequency are considered [7]. A common way to escape the problems raised by consequent gravitational effects, exploits the fact that only differences of energy are involved in all other interactions. Vacuum energy is set to zero by definition, a prescription which is embodied in normal ordering of quantum fields. In such a scheme, variations of vacuum energy, like the energy associated with Casimir forces [4], hardly give rise to inertia and gravitation. Furthermore, normal ordering cannot be implemented as a covariant prescription and leads to ambiguities in defining the gravitational effects of quantum fields [8]. Then, the question naturally arises of the compatibility of the mechanical effects induced by quantum field fluctuations with the general principles which govern the laws of mechanics.

---

[1]Unité propre du Centre National de la Recherche Scientifique,
associée à l'Ecole Normale Supérieure et à l'Université de Paris Sud.
[2]Unité de l'Ecole Normale Supérieure et de l'Université Pierre et Marie Curie,
associée au Centre National de la Recherche Scientifique.

As a new approach to this question, we discuss the mechanical effects of quantum field fluctuations on two mirrors building a Fabry-Perot cavity. We first put into evidence that the energy related with Casimir forces is an energy stored on field fluctuations as a result of scattering time delays. We then discuss the forces felt by the mirrors when they move within vacuum field fluctuations, and in particular the contribution of Casimir energy to inertia.

CASIMIR ENERGY

As a result of the radiation pressure of field quantum fluctuations in which they are immersed, two mirrors at rest in vacuum feel a mean Casimir force which depends on their distance q. For partially transmitting mirrors, characterised by their frequency dependent reflection coefficients (r_1 and r_2), the Casimir force takes a simple form (written here for a cavity in two-dimensional space-time immersed in the vacuum of a scalar field; similar expressions hold in four-dimensional space-time, and for electromagnetic and also thermal fields) [3]:

$$F_c = \int_0^\infty \frac{d\omega}{2\pi} \frac{\hbar\omega}{c} \{1 - g[\omega]\}$$

$$g[\omega] = \frac{1 - |r[\omega]|^2}{|1 - r[\omega]e^{2i\omega q/c}|^2} \qquad r[\omega] = r_1[\omega]r_2[\omega]$$

The first part of this expression corresponds to the energy-momentum of incoming vacuum field fluctuations ($\hbar$ is Planck constant, and c the light velocity). The second part describes the effect of the cavity on the modes: $g[\omega]$ describes an enhancement of energy density for modes inside the resonance peaks of the cavity, and an attenuation for modes outside.

This mean force can be seen as the variation of a potential energy, more precisely, as the length dependent part of the energy of the cavity immersed in field fluctuations:

$$dE_c = F_c dq$$

One easily derives the well-known phase-shift representation of Casimir energy [4], whose expression in the present case takes the simple following form:

$$E_c = \int_0^\infty \frac{d\omega}{2\pi} \hbar\{-\delta[\omega]\}$$

$$2\delta[\omega] = i Log \frac{1 - r[\omega]e^{2i\omega q/c}}{1 - r[\omega]^* e^{-2i\omega q/c}}$$

$$det S = det S_1 det S_2 e^{2i\delta}$$

$\delta[\omega]$ is the frequency dependent phase-shift introduced by the cavity on the propagation of field modes, as given by the scattering matrix (S) of the cavity (more precisely, its definition divides by the individual scattering matrices of the mirrors, whose contributions to the total energy are length independent).

The frequency dependent phase-shift corresponds to time delays in the propagation of fields through the cavity:

$$\tau[\omega] = \partial_\omega \delta[\omega]$$

This time delay [9] describes the time lag undergone by a wave packet around frequency ω and is the sum of several contributions:

$$\tau[\omega] = -\{1 - g[\omega]\}\{\frac{q}{c} + \frac{1}{2}\partial_\omega\varphi\}$$
$$+ g[\omega]sin(2\omega\frac{q}{c} + \varphi)\frac{\partial_\omega\rho}{1 - \rho^2}$$
$$r[\omega] = \rho[\omega]e^{i\varphi[\omega]} \qquad (1)$$

The main contribution identifies with the length of the cavity (divided by c), modified by the function g describing energy densities within the cavity. Other contributions are corrections due to the frequency dependence of the mirrors' reflection coefficients, i.e. delays introduced during reflection on the mirrors themselves.

Casimir energy can be rewritten in terms of these scattering time delays, integrating by parts and noting that boundary terms vanish in particular because of high frequency transparency:

$$E_c = \int_0^\infty \frac{d\omega}{2\pi}\hbar\omega\tau[\omega]$$

The result takes a simple form, as an integral over all modes of the product of the spectral energy density of quantum field fluctuations by the corresponding time delay. In particular, the length dependent part of Casimir energy is negative, corresponding to a binding energy, so that negative time delays contribute in majority [10]. As time delays are indeed relative to free propagation, i.e in abscence of cavity, the retardation effect of the cavity on resonant modes is thus dominated by the opposite effect on modes outside resonance peaks.

It can be shown that the same expressions remain valid for Casimir force and energy of a cavity immersed in thermal fields, provided the spectral energy density for thermal quantum fluctuations is substituted (T is the temperature) [3]:

$$F_c = \int_0^\infty \frac{d\omega}{2\pi}2\frac{\hbar\omega}{c}\{\frac{1}{2} + \frac{1}{e^{\hbar\omega/T} - 1}\}\{1 - g[\omega]\}$$

$$E_c = \int_0^\infty \frac{d\omega}{2\pi}2\hbar\omega\{\frac{1}{2} + \frac{1}{e^{\hbar\omega/T} - 1}\}\tau[\omega]$$

To the contribution of zero-point fluctuations, one must add the contribution due to the mean number of photons as given by Planck's formula. In all cases, Casimir energy appears as part of the energy of quantum field fluctuations which is stored inside the cavity, as a consequence of scattering time delays.

MOTIONAL CASIMIR FORCES

The Casimir forces felt by two mirrors at rest result from the radiation pressure exerted by the fluctuating quantum fields in which they are immersed. Hence, these forces also fluctuate and their fluctuations can be characterised by their correlations ($i, j = 1, 2$ label the two mirrors):

$$< F_i(t)F_j(0) > - < F_i >< F_j >= C_{F_iF_j}(t)$$

For a stationary state of the field, correlations are equivalently characterised by spectral functions [11]:

$$C_{F_iF_j}(t) = \int_{-\infty}^\infty \frac{d\omega}{2\pi}e^{-i\omega t}C_{F_iF_j}[\omega] \qquad (2)$$

The fluctuating forces induce random motions of the mirrors around their mean positions which can be described as quantum Brownian motions. As a consequence of general principles governing motion in a fluctuating environment [12], when set into motion mirrors feel additional forces which depend on their motions. For small displacements (δq_i), these forces are conveniently described by motional susceptibilities:

$$< \delta F_i[\omega] >= \sum_j \chi_{F_i F_j}[\omega] \delta q_j[\omega] \tag{3}$$

The motional forces can be obtained using motion dependent scattering matrices [13]. The scattering matrix of a mirror in its rest frame leads to a scattering matrix in the original frame which depends on the mirror's motion, and can easily be obtained up to first order in the mirror's displacement. Radiation pressures and forces exerted on the mirrors are thus obtained up to the same order [11]. (For perfect mirrors, forces have been obtained exactly for arbitrary motions of the mirrors [14]).

According to linear response theory [15], fluctuation-dissipation relations identify the imaginary (or dissipative) part of a susceptibility with the commutator of the corresponding quantity with the generator of the perturbation. In the case of mirrors' displacements, the generators are the forces exerted on the mirrors:

$$\chi_{F_i F_j}[\omega] - \chi_{F_j F_i}[-\omega] = \frac{i}{\hbar} \{ C_{F_i F_j}[\omega] - C_{F_j F_i}[-\omega] \}$$

Thus, fluctuation-dissipation relations provide a check for the results one obtains independently for force fluctuations (2) and for motional susceptibilities (3).

Although rather complex in their total generality, explicit expressions for motional forces induced by vacuum fluctuations on partially transmitting mirrors satisfy some general interesting properties [11]. As expected, the motional forces present mechanical resonances for frequencies which coincide with optical modes of the cavity:

$$\omega = n\pi \frac{c}{q}$$

Although motional Casimir forces are naturally small, much smaller than static Casimir forces, resonance properties might be used to compensate their smallness using cavities with very high quality factors, thus possibly leading to experimental evidence.

Other interesting properties of these forces appear at the quasistatic limit, i.e. at the limit of very slow motions [16]. For displacements which vary slowly in time, one can use a quasistatic expansion (expansion around zero frequency $\omega \sim 0$) of the expressions for motional susceptibilities (3) (a dot stands for time derivative):

$$< \delta F_i[\omega] >= \sum_j \{ \chi_{F_i F_j}[0] \delta q_j[\omega] + \frac{1}{2} \chi''_{F_i F_j}[0] \omega^2 \delta q_j[\omega] + \ldots \}$$

$$< \delta F_i(t) >= - \sum_j \{ \kappa_{ij} \delta q_j(t) + \mu_{ij} \delta \ddot{q}_j(t) + \ldots \}$$

The first term, described by κ_{ij} ($-\chi_{F_i F_j}[0]$), just reproduces the variations of the static Casimir force when the length of the cavity is changed. The further terms correspond to new forces which emerge when the mirrors are accelerated in vacuum and which exhibit peculiar features. These forces are proportional to the mirrors' accelerations and are conveniently expressed under the form of a mass matrix μ_{ij} ($\frac{1}{2} \chi''_{F_i F_j}[0]$). Diagonal terms are corrections to the mirrors' masses. They show that each mirror's mass is modified by the presence of the other mirror, with a correction which depends on the distance between the two mirrors. But non diagonal terms are also present, corresponding to the emergence of an inertial force for one mirror when the other mirror is accelerated. These properties of the inertial forces induced by vacuum fluctuations are reminiscent of Mach's principle of relativity of inertia. They indeed satisfy the requirements that Einstein [17] stated in his analysis of Mach's conception of inertia and in the context of gravity. They strongly suggest a relation between modifications of vacuum fields and gravitational effects [18].

Inertial forces acting on the cavity as a whole are related with global motions of the cavity, i.e. identical motions of the two mirrors (in linear approximation for displacements):

$$\delta \ddot{q}_1(t) = \delta \ddot{q}_2(t) = \delta \ddot{q}(t)$$

The total force acting on the cavity moving in vacuum fields then contains a component which dominates for slow motions and which is proportional to the cavity's acceleration:

$$< \delta F(t) >=< \delta F_1(t) + \delta F_2(t) >= -\{\mu \delta \ddot{q}(t) + \ldots\}$$

$$\mu = \sum_{ij} \mu_{ij}$$

Explicit computation [16] shows that the corresponding mass correction for the cavity is proportional to the length of the cavity and to the Casimir force between the two mirrors:

$$\mu c^2 = -2F_c q \tag{4}$$

This correction appears to be proportional to the contribution of the intracavity fields to the Casimir energy, i.e the energy stored on vacuum fluctuations due to the propagation delay inside the cavity (see (1)). For a cavity built with perfect mirrors in particular, this corresponds to Casimir energy:

$$E_c = -F_c q$$

Although not quite obvious at first sight, the factor 2 is in fact the correct one in the present case. Indeed, it was already shown by Einstein [19], that for a stressed rigid body Lorentz invariance implies a relation for the mass (μ), i.e the ratio between momentum and velocity, that not only involves the internal energy of the body (E_c) but also the stress (F_c) exerted on the body:

$$\mu c^2 = E_c - F_c q$$

When comparing the total momentum with the velocity of the center of inertia of the whole system, i.e. taking into account not only the masses of the two mirrors but also the energy stored in the fields inside the cavity, this relation leads to the usual equivalence between mass and energy. Thus, the energy of vacuum field fluctuations stored inside the cavity contributes to inertia in conformity with the law of inertia of energy.

However, for partially transmitting mirrors, the energy stored according to time delays due to reflection upon the mirrors (see (1)) is missing in the mass correction (4). The inertial forces obtained for a cavity moving in vacuum satisfy the law of inertia of energy for the energy of vacuum fluctuations stored inside the cavity, but not for the energy stored in the mirrors themselves. This result must be compared with a previous computation of the force exerted on a single, partially transmitting, mirror moving in vacuum fields, which appeared to vanish for uniformly accelerated motion [13]. This discrepancy with the general equivalence between mass and energy reflects a defect in the representation of the interaction of the mirror with the field. We shall now discuss, using an explicit model of interaction between mirror and field, how this representation can be improved.

POINTLIKE SCATTERER

We consider the case of a scalar field ϕ interacting with a pointlike mirror, located at q, in two-dimensional space-time $((x^\mu)_{\mu=0,1} = (t,x))$, described by the following manifestly relativistic Lagrangian (from now on, $c = 1$) [20]:

$$\mathcal{A} = \int \frac{1}{2}(\partial\phi)^2 d^2x - \int m\sqrt{1-\dot{q}^2}\,dt$$

$$\mathcal{L} = \frac{1}{2}(\partial\phi)^2 - m\sqrt{1-\dot{q}^2}\delta(x-q) \tag{5}$$

$$m = m_b + \Omega\phi(q)^2 \tag{6}$$

The two terms are the usual Lagrangians for a free scalar field and a free particle, except that the mass of the particle is assumed to also contain a contribution which depends on the field. Such contribution generally describes a relativistically invariant interaction term for the field and the sources which are located on the mirror. In order to facilitate comparison with the simplified representation in terms of a 2×2 scattering matrix, the interaction is further assumed to be quadratic in the field. Ω is the inverse of a proper time characterising field scattering. Equations for the field involve the scatterer's position and result in highly non linear coupling:

$$\partial^2\phi = -2\sqrt{1-\dot{q}^2}\,\Omega\phi\delta(x-q) \tag{7}$$

However, if one considers as a first approximation that the mirror remains at rest at a fixed position q, then (7) becomes a linear equation describing propagation in presence of a pointlike source. The field on both sides of the scatterer decomposes on two components which propagate freely in opposite directions and which can be identified with incoming and outcoming fields. The scattering matrix which relates outcoming and incoming modes is obtained from equation (7), and identifies with a simple symmetric 2×2 matrix determined by the following frequency dependent diagonal ($s[\omega]$) and non diagonal ($r[\omega]$) elements:

$$s[\omega] = 1 + r[\omega] \qquad r[\omega] = -\frac{\Omega}{\Omega - i\omega} \tag{8}$$

This corresponds to the simple model of partially transmitting mirror, with a reflection time delay having a Lorentzian frequency dependence:

$$\tau[\omega] = \frac{\Omega}{\Omega^2 + \omega^2} \tag{9}$$

Simple computation shows that the energy stored on field fluctuations due to this reflection time delay indeed identifies with the mass term describing the interaction with the field (6). The mean mass is determined by the correlations of the local field, which can be expressed in terms of incoming correlations and of the scattering matrix. For incoming fields in vacuum:

$$< \Omega\phi(q)^2 > = \int_0^\infty \frac{d\omega}{2\pi}\hbar\omega\tau[\omega] \tag{10}$$

Actually, the expression thus obtained for the mean value of the scatterer's mass in vacuum is infinite, as a result of a diverging contribution of high frequency fluctuations. In fact, the approximation of a scatterer staying at rest, on which expression (9) for the time delay relies, cannot remain valid for sufficiently high frequencies. At field frequencies which become comparable with the scatterer's mass, recoil of the scatterer cannot be neglected, so that the simplified 2×2 scattering matrix and its associated reflection time delay fail to be good approximations. Although consistent with the approximation which neglects the scatterer's recoil for all field frequencies, the result of an infinite stored energy does not correspond to the general case.

For a finite mass scatterer, the scatterer's recoil must be taken into account. This is described by the equations of motion for the scatterer which are derived from Lagrangian (5):

$$
\begin{aligned}
\frac{dp^\mu}{dt} &= F^\mu = 2\Omega\sqrt{1 - \dot{q}^2}\,\phi\partial^\mu\phi(q) \\
p^\mu &= (\frac{m}{\sqrt{1 - \dot{q}^2}}, \frac{m\dot{q}}{\sqrt{1 - \dot{q}^2}})
\end{aligned}
\tag{11}
$$

These correspond to Newton equation, with a force depending on the local field. Recalling the equations of motion for the field (7), the force identifies with the radiation pressure exerted by the scattered field. An important feature of the equations characterising the scatterer's recoil is that the mass involved in the relation between the force and the scatterer's acceleration includes the mass correction (6), that is the energy stored by the scatterer on incoming field fluctuations. As exemplified by this simple model, a correct treatment of the interaction between field and a partially transmitting mirror leads to an energy stored on vacuum field fluctuations due to reflection time delays which also satisfies the universal equivalence between mass and energy.

As shown by equations (11), the energy and momentum of the scatterer satisfy the usual relations:

$$
p_0^2 - p_1^2 = m^2 \qquad p^1 = p^0\dot{q}
$$

When submitted to the fluctuating radiation pressure of the field, the scatterer undergoes a relativistic stochastic process which remains causal, i.e with a velocity never exceeding the light velocity. When fields with frequencies much smaller than the scatterer's mass ($\hbar\omega \ll m >$) are reflected, recoil can be neglected and the scattering matrix is well approximated by the linear 2×2 matrix (8). However, for frequencies of the order of the scatterer's mass, recoil must be taken into account and the frequency dependence of scattering time delays differs significantly from the dependence at low frequencies (9). A complete and accurate treatment should then consistently provide a finite stored energy for a finite mass scatterer.

Integration of the stored energy in the inertial mass in a consistent way leads to interesting new consequences. It directly results from their expressions in terms of quantum field fluctuations (for instance (10)), that stored energies not only possess a mean value but also fluctuations. Hence, the inertial mass is a fluctuating quantity, with a characteristic noise spectrum:

$$
< m(t)m(0) > - < m >^2 = \int \frac{d\omega}{2\pi} e^{-i\omega t} C_{mm}[\omega]
$$

For the pointlike scatterer just described, the inertial mass correction is quadratic in the local field, and mass fluctuations are derived from incoming field fluctuations and the scattering matrix. For frequencies well below the scatterer's mass, recoil can be neglected and the mass noise spectrum in vacuum is readily obtained form (8):

$$C_{mm}[\omega] = 2\hbar^2\theta(\omega)\int_0^\omega \frac{d\omega'}{2\pi}\omega'\tau[\omega'](\omega-\omega')\tau[\omega-\omega'] \qquad (\hbar\omega \ll <m>)$$

This spectrum shows the characteristic positive frequency domain of vacuum fluctuations. It also corresponds to a convolution (a direct product in time domain) of two expressions equal to the mean mass correction, a consequence of the gaussian property of local field fluctuations (at this level of approximation). Inertial mass thus exhibits properties of a quantum variable.

As expected, mass fluctuations become extremely small for ordinary time scales, i.e. for low frequencies. For frequencies below the reflection cut-off Ω, the mass noise spectrum grows like ω^3:

$$C_{mm}[\omega] \simeq \frac{\hbar^2}{6\pi\Omega^2}\theta(\omega)\omega^3$$

The inertial mass remains practically constant in usual mechanical situations. For high frequencies however, mass fluctuations become important and cannot be neglected at very short time scales. As an illustration (of course, recoil should be accounted for at such frequencies), the same expression exhibits mass fluctuations which become comparable with the mean mass (for $m_b = 0$):

$$C_{mm}(t=0) = <m^2> - <m>^2 = 2<m>^2$$

CONCLUSION

Scattering time delays lead to a temporary storage of quantum field fluctuations by scatterers. Vacuum quantum field fluctuations induce stored energies and inertial masses which satisfy the universal equivalence between mass and energy, including for their fluctuations. Vacuum fluctuations result in mechanical effects which conform with general principles of mechanics. It can be expected that energies stored on quantum field fluctuations should also lead to gravitation, in conformity with the principle of equivalence. Moreover, mass fluctuations due to vacuum field fluctuations could play a significant role in a complete and consistent formulation of gravitational effects.

References

[1] F. Rohrlich. *Classical Charged Particles*, Addison-Wesley, Reading (1965).

[2] M.T. Jaekel and S. Reynaud, *Phys. Lett.* A 167:227 (1992).

[3] M.T. Jaekel and S. Reynaud, *J. Phys. I France* 1:1395 (1991).

[4] H.B.G. Casimir, *Proc. K. Ned. Akad. Wet.* 51:793 (1948).
G. Plunien, B. Müller and W. Greiner, *Phys. Rep.* 134:87 (1986).

[5] W. Nernst, *Verh. Deutsch. Phys. Ges.* 18:83 (1916).

[6] C.P. Enz, in: *Physical Reality and Mathematical Description*,
C.P. Enz and J. Mehra, eds., Reidel, Dordecht (1974).

[7] P.S. Wesson, *ApJ* 378:466 (1991).

[8] N.D. Birell and P.C.W. Davies. *Quantum Fields in Curved Space*,
University Press, Cambridge (1982).
S.A. Fulling. *Aspects of Quantum Field Theory in Curved Spacetime*,
University Press, Cambridge (1989).

[9] E.P Wigner, *Phys. Rev.* 98:145 (1955).

[10] E.L. Bolda, R.Y. Chiao, G.C. Garrison, *Phys. Rev.* A 48:3890 (1993).

[11] M.T. Jaekel and S. Reynaud, *J. Phys. I France* 2:149 (1992).

[12] A. Einstein, *Phys. Z.* 18:121 (1917).

[13] M.T. Jaekel and S. Reynaud, *Quantum Opt.* 4:39 (1992).

[14] G.T. Moore, *J. Math. Phys.* 11:2679 (1970).

[15] R. Kubo, *Rep. Prog. Phys.* 29:255 (1966).
L.D. Landau and E.M. Lifschitz. *Cours de Physique Théorique,
Physique Statistique, première partie*, Mir, Moscou (1984).

[16] M.T. Jaekel and S. Reynaud, *J. Phys. I France* 3:1093 (1993).

[17] A. Einstein. *The Meaning of Relativity*, University Press, Princeton (1946).

[18] A.D. Sakharov, *Doklady Akad. Nauk SSSR* 177:70 (1967)
[*Sov. Phys. Doklady* 12:1040 (1968)].

[19] A. Einstein, *Jahrb. Radioakt. Elektron.* 4:411 (1907); 5:98 (1908)
[translated by H.M. Schwartz, *Am. J. Phys.* 45:512, 811, 899 (1977)].

[20] M.T. Jaekel and S. Reynaud, *Phys. Lett.* A 180:9 (1993).

MECHANICAL EFFECTS OF RADIATION PRESSURE QUANTUM FLUCTUATIONS

Marc-Thierry Jaekel[1] and Serge Reynaud[2]

[1] Laboratoire de Physique Théorique
   de l'Ecole Normale Supérieure[1](CNRS),
   24 rue Lhomond, F-75231 Paris Cedex 05, France
[2] Laboratoire Kastler Brossel[2](UPMC-ENS-CNRS), case 74,
   4 place Jussieu, F-75252 Paris Cedex 05, France

INTRODUCTION

Lorentz electron theory [1] was an early unification of fields and particles, in that case electromagnetic fields and charged particles, in a common and universal description. This frame played a determinant role in a consistent development of classical field theory and relativistic mechanics [2]. This close connection was deeply perturbed by the advent of quantum formalisms, which ultimately emphasize the primary role of quantum fields. Within the framework of quantum electrodynamics, mechanical effects on charged particles, although obtainable in principle, are usually derived with difficulties [3].

When taking into account physical limits in space-time probing, field theory and mechanics emerge as complementary representations in space-time. Fields are measured by their mechanical effects on test particles, while particle positions are measured through probe fields. Space-time measurements are determined by energy-momentum exchanges between fields and particles. In the same way as radiation of test particles affects field measurements, mechanical effects of the probe field radiation pressure affect position measurements, and hence the determination of motions in space-time.

Quantum fluctuations impose ultimate limitations which affect both particle positions and field values in space-time [4]. As a consequence of Heisenberg inequalities, oscillators have fluctuations which subsist in their ground states. Quantum field fluctuations persist in vacuum, and those vacuum fluctuations result in fundamental limitations on the determination not only of fields, but also of positions and motions in space-time [5]. The presence of ultimate quantum fluctuations must then be taken into account in a consistent development of mechanics.

---

[1]Unité propre du Centre National de la Recherche Scientifique,
associée à l'Ecole Normale Supérieure et à l'Université de Paris Sud.
   [2]Unité de l'Ecole Normale Supérieure et de l'Université Pierre et Marie Curie,
associée au Centre National de la Recherche Scientifique.

Macroscopic objects feel the effects of quantum field fluctuations, even in vacuum. A well-known example is that of Casimir forces between macroscopic conductors or dielectrics [6]. Casimir forces can be seen as a mechanical signature of the radiation pressure of quantum field fluctuations. But radiation pressure itself presents quantum fluctuations, which result in further mechanical effects. When moving in a fluctuating environment, a scatterer radiates and experiences a radiation reaction force [7]. There results a typical quantum Brownian motion which is determined by radiation pressure fluctuations and which persists in vacuum. We shall here briefly survey these effects, and also discuss the fundamental consequences for stability of motion in vacuum and for ultimate fluctuations of position.

RADIATION PRESSURE OF QUANTUM FIELD FLUCTUATIONS

Reflectors which are immersed in a fluctuating field, like conductors or dielectrics in a fluctuating electromagnetic field for instance, feel the effects of field fluctuations. Even at the limit of zero temperature, i.e. in a field which does not contain any photon, that is in the vacuum state, effects of quantum field fluctuations persist. Vacuum fluctuations are responsible for a mean force, the so-called Casimir force, between reflectors. Two plane infinite perfect mirrors in the electromagnetic vacuum are submitted to a mean attractive pressure (force per unit area), which decreases like the fourth power of the mirrors' distance q ($\hbar$ is Planck constant, A the area, light velocity is equal to 1):

$$F_c = \frac{\pi^2}{240} \frac{\hbar}{q^4} A \qquad (1)$$

Although very weak, Casimir forces between macroscopic plates have effectively been observed [6]. As a useful simple illustration, we shall also discuss in the following the similar system made of two perfect mirrors in two-dimensional space-time, in the vacuum of a scalar field. In that case, the mean force decreases like the second power of the mirrors' distance:

$$F_c = \frac{\pi}{24} \frac{\hbar}{q^2} \qquad (2)$$

For perfect mirrors, the mean Casimir force can easily be derived, using the following simple argument [6]. The field can be considered as a set of free oscillators in each of the space domains delimited by the mirrors. Each of these oscillators possesses (zero-point) quantum fluctuations in its ground state, with a corresponding energy of $\frac{1}{2}\hbar\omega$ (ω the oscillator's frequency). The mirrors play the role of boundary conditions for the field, modifying the spectrum of mode frequencies allowed in the cavity they form. The total zero-point energy of the field ($\sum_n \frac{1}{2}\hbar\omega_n$) varies with the mirrors' distance, and results in the mean force (1) between the mirrors. Although it clearly exhibits the role of vacuum field fluctuations, this simple interpretation in terms of vacuum energy leads to problems. Because of its high frequency dependence, the mode spectrum results in a total zero-point energy which is infinite, or which at least corresponds to a high energy density that induces problematic gravitational and cosmological consequences [8]. On another hand, when comparing different physical situations, variations of vacuum energy (with the distance for instance) produce finite and observable effects. Moreover, realistic mirrors must clearly be transparent to sufficiently high field frequencies, whose fluctuations should then have no incidence on the effect.

An alternate and more consistent derivation of Casimir forces uses a local description in terms of the radiation pressure exerted by field fluctuations on the mirrors [9]. We briefly recall this derivation in the case of the simple model of mirrors in two-dimensional space-time $((x^\mu)_{\mu=0,1} = (t, x))$ (2). Each mirror determines two regions of space where the scalar field ϕ can be decomposed on two components which propagate freely (Figure 1):

$$\phi(t, x) = \varphi(t - x) + \psi(t + x)$$

A perfect mirror corresponds to a boundary condition for the field, saying that the field vanishes at the mirror's position in space (q):

$$\varphi_{out}(t - q) = -\psi_{in}(t + q) \qquad\qquad \psi_{out}(t + q) = -\varphi_{in}(t - q) \qquad (3)$$

From now on, we shall use the following notation for Fourier transforms:

$$f(t) = \int_{-\infty}^{\infty} \frac{d\omega}{2\pi} e^{-i\omega t} f[\omega]$$

In a more realistic model, the mirror partially reflects and transmits both components of the field and is described by a scattering matrix [10]:

$$\begin{pmatrix} \varphi_{out}[\omega] \\ \psi_{out}[\omega] \end{pmatrix} = S[\omega] \begin{pmatrix} \varphi_{in}[\omega] \\ \psi_{in}[\omega] \end{pmatrix} \qquad S[\omega] = \begin{pmatrix} s[\omega] & r[\omega]e^{-2i\omega q} \\ r[\omega]e^{2i\omega q} & s[\omega] \end{pmatrix} \qquad (4)$$

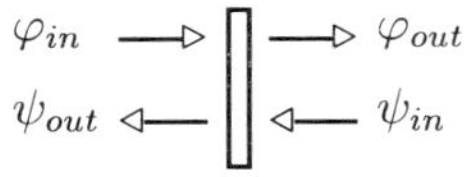

Figure 1. Fields scattered by a single mirror.

The mirror is assumed to be very heavy when compared with the field energy, so that under reflection momentum is transfered to the field while its energy is preserved (the mirror's recoil is neglected). s and r are frequency dependent transmission and reflection amplitudes, and must obey the general analyticity and unitarity conditions of scattering matrices which correspond to causality of field scattering and conservation of probabilities. In addition, we shall assume high frequency transparency, i.e. that reflection coefficients vanish sufficiently rapidly when frequency goes to infinity.

Two mirrors form a Fabry-Perot cavity which divides space into three domains where the field propagates freely (Figure 2).

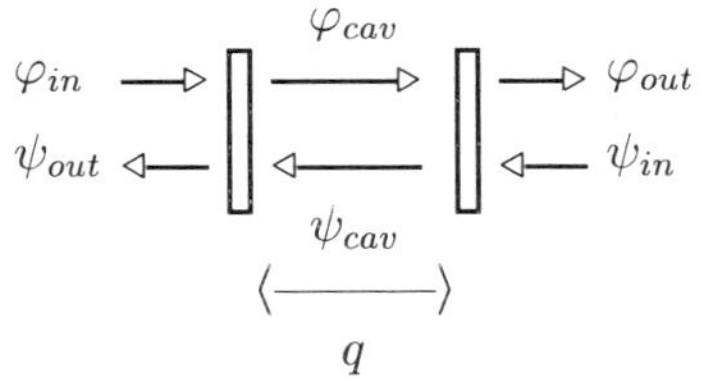

Figure 2. Fields scattered by a cavity.

All fields are determined by input fields and the mirrors' scattering matrices. On each side of each mirror, the radiation pressure exerted by the field is provided by the field stress tensor $(T^{\mu\nu})$:

$$\begin{aligned} T^{00} &= T^{11} = \frac{1}{2}(\partial_t\phi^2 + \partial_x\phi^2) \\ T^{01} &= T^{10} = -\partial_t\phi\partial_x\phi \end{aligned} \qquad (5)$$

The force is given by the flux of stress tensor component T^{11} through the mirror (difference between left and right sides; a dot stands for time derivative):

$$
\begin{array}{rcl}
F_1 & = & \{\dot\varphi_{in}(t-q_1)^2 + \dot\psi_{out}(t+q_1)^2\} - \{\dot\varphi_{cav}(t-q_1)^2 + \dot\psi_{cav}(t+q_1)^2\} \\
F_2 & = & \{\dot\varphi_{cav}(t-q_2)^2 + \dot\psi_{cav}(t+q_2)^2\} - \{\dot\varphi_{out}(t-q_2)^2 + \dot\psi_{in}(t+q_2)^2\}
\end{array}
$$

(6)

$< F_1 >= - < F_2 >= F_c$ is the mean radiation pressure, or Casimir force, felt by the mirrors [10].

In vacuum state, the two input field components are uncorrelated and have identical auto-correlations:

$$
< \varphi_{in}[\omega]\varphi_{in}[\omega'] >=< \psi_{in}[\omega]\psi_{in}[\omega'] >= \frac{2\pi}{\omega^2}\delta(\omega+\omega')\theta(\omega)\frac{1}{2}\hbar\omega
$$

(7)

(θ is Heaviside's step function). Although the corresponding vacuum mean energy density is infinite (the spectral energy density increases linearly with frequency), the mean forces exerted on the mirrors (6) are finite, as the reflection coefficients of the mirrors (r_1 and r_2) satisfy high frequency transparency ($q = q_2 - q_1$):

$$
F_c = \int_0^\infty \frac{d\omega}{2\pi}\hbar\omega\{1 - g[\omega]\}
$$

$$
g[\omega] = \frac{1 - |r[\omega]|^2}{|1 - r[\omega]e^{2i\omega q}|^2} \qquad r[\omega] = r_1[\omega]r_2[\omega]
$$

g describes the field spectral energy density inside the cavity. At the limit of perfect mirrors ($r = 1$), g becomes a sum of delta functions at frequencies equal to the modes of the cavity ($n\frac{\pi}{q}$) and the Casimir force between the two mirrors identifies with (2).

QUANTUM FLUCTUATIONS OF RADIATION PRESSURE

The radiation pressure exerted on a scatterer is related to the field stress tensor, so that it is a function of fields (in general a quadratic form, see (5) for instance). Consequently, quantum fluctuations of fields also induce quantum fluctuations of stress tensors and radiation pressures. The fluctuations of Casimir forces exerted on mirrors, due to quantum fluctuations of electromagnetic fields, have recently been studied [11]. We shall just discuss some general properties of stress tensor and radiation pressure fluctuations in vacuum.

Electromagnetic fields $F_{\mu\nu}$ and stress tensor components $T_{\mu\nu}$ can be derived from the electromagnetic potentials A_μ ($F_{\mu\nu} = \partial_\mu A_\nu - \partial_\nu A_\mu$, and $\eta_{\mu\nu}$ is Minkowski metric $diag(1,-1,-1,-1)$):

$$
T_{\mu\nu} = F_\mu{}^\lambda F_{\nu\lambda} - \frac{1}{4}\eta_{\mu\nu}F^{\rho\lambda}F_{\rho\lambda}
$$

Vacuum correlations of electromagnetic potentials are determined from propagation equations (in Feynman gauge):

$$
\begin{array}{rcl}
< A_\mu(x)A_\nu(0) > & = & \int \frac{d^4k}{(2\pi)^4}e^{-ik.x}C_{A_\mu A_\nu}[k] \\
C_{A_\mu A_\nu}[k] & = & 2\pi\hbar\theta(k_0)\delta(k^2)\eta_{\mu\nu}
\end{array}
$$

These expressions exhibit translation and Lorentz invariances, and a spectrum limited to light-like momenta with positive frequencies (as vacuum is the state of minimum energy, transitions only have positive frequencies). In vacuum, correlations of stress tensors are determined from field correlations using Wick'rules:

$$< T_{\mu\nu}(x)T_{\rho\sigma}(0) > - < T_{\mu\nu}(x) >< T_{\rho\sigma}(0) >= \int \frac{d^4k}{(2\pi)^4}e^{-ik.x}C_{T_{\mu\nu}T_{\rho\sigma}}[k]$$

$$C_{T_{\mu\nu}T_{\rho\sigma}}[k] = \frac{\hbar^2}{40\pi}\theta(k_0)\theta(k^2)(k^2)^2\pi_{\mu\nu\rho\sigma}$$

$$\pi_{\mu\nu\rho\sigma} = \frac{1}{2}(\pi_{\mu\rho}\pi_{\nu\sigma} + \pi_{\mu\sigma}\pi_{\nu\rho}) - \frac{1}{3}\pi_{\mu\nu}\pi_{\rho\sigma}$$

$$\pi_{\mu\nu} = \eta_{\mu\nu} - \frac{k_\mu k_\nu}{k^2}$$

Stress tensor correlations are gauge independent and are in fact completely determined, up to a numerical factor, by general symmetries that are satisfied by correlation functions in vacuum. Translation and Lorentz invariances imply that correlations in momentum domain are tensors built from k_μ and $\eta_{\mu\nu}$ only. Correlations of stress tensors $T_{\mu\nu}$ and $T_{\rho\sigma}$ decompose on tensors which are symmetric in indices (μ, ν), (ρ, σ) and exchange of these pairs, and which are transverse because of energy-momentum conservation ($\partial^\mu T_{\mu\nu} = 0$). As Maxwell stress tensor is moreover traceless, this leaves only one such tensor $\pi_{\mu\nu\rho\sigma}$. In momentum domain, stress tensor correlations are obtained as convolutions of field correlation functions in vacuum, so that they only contain time-like momenta (given by adding two light-like momenta of positive frequencies), and thus a factor $\theta(k_0)\theta(k^2)$. Quite generally, as discussed at the end of this section, such factor can also be seen as a consequence of fluctuation-dissipation relations characteristic of vacuum, and of Lorentz invariance. By dimensionality, these correlations are proportional to $(k^2)^2$. Explicit computation provides the remaining numerical factor.

When seen as a boundary condition for the field, a perfect mirror determines a relation between outcoming and incoming fields, which allows one to derive the fluctuations of radiation pressure exerted on the mirror from those of incoming fields. Explicit computations have been performed for mirrors of different shapes [11, 12].

We shall briefly discuss the model of partially transmitting mirror in two-dimensional space-time. The force exerted on the mirror is given by the difference of field energy densities between the two sides of the mirror (6). For a single mirror and vacuum input fields, its mean value vanishes. However, the radiation pressures exerted on both sides of the mirror have independent fluctuations, and the resulting force still fluctuates. Force fluctuations on a mirror at rest are stationary and determined by the mirror's scattering matrix (4) and input field correlations (7) [13]:

$$< F(t)F(t') >= C_{FF}(t - t')$$

$$C_{FF}[\omega] = 2\hbar^2\theta(\omega)\int_0^\omega \frac{d\omega'}{2\pi}\omega'(\omega - \omega')\text{Re}\{1 - s[\omega']s[\omega - \omega'] + r[\omega']r[\omega - \omega']\} \quad (8)$$

Force correlations are positive and always finite. They vanish as ω^3 around zero frequency (energy-momentum is conserved in vacuum). In particular for a perfect mirror, they are directly related to correlations of momentum densities of incoming free fields (see (3)):

$$C_{FF}[\omega] = \frac{\hbar^2}{3\pi}\theta(\omega)\omega^3 \quad (9)$$

Correlation spectra in vacuum contain positive frequencies only. Force correlations can similarly be derived for thermal fields. In thermal equilibrium, a fluctuation-dissipation relation relates commutators and anticommutators [14]:

$$2\hbar\xi_{FF}(t) = <[F(t), F(0)]>$$
$$2\hbar\sigma_{FF}(t) = <\{F(t), F(0)\}>$$

$$2\hbar\xi_{FF}[\omega] = (1 - e^{-\frac{\hbar\omega}{T}})C_{FF}[\omega] \tag{10}$$

$$\sigma_{FF}[\omega] = \coth\frac{\hbar\omega}{2T}\xi_{FF}[\omega]$$

(T is the temperature). This fluctuation-dissipation relation, which was first studied for Nyquist noise in electric circuits [15], allows to determine fluctuations in a state of thermal equilibrium from the commutator only. As discussed in next section, the commutator also identifies with the dissipative part of the linear response of the system to an external perturbation [14]. At the limit of zero temperature, that is in vacuum, the fluctuation-dissipation relation leads to a fluctuation spectrum which is limited to positive frequencies:

$$C_{FF}[\omega] = 2\hbar\theta[\omega]\xi_{FF}[\omega] \tag{11}$$

The factor $\theta[\omega]$ explicitly shows that correlations (8) are not symmetric under exchange of time arguments and thus exhibits the non-commutative character (i.e. the quantum nature) of fluctuations in vacuum.

RADIATION REACTION FORCE

As first discussed by Einstein, a scatterer immersed in a fluctuating field undergoes a Brownian motion [7]. The fluctuating force exerted by the field leads to a diffusion process for the scatterer's momentum P, which spreads in time according to:

$$< \Delta P^2 >\sim 2D\Delta t$$

The momentum distribution being constant at thermal equilibrium, the effect of the fluctuating force must be exactly compensated by the effect of a further cumulative force which appears when the scatterer moves (with velocity $\delta\dot{q}$):

$$< \delta F >\sim -\gamma\delta\dot{q}$$

This implies a relation between the momentum diffusion coefficient D which characterises force fluctuations and the friction coefficient γ which characterises the mean dissipative force:

$$D = \gamma T \tag{12}$$

(T is the temperature). In the general quantum case, when small displacements of the scatterer are considered, a relation still holds between force fluctuations exerted on the scatterer at rest and the mean dissipative force exerted on the moving scatterer, i.e. the imaginary part of the force susceptibility χ_{FF}:

$$< \delta F[\omega] >= \chi_{FF}[\omega]\delta q[\omega] \tag{13}$$

Indeed, according to linear response theory [14], the susceptibility of a quantity under a perturbation is related to the correlations, in the unperturbed state, of this quantity with the generator of the perturbation. More precisely, fluctuation-dissipation relations identify the imaginary (or dissipative) part of the susceptibility of a quantity with its commutator with the generator of the perturbation. In the case of displacements, the generator is given by the force itself:

$$\text{Im}\chi_{FF}[\omega] = \xi_{FF}[\omega] \tag{14}$$

In a thermal state, according to fluctuation-dissipation relations (10) and (14), force fluctuations are connected with the mean motional force:

$$2\text{Im}\chi_{FF}[\omega] = \frac{1}{\hbar}(1 - e^{-\frac{\hbar\omega}{T}})C_{FF}[\omega]$$

At the limit of high temperature, Einstein's relation is recovered (see 12):

$$2\text{Im}\chi_{FF}[\omega] = \frac{\omega}{T}C_{FF}[\omega]$$

At the limit of zero temperature, force fluctuations subsist on a scatterer at rest and imply that, when moving in vacuum, a scatterer is submitted to a dissipative force. The motional susceptibility of the force is then determined by force correlations in vacuum (8) and fluctuation-dissipation relation (14). For a perfect mirror, at first order in the mirror's displacement, the mean dissipative force is proportional to the third time derivative (see (9)):

$$< \delta F(t) > = \frac{\hbar}{6\pi} \dddot{q}(t) \tag{15}$$

More recently, effects of moving boundaries on quantum fields have been discussed to study the influence of classical constraints on quantum fields, and in particular as an analogy for quantum fields in a classical curved space [16]. A perfect mirror moving in vacuum has been shown to radiate and hence to undergo a radiation reaction force. In two-dimensional space-time, and for scalar fields in vacuum, the radiation reaction force is proportional to the two-dimensional version of Abraham vector [17], and identifies with (15) in linearised and non relativistic limits. The radiation reaction force for partially transmitting mirrors can also be obtained following this approach. A moving mirror induces a motional modification of the field scattering matrix (4). Using a coordinate transformation to the mirror's proper frame, the modification of the scattering matrix is easily obtained up to first order in the mirror's displacement [13]:

$$\begin{pmatrix} \delta\varphi_{out}[\omega] \\ \delta\psi_{out}[\omega] \end{pmatrix} = \int \frac{d\omega'}{2\pi} \delta S[\omega, \omega'] \begin{pmatrix} \varphi_{in}[\omega'] \\ \psi_{in}[\omega'] \end{pmatrix} \tag{16}$$

For general motions of the mirror, vacuum incoming fields are transformed into outcoming fields whose correlations do not correspond to vacuum. Then, energy-momentum is radiated by the moving mirror. As the mean stress tensor of outcoming fields differs from that of incoming fields, the radiation pressure exerted by scattered fields does not vanish and the mirror is submitted to a radiation reaction force (13). The radiation pressure exerted by scattered fields (see (6)) can be obtained using the modified scattering matrix (16) [13]:

$$\chi_{FF}[\omega] = i\hbar \int_0^\omega \frac{d\omega'}{2\pi}\omega'(\omega - \omega')\{1 - s[\omega']s[\omega - \omega'] + r[\omega']r[\omega - \omega']\} \tag{17}$$

As expected, this expression and (8) satisfy fluctuation-dissipation relation (14). As a consequence of Lorentz invariance of vacuum, the radiation reaction force vanishes for uniform motion; expression (17) leads to (15) for a perfect mirror and also vanishes for uniformly accelerated motion:

$$\chi'_{FF}[0] = \chi''_{FF}[0] = 0$$

The radiation reaction force felt by a plane perfect mirror in four-dimensional space-time has been obtained for motions in scalar field [18] and in electromagnetic [19] vacua. Direct comparison between radiation pressure fluctuations and the dissipative force shows that fluctuation-dissipation relations are satisfied in these cases.

STABILITY OF MOTION AND POSITION FLUCTUATIONS

It is well-known from classical electron theory [20], that a radiation reaction force proportional to Abraham-Lorentz vector (15) leads to motions which are either unstable (the so-called "runaway solutions") or violate causality. Perfect mirrors, and in particular mirrors treated as field boundaries, have motions in vacuum which are affected by the same stability and causality problems. However, as we briefly show in the following, partially transmitting mirrors can avoid these difficulties [21].

When a scatterer is submitted to an applied external force $F(t)$, its motion is determined by an equation which also takes into account the force induced by motion (13). It will be sufficient for our purpose to consider small displacements only, so that the equation of motion reads (we generally consider a mirror of mass m_0, bound with a proper frequency ω_0; a free mirror is recovered for $\omega_0 = 0$):

$$m_0(\ddot{q}(t) + \omega_0^2 q(t)) = F(t) + \int_{-\infty}^{\infty} dt' \chi_{FF}(t - t') q(t') \tag{18}$$

The motional response is best characterised in the frequency domain, by a mechanical impedance $Z[\omega]$ which relates the resulting velocity to the applied force:

$$-i\omega Z[\omega] q[\omega] = F[\omega] \qquad Z[\omega] = -im_0\omega + i\frac{m_0\omega_0^2}{\omega} + \frac{\chi_{FF}[\omega]}{i\omega} \tag{19}$$

From analytic properties of the scattering matrix (4), it is easily derived that χ_{FF} is also analytic in the upper half plane $(\text{Im}(\omega) > 0)$, showing that the induced force is a causal function of the scatterer's motion. The mechanical impedance of a perfect mirror is obtained from (15):

$$Z[\omega] = -im_0\omega + i\frac{m_0\omega_0^2}{\omega} + \frac{\hbar}{6\pi}\omega^2 \tag{20}$$

As easily seen, the mechanical admittance $(Y[\omega] = Z[\omega]^{-1})$ is no longer analytic but has a pole in the upper half plane. Such a pole corresponds to unstable motions, i.e. "runaway solutions" in the free mirror case. If supplementary boundary conditions are imposed (at large time for instance) to exclude such solutions, then the resulting motions violate the causal dependence on the applied force.

A partially transmitting mirror is characterised by a force susceptibility of the general form:

$$\chi_{FF}[\omega] = i\frac{\hbar}{6\pi}\omega^3 \Gamma[\omega]$$

High frequency transparency and analyticity lead to a better behaviour of the susceptibility [21]:

$$\Gamma[\omega] \sim i\frac{\omega_c}{\omega} \qquad \text{for} \qquad \omega \to \infty$$

where ω_c defines a high frequency cut-off. Characteristic relations in vacuum derived from (11) and (14) furthermore imply:

$$\text{Im}(\frac{\chi_{FF}[\omega]}{\omega}) \geq 0 \qquad \text{for} \qquad \text{Im}(\omega) \geq 0 \qquad (21)$$

This property means that the force susceptibility is related to a positive function [22]. The mechanical impedance defined by (19) corresponds to a positive function as long as its pole at infinity satisfies:

$$m_\infty = m_0 - \frac{\hbar\omega_c}{6\pi} \geq 0 \qquad (22)$$

The inverse of a positive function is still a positive function, and furthermore causality follows from positivity [22]. When inequality (22) is satisfied, the mechanical admittance is also a causal function, and no "runaway solutions" can appear (see (19)). m_0 and m_∞ can be seen as describing a low and a high frequency mass for the mirror, their difference being a mass correction induced by the field interacting with the mirror. Stability in vacuum is insured as long as the free mass of the mirror m_∞ is positive, or else as long as the quasistatic mass of the mirror m_0 is larger than the induced mass. Realistic mirrors certainly satisfy this condition, contrarily to perfect mirrors. This discussion also shows the incompatibility between stability in vacuum and mass renormalisation, since m_∞ is infinitely negative in that case (see [23] for a similar discussion for electrons).

While motion of the mirror modifies the scattered field, the radiation pressure exerted by the field also perturbs the mirror's motion. Equations of the coupled system, mirror and vacuum field radiation pressure, when treated linearly, can be solved to obtain fluctuations for the interacting system in terms of input fluctuations only [24]. In particular, equation (18) provides the mirror's position fluctuations in terms of input force fluctuations. As a consequence of consistency relations satisfied by linear response formalism, position fluctuations of the mirror coupled to vacuum fields also satisfy fluctuation-dissipation relations, which relate the position commutator to the dissipative part of the mechanical admittance (the mirror's response to an applied force):

$$\begin{aligned} Y[\omega] &= \frac{-i\omega}{m_0(\omega_0^2 - \omega^2) - \chi_{FF}[\omega]} \\ \xi_{qq}[\omega] &= \text{Re}Y[\omega]/\omega \end{aligned} \qquad (23)$$

Fluctuations of the coupled system also satisfy the relations characteristic of vacuum:

$$C_{qq}[\omega] = 2\hbar\theta(\omega)\xi_{qq}[\omega]$$

The mechanical admittance completely determines position fluctuations in vacuum. As shown by expression (23), position fluctuations consist of two main parts. One part corresponds to a resonance peak at the oscillator's proper frequency ω_0. These fluctuations subsist at the limit of decoupling between mirror and field and can be seen as proper position fluctuations of the mirror; they identify with the fluctuations associated with Schrödinger equation for a free oscillator. The other part is a background noise spreading over all frequencies. These are position fluctuations induced by the fluctuating radiation pressure of vacuum fields and are dominant outside resonance peaks; they describe ultimate position fluctuations and correspond to a quantum limit on the sensitivity of an optimal position measurement performed at such frequencies [5, 24].

CONCLUSION

Quantum field fluctuations in vacuum exert a fluctuating radiation pressure on scatterers. In agreement with fluctuation-dissipation relations, a scatterer moving in vacuum experiences an additional force depending on its motion. Like in classical electron theory, it generally modifies the scatterer's motional response to an applied force. For realistic mirrors, which are transparent to high frequencies, motions can be shown to remain stable and causal in vacuum, a property which is violated by mass renormalisation.

The quantum Brownian motion induced on the mirror's position by its coupling to vacuum field radiation pressure can be described consistently within linear response formalism. Position fluctuations are determined from the mechanical susceptibility of the mirror through fluctuation-dissipation relations. Vacuum field fluctuations then lead to ultimate quantum fluctuations for positions in space-time.

References

[1] H.A. Lorentz. *The Theory of Electrons*, Leipzig (1915) [reprinted by Dover, New York (1952)].

[2] A. Einstein, *Ann. Physik* 18:639 (1905), 20:627 (1906); *Jahrb. Radioakt. Elektron.* 4:411 (1907), 5:98 (1908) [translated by H.M. Schwartz, *Am. J. Phys.* 45:512, 811, 899 (1977)].

[3] H. Groch and E. Kazes, in: *Foundations of Radiation Theory and QED*, A.O. Barut, ed., Plenum Press, New York (1980).

[4] N. Bohr and L. Rosenfeld, *Phys. Rev.* 78:794 (1950).

[5] V.B. Braginsky and F.Ya. Khalili. *Quantum Measurement*, University Press, Cambridge (1992). M.T. Jaekel and S. Reynaud, *Europhys. Lett.* 13:301 (1990); *Phys. Lett.* A 185:143 (1994).

[6] H.B.G. Casimir, *Proc. K. Ned. Akad. Wet.* 51:793 (1948). G. Plunien, B. Müller and W. Greiner, *Phys. Rep.* 134:87 (1986).

[7] A. Einstein, *Ann. Physik* 17:549 (1905); *Phys. Z.* 10:185 (1909), 18:121 (1917).

[8] P.S. Wesson, *ApJ* 378:466 (1991).

[9] L.S. Brown and G.J. Maclay, *Phys. Rev.* 184:1272 (1969).

[10] M.T. Jaekel and S. Reynaud, *J. Phys. I France* 1:1395 (1991).

[11] G. Barton, *J. Phys. A: Math. Gen.* 24:991, 5533 (1991); in: *Cavity Quantum Electrodynamics (Supplement: Advances in Atomic, Molecular and Optical Physics*, P. Berman, ed., Academic Press (1994).

[12] C. Eberlein, *J. Phys. A: Math. Gen.* 25:3015, 3039 (1992).

[13] M.T. Jaekel and S. Reynaud, *Quantum Opt.* 4:39 (1992).

[14] R. Kubo, *Rep. Prog. Phys.* 29:255 (1966).
L.D. Landau and E.M. Lifschitz. *Cours de Physique Théorique,*
Physique Statistique, première partie, Mir, Moscou, (1984).

[15] H.B. Callen and T.A. Welton, *Phys. Rev.* 83:34 (1951).

[16] B.S. de Witt, *Phys. Rep.* 19:295 (1975).
N.D. Birell and P.C.W. Davies. *Quantum Fields in Curved Space,*
University Press, Cambridge (1982).
S.A. Fulling. *Aspects of Quantum Field Theory in Curved Spacetime,*
University Press, Cambridge (1989).

[17] S.A. Fulling and P.C.W, Davies, *Proc. R. Soc.* A 348:393 (1976).

[18] L.H. Ford and A. Vilenkin, *Phys. Rev.* D 25:2569 (1982).

[19] P.A. Maia Neto, *J. Phys. A Math. Gen.* 27:2167 (1994).

[20] F. Rohrlich. *Classical Charged Particles,* Addison-Wesley, Reading (1965).

[21] M.T. Jaekel and S. Reynaud, *Phys. Lett.* A 167:227 (1992).

[22] J. Meixner, in: *Statistical Mechanics of Equilibrium and Non-Equilibrium,*
J. Meixner, ed., North-Holland, Amsterdam (1965).

[23] H. Dekker, *Phys. Lett.* A 107:255 (1985);
Physica 133 A:1 (1985).

[24] M.T. Jaekel and S. Reynaud, *J. Phys.I France* 3:1 (1993).

VACUUM STRUCTURES FOR THE CHARGED FIELDS IN PRESENCE OF E-M POTENTIALS AND BOUNDARIES

I. H. Duru

Trakya University
Department of Mathematics
P.K. 126, Edirne, Turkey
and
TUBITAK, Marmara Research Center,
P.O. Box 21, 41470 Gebze, Turkey

INTRODUCTION

In 1951 Schwinger discovered that in constant electric field pairs of charged particles are spontaneously produced [1]. This phenomena has extensively been studied in time dependent gauge and in the path integral formalism [2]. It is possible that some other electric fields, which are time dependent also produces pairs [3]. Coulomb potential of a charge exceeding the value 137/2 is another example to the particle producing fields [4].

In this work we first present an investigation on the possible pair production of the electrically charged particles by magnetic monopoles. It is observed that in the field of magnetic monopole g, electrically charged pairs are produced with total

angular momenta ℓ, satisfying $(l+1/2)^2 < e^2 g^2$ and with azimuthal component $m=0$.

In the second part we discuss another effect involving the nontrivial vacuum structure of charged particle fields: We propose the existence of an attractive Casimir force between the Aharonov-Bohm (A-B) solenoids [5]. The usual Casimir effect, i.e., the attraction between the perfectly conductive parallel plates, is due to the non zero vacuum expactation value of the electromagnetic field [6]. The perfectly conductive plates constitute boundaries for the radiation field. If one can realise boundaries for another field one may well expect a Casimir type effect resulting from the vacuum expectation value of that field. In fact A-B solenoids create a non-trivial geometry, i.e., a multiply connected topology for the electrically charged particles.

In the final section we mention a possible observation of Casimir effect in the solid state structures in disordered materials.

Electron Theory and Quantum Electrodynamics: 100 Years Later
Edited by Dowling, Plenum Press, New York, 1997

PARTICLE CREATION BY MAGNETIC MONOPOLES

Potential of a magnetic monopol g, located at the origin is expressed in terms of the spherical coordinates is $(\hbar = c = 1)$

$$A = \frac{g(1-\cos\theta)}{r\sin\theta} \hat{\phi} \tag{1}$$

Klein-Gordon equation* for a scalar particle with charge e and mass μ in the field of the above potential given by

$$[(i\partial_\mu - eA_\mu)^2 - \mu^2] f(x) = 0 \tag{2}$$

can be solved by

$$f_{Elm}(x) = e^{-iEt} \frac{e^{-im\varphi}}{\sqrt{2\pi}} \sqrt{l+1/2} \; P_{\alpha\beta}^l(\cos\theta) R(r) \tag{3}$$

where

$$\alpha = m - eg \quad , \quad \beta = eg \quad . \tag{4}$$

and $P_{\alpha\beta}^l$ is the Jacobi function. The radial wave function R is given either by

$$R_>(r) \sim \frac{1}{\sqrt{r}} \; H_{\pm\gamma}^{(2)} (\sqrt{E^2 = \mu^2} \; r) \tag{5}$$

which is regular at $r \to \infty$, or by

$$R_<(r) \sim \frac{1}{\sqrt{r}} \; J_{\pm\gamma} (\sqrt{E^2 - \mu^2} r) \tag{6}$$

which can be regular at $r \to 0$. The index γ is defined as

$$\gamma = \sqrt{(l+1/2)^2 - e^2 g^2}. \tag{7}$$

Consider the behaviour of the Bessel function $J_{\pm\gamma}$ near the origine:

$$J_{\pm\gamma} \sim r^{\pm\gamma/2} \quad . \tag{8}$$

If γ is real and positive (or negative), the positive (or negative) sign for the index of

---

*For simplicity we discuss the creations of scalar particles. Case involving the Dirac particles is essentially the same.

the Bessel function (6) is chosen to have finite solution at the origin. Thus for $(l+1/2)^2 > e^2g^2$ we have a unique solution at the origine. For imaginary values of γ we do not have any physical criteria to distinguish between J_γ and $J_{-\gamma} = J_\gamma^*$. Therefore recalling the relation between the Hankel and Bessel functions [6]

$$H_\gamma^{(2)} = \frac{1}{\sin i\pi\gamma} \, (J_\gamma(z) \, e^{i\pi\gamma} - J_{-\gamma}(z)) \, , \tag{9}$$

for imaginary γ's it is possible to establish a correlation between the states describing the particles at $r \to \infty$ and $r \approx 0$.

Choose the wave functions describing particles at $r \to \infty$ and $r \approx 0$ are as

$$\vec{f}_{Elm} \approx e^{-iEt} \frac{e^{-im\varphi}}{\sqrt{2\pi}} \, \sqrt{l+1/2} \; P_{\alpha\beta}^l \, (\cos\theta) H_\gamma^{(2)} \, (\sqrt{E^2-\mu^2} \, r) \tag{10}$$

and

$$\vec{f}_{Elm}(x) \approx e^{-iEt} \frac{e^{-im\varphi}}{\sqrt{2\pi}} \, \sqrt{l+1/2} \; P_{\alpha\beta}^l(\cos\theta) \, J_\gamma \, (\sqrt{E^2-\mu^2} \, r). \tag{11}$$

Then definition (4) and relation (9) suggest the following relation for $m=0$:

$$\vec{f}_{Elo}(x) \approx \frac{-1}{\sin\pi\,|\gamma|} \, [-e^{-\pi|\gamma|} \, \vec{f}_{Elo}(x) + (\vec{f}_{Elo}(x))^* \,] \quad . \tag{12}$$

This means that for certain values of quantum numbers (i.e. for $(l+1/2)^2 < e^2g^2$ and $m=0$) the particle states at $r \approx 0$ is equal to the linear combination of the particle and antiparticle states of $r \to \infty$.

The asymptotic vacuum is equivalent to the vacuum at the origin only with the probability $(e^{-\pi|\gamma|})^2$. On the other hand with the relative probability

$$\omega = 1 - e^{-2\pi|\gamma|} \tag{13}$$

charged pairs are created with energies E, $-E$; having the total angular momenta $l < |eg| - 1/2$ and azimuthal angular momenta $m=0$.

CASIMIR FORCE BETWEEN AHARANOV-BOHM SOLENOIDS

Consider two, infinitely long cylindrical solenoids confining the fluxes ϕ_1 and

ϕ_2 ; parallel to the z-axis with their centers located at

$$x_1 = 0 \quad \text{and} \quad x_2 = (a,0,0) \ .$$

To calculate the vacuum expectation value of the charged scalar field in the region outside the solenoids, we first write down the symmetric Green function:

$$G(x_\mu, x_\mu') = [\cos \frac{e}{2\pi} \phi_1 (\theta - \theta') \cos \frac{e}{2\pi} \phi_2 (\theta - \theta_2') - 1] G_M(x_\mu, x_\mu') \tag{14}$$

Here G_M is the Minkowski space Green function. The angle θ_2 is given by

$$\sin \theta_2 = y/\sqrt{(x-a)^2 + y^2} \ . \tag{15}$$

The vacuum expectation value of the energy density for the massive charged field is calculated by the coincidence limit formula:

$$<E> = \lim_{x_\mu \to x_\mu'} \frac{1}{4} (\partial_t \partial_t' + \nabla \cdot \nabla' + \mu^2) G(x_\mu, x_\mu') \ . \tag{16}$$

In the above formula there are "self-energy" terms involving only ϕ_1^2 or ϕ_2^2 , which have infinities. However, the interaction term depending on $(\phi_1 \phi_2)^2$ is finite:

$$<E>_{int} = \frac{e^4}{(2\pi)^4} \frac{\phi_1^2 \phi_2^2}{16\pi^2} \frac{1}{\rho^2 \rho_2^2} (1 - \frac{(\rho^2 - ax)^2}{\rho^2 \rho_2^2}) \ . \tag{17}$$

Here ρ is the polar coordinate in xy-plane and ρ_2 is given by

$$\rho_2 = \sqrt{(x-a)^2 + y^2} \ .$$

Integrating (17) over xy-plane we obtain the total interaction energy in the slice of space with unit thickness having the normal direction in the z-direction; which in thin solenoid limit, i.e., for solenoids with radii small compared to a is given by

$$E_{int} = - \frac{\hbar c e^4}{32\pi^6} \frac{\phi_1^2 \phi_2^2}{a^2} \ . \tag{18}$$

Here we have introduced $\hbar c$ to give dimensions. The above interaction energy implies an attractive force per unit length of the solenoids:

$$F_{int} = - \frac{\partial E_{int}}{\partial a} = - \frac{\hbar c e^4}{16\pi^6} \frac{\phi_1^2 \phi_2^2}{a^3} \ . \tag{19}$$

This result should be compared to the usual Casimir force per unit area between parallel plates [7]:

$$F_{plate} = -\frac{\pi}{240}\frac{\hbar c}{a^4} \quad .$$ (20)

For the typical experimental distance $a = 0.5\,\mu m$, this force is 0.2 dyn/cm^2 [8]. For the same distance, the magnitude of our force is

$$F_{int} \approx (\frac{e}{2\pi})^4 (\phi_1 \phi_2)^2 \times 10^{-5} dyn/cm \quad .$$ (21)

CASIMIR ENERGIES FOR LOCALIZED FIELDS
IN DISORDERED MATERIALS [9]

Recent achievements in growth technologies made it possible to confine electrons and light in small cavities in the solid state structures. Therefore we can talk of Casimir effect for these cavities.

Theoretically for a spherical cavity of radius R, with perfectly conductive walls the vacuum energy of the electromagnetic field is $(\hbar = c = 1)$ [8]:

$$\varepsilon = 0.0923/2R \quad .$$ (22)

For massive fields the above formula to be multiplied by

$$e^{-2mR} \quad .$$ (23)

Typical experimental size of the confinement is of the order of nanometers, i.e., $R \approx 10^{-7} cm$. For this size of cavity the Casimir energy (22) for light is

$$\varepsilon \approx 0.5 \times 10^6 cm^{-1} \approx 10\,eV \quad .$$ (24)

For electrons, using $m \approx 0.5\,MeV \approx 2.5 \times 10^{10}\,cm^{-1}$ and $R \approx 10^{-7} cm$, the exponantial factor (23) is

$$2mR \approx 5 \times 10^3 \quad .$$ (25)

Thus, the Casimir energy for electrons is very small.

REFERENCES

1) J. Schwinger, Phys.Rev. 82, 664 (1951).
2) A.I. Nikishov, Zh.Eksp.Teor.Fiz. 57, 1210 (1969) [Sov.Phys.JETP 30, 660 (1970)]; N.B. Narozhnyi and A.I. Nikishov, Yad.Fiz. 11, 1072 (1970) [Sov.J.Nucl.Phys. 11, 596 (1970)]. A.O. Barut and I.H. Duru, Phys.Rev. D41, 1312 (1990).

3) S.A. Baran and I.H. Duru, J. of Sov. Laser Research, <u>13</u>, 241 (1992).

4) See for example A.S. Lapedes, Phys.Rev. <u>D17</u>, 2556 (1978).

5) I.H. Duru, Foundations of Phys. <u>23</u>, 809 (1993).

6) W. Magnus, F. Oberhettinger and R.P. Soni, "Formulas and Theorems for the Special Functions of Mathematical Physics" 3<u>rd</u> edn. (Springer, Berlin, 1966).

7) See for example N.D. Birrell and P.C. W. Davies, "Quantum Fields in Curved Space" (Cambridge University Press, Cambridge, 1982).

8) See for example V.M. Mostepanenko and N.N. Trunov, Sov.Phys.Usp. <u>31</u>, 965 (1988).

9) I.H. Duru and M. Tomak, Phys.Letters <u>A176</u>, 265 (1993).

QUANTUM ELECTRODYNAMICS
IN SUPERCONDUCTING CIRCUITS

J.F. Ralph,[1,2] R. Whiteman,[2] J. Diggins,[2] R.J. Prance,[2]
T.D. Clark,[2] H. Prance,[2] T.P. Spiller,[2] and A. Widom [3]

[1]Scuola Normale Superiore, Piazza dei Cavalieri 7,
56126 Pisa, Italy.
[2]Mathematical and Physical Sciences, University of Sussex,
Falmer, BN1 9QH, U.K.
[3]Physics Department, Northeastern University, Boston,
MA 02115, U.S.A.

Abstract

We discuss the appearence of quantum electrodynamic effects in superconducting circuits. We concentrate on one simple system; a thick superconducting ring containing a single Josephson weak link. We construct a Hamiltonian for the weak link ring and solve the time-independent Schrödinger equation to find its energy eigenstates. We then describe a method of probing the behaviour of this macroscopic quantum circuit through its response to an applied magnetic flux, giving experimental and theoretical results.

Introduction

It is possibly not suprising that quantum electrodynamic effects can appear in electrical circuits. After all, we tend to assume that all matter is essentially quantum mechanical at some fundamental level. The question is whether such effects are important when determining the behaviour of a given circuit. The problem is that most electrical circuits are far too complicated to deal with from a microscopic level because there are far too many degrees of freedom. Apart from calculating the energy band structure of the constituent materials, very little quantum mechanics is used; the gross behaviour of the system is generally considered to be entirely classical. In most cases this is a perfectly justifiable approximation.

In this paper, however, we will deal with one case where quantum mechanical effects are important, simple superconducting circuits (for recent reviews see [1, 2,

Electron Theory and Quantum Electrodynamics: 100 Years Later
Edited by Dowling, Plenum Press, New York, 1997

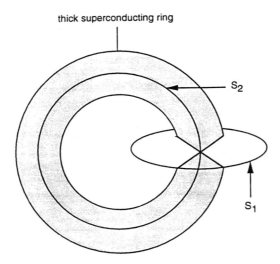

thick superconducting ring

S_2

S_1

Figure 1. A schematic diagram of a thick superconducting ring with a weak link constriction, indicating the open surfaces S_1 and S_2 used to define the variables Q and Φ.

3, 4]). In a bulk superconductor, the appearence of a coherent macroscopic phase means that an otherwise very complex system can be treated as a single quantum mechanical degree of freedom [5, 6]. The superconducting energy gap ensures that it is the excitations of this macroscopic phase which form the lowest lying energy states rather than any microscopic excitations.

We will concentrate on one of the simplest superconducting circuits; a thick superconducting ring containing one Josephson weak link [7] (the rf-SQUID ring [8]). We will introduce a simple lumped circuit model for the weak link ring and derive a Hamiltonian for this circuit. By imposing the usual canonical commutation relations, we can solve the time-independent Schrödinger equation for the energy eigenstates corresponding to the macroscopic excitations of the ring. (In the experments discussed here, the ring will be truely macroscopic, with typical dimensions being of the order of $\sim$ 1cm.)

By coupling such a ring to a classical LC oscillator, we show that it is possible to identify which energy state the ring is in through its response to an external applied magnetic flux. Experimental and theoretical results are given which are in remarkably good agreement for the ground state and first two excited states. We will then discuss some limitations of this lumped circuit approach.

Theoretical model

In this section we will introduce our theoretical model for the superconducting weak link ring (c.f. Fig. 1). The simplest (and most common) way to describe this system is by replacing the ring by an equivalent lumped circuit (c.f. Fig. 2) [8]. In this model, the thick ring is replaced by an inductor Λ and the weak link is replaced

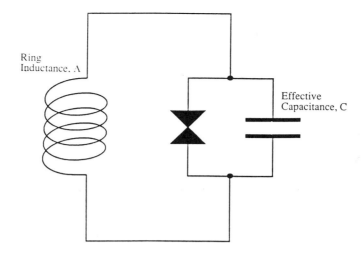

Figure 2. Simple lumped circuit model of superconducting weak link ring. The thick ring is replaced by an inductor Λ and the weak link is replaced by a Josephson junction and an effective parallel capacitance C.

by a Josephson junction circuit element [7, 8] and an effective parallel capacitance C. From the conservation of electrical current around the ring, we can derive an effective Hamiltonian,

$$H = \frac{Q^2}{2C} + \frac{\Phi^2}{2\Lambda} - \hbar\nu \cos\left(\frac{2\pi\Phi}{\Phi_0}\right) \tag{1}$$

where $C\frac{d\Phi}{dt} = -Q$ is the conjugate momentum to the magnetic flux Φ threading the circuit. The final term is simply the Josephson energy, coming from the Josephson current [7, 8],

$$I_J = -2e\nu \sin\left(\frac{2\pi\Phi}{\Phi_0}\right) \tag{2}$$

where ν is the angular frequency for pair charges to tunnel coherently across the weak link.

We can then impose the usual canonical commutation relation between conjugate variables [1, 2, 3, 4, 9, 10],

$$[Q, \Phi] = i\hbar \tag{3}$$

before solving the Schrödinger equation to find the energy eigenstates of the system.

We note that this commutation relation can easily be derived from the commutation relation between the electric and magnetic fields [10, 11],

$$[E_i(\mathbf{r}), B_j(\mathbf{s})] = \frac{i\hbar}{\epsilon_0}\epsilon_{ijk}\partial_k\delta(\mathbf{r} - \mathbf{s}) \tag{4}$$

(where ϵ_{ijk} is the Levi-Civita tensor and summation is assumed), by defining two quantities: the electric flux Q crossing an open surface S_1 and the magnetic flux Φ

contained within the ring (∂S_2) (c.f. Fig. 1).

$$Q = \epsilon_0 \int_{S_1} d\mathbf{S} \cdot \mathbf{E} \tag{5}$$

$$\Phi = \int_{S_2} d\mathbf{S} \cdot \mathbf{B} \tag{6}$$

The choice of the open surface S_2 is obvious, its boundary ∂S_2 simply follows the ring, as expected. The choice of S_1 is less obvious. However, it is reasonable to define S_1 such that it is crossed by all of the field lines once, and once only. Note that, whilst we have denoted the electric flux by Q and it does have units of charge, the variable Q is only a pseudo-charge and is not necessarily the actual charge in or around the neighbourhood of the weak link [10].

Of course, the commutation relation between the electric and magnetic fields is quite general. It must hold for any system, not just a superconductor. The important point about the superconducting circuit is that the currents and the fields that they generate are coherent. This enables us to define the conjugate quantities Q and Φ over macroscopic length scales ($\sim$ 1cm) and provided that the critcal current of the superconductor is not exceeded, the fields should remain coherent.

In the presence of an external applied magnetic flux Φ_{ext}, the Hamiltonian (which we have introduced for convenience later) becomes,

$$H = \frac{Q^2}{2C} + \frac{(\Phi - \Phi_{ext})^2}{2\Lambda} - \hbar\nu\cos\left(\frac{2\pi\Phi}{\Phi_0}\right) \tag{7}$$

The quantum mechanical behaviour of this Hamiltonian is well documented [4, 12, 13]. Solving the Schrödinger equation for the system [12, 13], we can find the energy eigenstates of the ring $\Psi_\kappa(\Phi, \Phi_{ext})$, corresponding to the macroscopic exciations of the ring. The energy eigenvalues $E_\kappa(\Phi_{ext})$ are even, periodic functions of the applied magnetic flux (period Φ_0) and are different for each state κ (c.f. Fig. 3).

It is this energy level spectrum which we wish to investigate, experimentally and theoretically.

Experimental system

Now that we have a model for our superconducting weak link ring, with a Hamiltonian and an energy eigenvalue spectrum, we need to consider how such a system can be investigated without introducing significant environmental dissipation, which would tend to hide its quantum mechanical nature [14, 15, 16]. One way of achieving this is to probe the local environment of the quantum object rather than the quantum object itself [4, 17]. In this way it may be possible to infer the behaviour of the quantum object by its reactive effect on its immediate environment. If the quantum mechanical system were microscopic, the back-reaction of the system on its environment would be negligible and can generally be ignored. If the system of interest is macroscopic, as it is here, this back-reaction may be significant. (Of course, great care is needed to ensure that all couplings to additional, unprobed degrees of freedom are negligible).

In our case, the environment will be a simple LC oscillator circuit. This linear oscillator is assumed to be essentially classical and will act as a source of external magnetic flux for the weak link ring. The flux generated in this classical circuit is coupled to the ring via a mutual inductance M (c.f. Fig. 4) to the ring, where it will

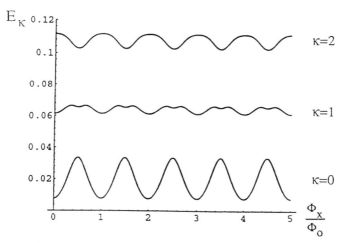

Figure 3. Lowest few energy eigenvalues $E_\kappa(\Phi_{ext})$ for the weak link ring Hamiltonian using $\hbar\omega_0 = 0.043\Phi_0^2/\Lambda$ and $\hbar\nu = 0.020\Phi_0^2/\Lambda$. ($\omega_0 = (\Lambda C)^{-1/2}$).

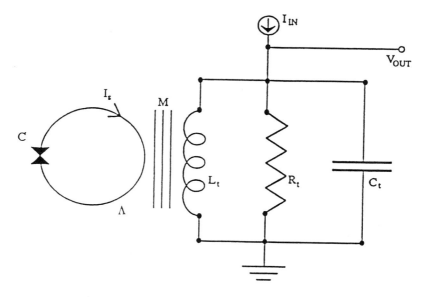

Figure 4. Schematic diagram of coupled SQUID ring-linear oscillator system.

give rise to a screening current, which is then coupled back to the circuit. In this way, the response of the quantum mechanical SQUID ring can alter the otherwise linear behaviour of the classical circuit in a nonlinear way. By using an analogue of the Born-Oppenheimer approximation [18] to integrate out the quantum degrees of freedom, it is possible to derive a classical nonlinear equation for the coupled system which contains the expectation value of the screening current for a given energy eigenstate [19].

$$C_t \frac{d^2 \Phi_t}{dt^2} + \frac{1}{R_t} \frac{d\Phi_t}{dt} + \frac{\Phi_t}{L_t} = I_{in}(t) + \mu \langle \kappa | \hat{I}_s(\mu \Phi_t + \Phi_{dc}) | \kappa \rangle \tag{8}$$

where $\hat{I}_s(\Phi_{ext}) = (\hat{\Phi} - \Phi_{ext})^2 / (\Lambda(1 - K^2))$, Φ_{dc} is an applied quasi-static DC magnetic flux, $K = M/\sqrt{L_t \Lambda}$, $\mu = M/L_t$ and $I_{in}(t)$ is a time-dependent external current source.

To ensure that the ring remains adiabatically in an energy eigenstate, we require that the energy level separations are large compared to $k_B T$ (where $T \simeq 4K$) or any other environmental fluctuations. This means that the energy level separations must be much larger than about 100GHz (in frequency units), although typical separations are more likely to be of the order of $500 - 1000$GHz.

We can see from the modified dynamical equation for the coupled system that the response of the system will be different for each energy eigenstate, since each eigenstate will tend to have a different screening current response. In the experiment detailed here, a very small coherent current is fed into the oscillator at a particular frequency and the output voltage across the circuit is measured. The average amplitude of the ouput voltage will be related to the frequency of the drive, the DC magnetic flux value and the energy level of the ring. By changing the drive frequency the frequency response of the system can be found, and the 'resonant' frequency at a particular value of Φ_{dc} and κ can be determined. (The fact that the system is non-linear means that a true resonant frequency cannot the defined but, for very small oscillations, a pseudo-resonant frequency can be defined by linearising the equation [19]). These experimental frequencies shifts can then be compared with calculations based on the theoretical model (with small coherent drive terms and noise terms corresponding to thermal noise at $T = 4K$.)

Results

In Fig. 5 we present experimental and theoretical frequency shift data for the lowest three energy states ($\kappa = 0, 1, 2$) taken on the same SQUID ring with the same weak link contact, corresponding to the energy levels shown in Fig. 3. The experimental data was obtained from a system similar to that discussed above (experimental details can be found in ref. [20]). The frequency shift measurements were taken using a coherent current source from a spectrum analyzer (Rohde and Schwartz FSAS model) to drive the oscillator circuit, which was also used to plot the resonant peak of the coupled system. The signal was arranged to be very small, such that $\mu \Phi_t \ll \Phi_0$. Typically, $\mu \Phi_t$ was less than $0.05\Phi_0$ (including both the coherent signal and total integrated flux noise due to the oscillator coil at 4K). The pseudo-resonant frequency was then found for different values of the static DC flux, so that the response of the system could be found over a whole period of Φ_0.

The weak link ring used was a Zimmerman-type two hole niobium SQUID ring with a niobium point screw-post contact weak link, with a geometric inductance of $\Lambda = 1.5 \times 10^{-10}$H. The oscillator circuit (coupled to the ring with no contact) resonates at approximately $\omega = (L_t C_t)^{-1/2} = 20$MHz and has a quality factor $Q_t =$

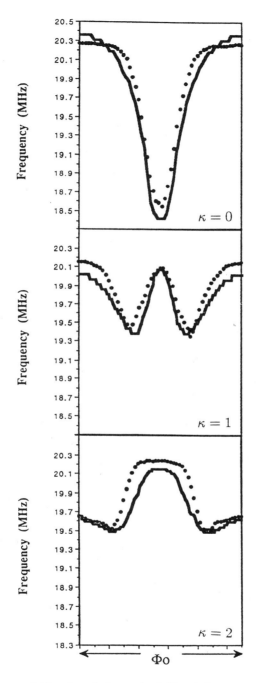

Figure 5. Experimental (dots) and theoretical (lines) frequency response of a coupled ring-oscillator system for the first three energy states ($\kappa = 0, 1, 2$). (The parameters are given in the text.)

$\omega R_t C_t = 200$. The strength of the coupling can be estimated to be around $K^2 = 0.15$, based on the known periodicity of the ring to applied magnetic flux and the values of the inductances L_t and Λ.

Using these experimental values and solving the nonlinear equation (16), we can fit a theoretical frequency response to the experimental data (c.f. Fig. 5) and estimate the two unknown quantities ω_0 and ν. In the data shown in Fig. 5, the best fit corresponds to the energy spectrum calculated for Fig. 3, $\hbar\omega_0 = 0.043\Phi_0^2/\Lambda$ and $\hbar\nu = 0.020\Phi_0^2/\Lambda$. This gives an estimate for the effective capacitance of the weak link of $C = 1 \times 10^{-16}$F.

Although the match is not exact, the agreement is very good for all three levels shown, in both the relative amplitude of the frequency shifts and their functional form. A change in either of the two unknown parameters of a few percent can dramatically alter both the shapes and the relative strengths of the frequency shifts of each level differently, so that the estimated values can be quoted with some confidence.

Lifetimes of states

The experimental frequency shifts corresponding to the excited states ($\kappa = 1, 2$) are found to be stable on timescales of many minutes. This is far longer than one might expect. Estimates based on classical circuit parameters (typically of the order of the RC or L/R time constants [8]) would indicate much shorter lifetimes. However, these estimates are somewhat naïve, they make no allowance for the quantum mechanical nature of the object. Such estimates would not be expected to hold for any other quantum mechanical system, and they do not hold for the quantised magnetic flux states of a thick superconducting ring. Any model which is used to estimate the lifetimes of the macroscopic excited quantum states of a weak link ring must also be able to explain the extremely long lifetimes of thick superconducting ring states.

To consider the problem of apparent lifetimes in more detail, we need to look at possible decay mechanisms. One source of dissipation which is frequently discussed in relation to superconducting circuits is the normal quasi-particle current [2, 3]. From the SQUID ring Hamiltonian (15), we can use Faraday's Law to find the quantum mechanical voltage operator,

$$V = -\frac{i}{\hbar}[H(\Phi_{ext}), \Phi] = \left(\frac{i\hbar}{C}\right)\frac{\partial}{\partial\Phi} \tag{9}$$

The mean value of the Faraday Law voltage operator in an energy eigenstate then vanishes,

$$\langle\kappa|V|\kappa\rangle = \frac{i\hbar}{C}\int d\Phi\ \Psi_\kappa^*(\Phi, \Phi_{ext})\frac{\partial\Psi_\kappa(\Phi, \Phi_{ext})}{\partial\Phi} = 0 \tag{10}$$

which follows most easily by choosing real wavefunctions for the energy eigenstates $\Psi_\kappa(\Phi, \Phi_{ext})$. The normal quasi-particle currents around the ring $I_n(V_\kappa, T)$ vanish in equilibrium because there is no voltage around the ring in an equilibrium state, i.e. $\langle\kappa|V|\kappa\rangle = 0$ implies $I_n = 0$.

This physical situation is well known for thick superconducting rings. If magnetic flux is inserted through the bulk ring, there will be a rapid change in the normal quasi-particle current until equilibrium is achieved. After equilibrium is achieved, the normal current will vanish, and the trapped flux supercurrent will remain constant for many years. Even if one were to add some shunt resistors to the ring, they would have some effect on the time constant required for the initial relaxation of the quasi-

90

particle normal currents to zero. But once equilibrium has been re-established, the shunt resistors would be shorted out by the supercurrents.

Similarly, for the weak link ring, provided there is no voltage generated around the ring (i.e. the system remains in an energy eigenstate) there should be no quasi-particle current at equalibrium. The persistent supercurrents may live for a very long time without dissipation, whilst any quasi-particle currents decay to zero.

Given that the energy eigenstates of our macroscopic Hamiltonian are stable relative to the normal quasi-particle currents when in equilibrium, the question arises as to what sort of radiative transitions are possible between these energy states. Obviously, if the macroscopic excitations are genuine quantum mechanical states then they should radiate in exactly the same way as atomic or nuclear systems. As we have already stated, the energy level separations correspond to frequencies of the order of $\sim 500 - 1000\,$GHz (for single photon processes), equivalent to wavelengths of the order of $\sim 0.5 - 0.1$mm. We have a situation where the wavelength for single photon emission is significantly smaller than the dimensions of the quantum object. We can no longer use simple multipole expansions to predict the expected quantum mechanical transition rates, as one might in atomic physics. Certainly, if one were to use such methods, it is the higher multipole transitions which would tend dominate; and since these involve the correlated emission of several photons, one might expect a rather long lifetime. (However, it should also be noted that if transitions do occur by emission of several lower frequency photons, the actual temperature and the electronic noise temperature of the environment may become important, 4K$\sim 90\,$GHz.)

What is clear is that whilst the lumped circuit model may provide approximate energy eigenstates for the system in certain regimes, a more sophisticated model is required to deal with radiative transitions, and this model must include spatial variations of the electromagnetic field and the quantum object [21].

Conclusions

In this paper we have discussed the behaviour of a simple superconducting circuit (a superconducting weak link ring) coupled to linear oscillator. By treating the ring as a macroscopic quantum electrodynamic object, we have taken the conventional lumped circuit model and quantised it using the usual canonical commutation relation. We then solved the time-independent Schrödinger equation to find the appropriate energy eigenstates in the presence of an external applied magnetic flux.

Using the fact that the response of the ring to an applied magnetic flux should be different for each energy eigenstate, we showed that it is possible to identify the quantum state of the ring simply through its reactive coupling to the linear oscillator. In Fig. 5 we compare experimental results with the predictions of our theoretical model. The agreement between the two is excellent. The experimental results show the first three energy states of the same ring with the same weak link contact. These three states provide a good match between experiment and theory and (most importantly) they do so consistently.

The authors would like to thank the Engineering and Physical Science Research Council, The European Community, The National Physical Laboratory and The Royal Society for their generous funding of this work.

References

[1] A. Widom, Y. Srivastava, *Phys. Rep.* **148**, 1 (1987).

[2] G. Schön, A.D. Zaikin, *Phys. Rep.* **198**, 237 (1990).

[3] D.V. Averin, K.K. Likharev, *Mesoscopic Phenomena in Solids*, Eds. B.L. Altshuler, P.A. Lee, R.A. Webb (Elsevier, 1991) Ch. 6.

[4] T.P. Spiller, T.D. Clark, R.J. Prance, A. Widom, *Prog. Low Temp. Phys. XIII*, Ch. 4, 219 (Elsevier, 1992).

[5] R.P. Feynman, *The Feynman Lectures on Physics* (Addison-Wesley, Reading, Massachusetts, 1966) Vol.III, Chap.21.

[6] A. Widom, T.D. Clark, G. Megaloudis, G. Sacco, *Il Nuovo Cimento* **61B**, 112 (1981).

[7] B.D. Josephson, *Phys. Lett.* **1**, 251 (1962).

[8] e.g. A. Barone, G. Paterno, *'Physics and Applications of the Josephson Effect'*, Wiley (1982).

[9] A. Widom, *J. Low Temp. Phys.* **37**, 449 (1979).

[10] A. Widom, *Macroscopic Quantum Phenonmena, Proc. LT-19 Satellite Workshop*, Eds. T.D. Clark, H. Prance, R.J. Prance, T.P. Spiller (World Scientific, Singapore, 1991) p. 55.

[11] e.g. H. Mandl and G. Shaw, *Quantum field theory* (Wiley, 1984) Ch. 5.

[12] J.E. Mutton, R.J. Prance, T.D. Clark, *Phys. Lett.* **37**, 449 (1984).

[13] A.J. Leggett, *Contemp. Phys.* **25**, 583 (1984).

[14] R.P. Feynman, F.L. Vernon Jr., *Ann. Phys.* **24**, 118 (1968).

[15] A.O. Caldeira, A.J. Leggett, *Phys. Lett.* **46**, 211 (1981).

[16] A.O. Caldeira, A.J. Leggett, *Ann. Phys.* **149**, 375 (1983).

[17] T.P. Spiller, T.D. Clark, R.J. Prance, H. Prance, D.A. Poulton, *Il Nuovo Cimento* **105B**, 749 (1990).

[18] M. Born, *Z. Physik* **38**, 803 (1926).

[19] J.F. Ralph, T.P. Spiller, T.D. Clark, R.J. Prance, H. Prance, *Int. J. Mod. Phys.* **B**, August (1994).

[20] R.J. Prance, T.D. Clark, R. Whiteman, J. Diggins, H. Prance, J.F. Ralph, T.P. Spiller, A. Widom, Y. Srivastava, *'Reactive probing of macroscopically quantum mechanical SQUID rings'* (to appear in Physica, Proc. NATO ARW on Mesoscopic Superconductivity, Karlsruhe, May 1994).

[21] J.F. Ralph, G.J. Colyer, T.D. Clark, T.P. Spiller, R.J. Prance, H. Prance, *'A quantum electrodynamic model for thick superconducting rings'* (to appear in Physica, Proc. NATO ARW on Mesoscopic Superconductivity, Karlsruhe, May 1994).

DYNAMICS OF THE MICROMASER FIELD

G. Raithel, O. Benson, and H. Walther

Max-Planck-Institut für Quantenoptik and Sektion Physik der Universität München
D-85748 Garching, Fed. Rep. of Germany

1 INTRODUCTION

In this contribution a cavity-QED system is investigated in which single atoms interact with a single mode of a high-Q microwave cavity. One can distinguish between two limiting cases of resonantly interacting atom-field systems. In one of those regimes, the perturbative regime, the photon storage time τ_{cav} of the cavity device is much shorter than the decay time of the excited atoms in the cavity field. This is the usual experimental situation when considering the interaction between single atoms and quantized electromagnetic fields. Important observations on perturbative cavity-QED systems include enhanced[1] or inhibited[2] spontaneous decay of the atoms, as well as radiative shifts of the atomic energy levels [3,4,5].

The one-atom maser or micromaser [6,7] belongs to the class of non-perturbative or strongly coupled cavity-QED systems. During the interaction events between the pump atoms and the maser field all interactions with field modes except the maser field mode can be neglected. The Rabi flopping period of the pump atoms in a ground state cavity field is of the order of $100\mu s$, this being much less than the photon storage time of the cavity which can be as long as $\tau_{cav}=200ms$[8]. The micromaser operates with only a few photons and less than one pump atom on the average in the maser field. It is thus an ideal device to study quantum phenomena in atom-radiation interaction. The dynamics of the atom-field interaction events is governed by the Jaynes-Cummings Hamiltonian:

$$H_{\mathrm{JCM}} = \hbar\omega_0|e\rangle\langle e| + \hbar\omega\, a^+ a + \hbar\Omega\, a\, |e\rangle\langle g| + \hbar\Omega\, a^+\, |g\rangle\langle e| \qquad (1)$$

Here, $\hbar\omega_0$ is the energy difference between the atomic levels $|e\rangle$ and $|g\rangle$ which are coupled by the maser field mode. The cavity frequency is denoted by $\omega/(2\pi)$, and the atom-field coupling constant by Ω.

In the micromaser experiment one mode of a superconducting niobium cavity is tuned to a microwave transition between highly excited states (Rydberg states) of the rubidium atom. The cavity field is pumped by a beam of velocity-selected Rydberg atoms entering the cavity in the upper maser level $|e\rangle$. The pumping occurs via the interaction of single atoms with the cavity field, the interaction approximately following the Hamiltonian Eq.(1). The duration of the pumping events t_{int} follows from the atomic velocity. The dynamics of the atom-field interaction displays features which are purely quantum, such as collapse and revival in the Rabi flopping process[7,9]. In the time intervals in between the pumping events the maser field is damped, the damping being described by the master equation of the damped quantum harmonic oscillator. The pumping-damping equilibrium corresponds to a steady-state photon number distribution $P(n)$ of the micromaser field which can be sub-Poissonian[8,10,11,12]. Thus, the micromaser is suited for the generation of non-classical radiation.

Electron Theory and Quantum Electrodynamics: 100 Years Later
Edited by Dowling, Plenum Press, New York, 1997

In the following we particularly deal with the *dynamical* properties of the maser field. In Sec.2 a brief review of some theoretical results on the micromaser is given. Various approaches such as the quantum micromaser model and the semiclassical micromaser description are discussed. After a brief presentation of the experimental setup in Sec.3 we discuss experimental results on the field dynamics in Sec.4. The experimental investigations reveal quantum jumps, metastability and hysteresis of the maser field [13], as predicted by the micromaser theory [10,14]. In Sec.5 experimental and theoretical results are compared. One of the conclusions is that we expect weak DC electric fields in the vicinity of the cavity holes. Those fields lead to a considerable red-detuning of the atomic resonance frequency close to the cavity holes, this giving rise to interference phenomena in the micromaser resonance lines. The novel interference structures presented in Sec.6 are explained in Sec.7 on the basis of a model assuming Ramsey-type atom-field interaction. In Sec.8 it is shown that the interferences are also related to metastability and hysteresis. Sec.9 concludes the contribution.

2 THE MICROMASER FIELD - VARIOUS MODELS

2.1 Monte-Carlo Simulations

A quite intuitive view of the dynamics of the maser field can be gained by Monte-Carlo simulations [12,15,16]. In general, the field of a micromaser is characterized by the field density matrix which is usually written in the Fock state representation of the field. The diagonal elements of the field density matrix correspond to the photon number distribution of the field. When the pump atoms enter the cavity in the upper state $|e\rangle$ or in an incoherent mixture of upper and lower state and when the non-diagonal elements of the field density matrix are initially zero, the non-diagonal elements of the field density matrix are always identical zero, i.e. the incoherently pumped micromaser does not build up any phase information [17]. Since the mentioned conditions apply to our experimental situation, the photon number distribution is sufficient for the field description.

In a Monte-Carlo simulation the evolution of the photon number distribution is calculated using various random numbers per cycle. The number of pump atoms per cavity decay time, N_{ex}, the atom-field coupling constant Ω and interaction time t_{int} as well as quantities such as the experimental uncertainty of t_{int} are entered as parameters into the simulation. The result of a simulation corresponds to an individual, possible realization of the actual field evolution. Thus, a single simulation should directly resemble the system evolution observed in the experiment. Since in the Monte-Carlo simulations the photon number distribution depends on time, one directly obtains information on dynamical properties of the maser field. It is possible, for instance, to simulate the field jumps discussed in Sec. 4.1: The simulations reveal that under conditions such that $\sqrt{N_{\mathrm{ex}}}\,\Omega\,t_{\mathrm{int}} \approx m\,2\pi$ with integer m the field statistically hops between two metastable states which can be identified with two well defined maxima of the steady-state photon number distribution $P(n)$ (see also Sec. 2.1 and Fig.1). The jumps of the maser field are a consequence of the quantum measurement process on the pump atoms [14,15]. When the simulation is run with zero detection efficiency, the photon number distribution does not display any jumps. The mechanism of the measurement-induced projection of the field onto one of the two peaks in the steady-state photon number distribution is related to the atom-field entanglement in the micromaser: a measurement performed on the atom leads to a change of the field state. The field jumps occur with typical hopping time constants τ_{field}. The number of detection events per τ_{field} required for a successful projection of the actual photon number distribution onto one of the two possible metastable field states is of the order of ten.

Properties of the ensemble-averaged field evolution are obtained by averaging the results of many simulations with different seeds for the random number generator and equal macroscopic parameters such as N_{ex}, Ω and t_{int}. The steady-state photon number distribution $P(n)$ can be evaluated as well. It is, however, found more readily using the method described next.

2.2 Steady-State Photon Number Distribution

The central result of the micromaser theory developed in [10,11] is the steady-state photon number distribution $P(n)$. It is obtained from the recursion formula

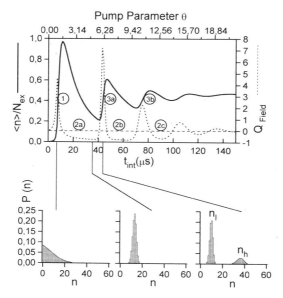

Figure 1. In the upper part the average photon number $\langle n \rangle$ normalized by N_{ex} and the Mandel Q-parameter of the steady state photon number distribution $P(n)$ are displayed as a function of the atom-field interaction time t_{int}. The upper scale shows the corresponding values of the pump parameter Θ. The distribution $P(n)$ is calculated using Eq.2 with $N_{ex} = 50$, $T = 0.2K$, $\Omega = 20 krad/s$, an atomic lifetime of $\tau_{at} = 400\mu s$, and a Gaussian interaction time distribution with a RMS deviation of 3% of the mean value t_{int}. The labels $1, 2a, 3a, ...$ are explained in the text.

In the lower part of the figure, the steady state photon number distributions $P(n)$ are shown for three values of t_{int}. The distribution to the left corresponds to the maser threshold, the one to the right gives an example for the double-peaked distributions associated with the dynamical phenomena discussed in Secs.4 and 5.

$$P(n) = \frac{n_b}{n_b + 1}(1 + \frac{N_{ex}\langle \beta_n \rangle}{n\, n_b}) P(n-1) \qquad (2)$$

The quantity n_b stands for the number of thermal photons. The β_n denote the probabilities that an atom emits a photon in a field of $n-1$ photons prior to the interaction. The values of β_n follow from the Jaynes-Cummings dynamics described by the Hamiltonian Eq.(1). The $\langle \ \rangle$ indicate that the β_n have to be averaged over all statistical fluctuations such as the spread of t_{int} caused by the velocity distribution of the pump atoms. The overall behavior of the steady-state photon number distribution $P(n)$ is essentially determined by the pump parameter, $\Theta = \sqrt{N_{ex}}\, \Omega\, t_{int}$.

Evaluating Eq.(2) one finds the following generic behavior of the quantity $\langle v \rangle = \langle n \rangle / N_{ex}$ as a function of the pump parameter Θ (see Fig.1): $\langle v \rangle$ suddenly increases at the maser threshold value $\Theta = 1$ (denoted by 1 in Fig.1), and reaches a maximum for $\Theta \approx 2$. For N_{ex} somewhat larger than 20, the maser threshold displays the characteristics of a continuous phase transition [10]. As Θ further increases, $\langle v \rangle$ decreases (region $2a$ in Fig.1) and reaches a minimum at $\Theta = 2\pi$, and then abruptly increases to a second maximum ($3a$ in Fig.1). This general type of behavior recurs roughly at integer multiples of 2π, but becomes less pronounced with increasing Θ. The Mandel Q-parameter of the field, $Q = \{(\langle n^2 \rangle - \langle n \rangle^2)/\langle n \rangle\} - 1$, which is also shown in Fig.1, drops below zero in the regions $2a$, $2b$, etc.. The Q-value equals zero for Poissonian photon number distributions. Thus, Fig.1 demonstrates the highly sub-Poissonian character of the maser field obtained in wide parameter ranges.

The reason for the periodic maxima of $\langle v \rangle$ close to the locations $3a$, $3b$, etc. in Fig.1 is that for integer multiples of $\Theta = 2\pi$ the pump atoms perform an almost integer number of full Rabi flopping

95

cycles, and start to flip over at a slightly larger value of Θ, this leading to enhanced photon emission. The periodic maxima in $\langle v \rangle$ occuring at $\Theta = 2\pi$, 4π,...can be interpreted as first-order phase transitions [10]. The field strongly fluctuates for all phase transitions (*1,3a,3b*, etc. in Fig.1), the large photon number fluctuations for $\Theta = 2\pi$ and multiples thereof being caused by the presence of two maxima in the photon number distribution *P(n)* at photon numbers n_l and n_h ($n_l < n_h$). This is demonstrated by the photon number distribution *P(n)* displayed on the lower right of Fig.1. For n_l and n_h the atoms perform almost integer numbers m or $m+1$ of full Rabi flopping cycles, respectively. When the pump parameter is scanned across $\Theta = m\,2\pi$, the maximum of *P(n)* at n_l completely dies out, and is replaced by the new peak at the higher photon number n_h. The spontaneous field jumps mentioned in Sec. 2.2 occur if *P(n)* displays two well separated maxima.

2.3 Fokker-Planck Model

The phenomenon of the two coexisting maxima in *P(n)* was also studied in a semi-heuristic Fokker-Planck (FP) approach [10] which is suited for the description of the ensemble-averaged field dynamics. The FP model is appropriate for photon numbers $n \gg 1$ and $\Omega t_{int} \ll \sqrt{n}$ [10]. Those conditions apply for the experiments discussed in Secs.4 and 5. The photon number distribution *P(n)* is replaced by a probability function $P(v, \tau)$ with continuous variables $\tau = t / \tau_{cav}$ and $v = n / N_{ex}$, the latter replacing the photon number n. The steady-state solution obtained for $P(v, \tau)$, $\tau \rightarrow \infty$, can be constructed by means of an effective potential $V(v)$ showing minima at positions where maxima of $P(v, \tau)$, $\tau \rightarrow \infty$, are found. Close to $\Theta = 2\pi$ and multiples thereof, the effective potential $V(v)$ exhibits two equally attractive minima located at stable gain-loss equilibrium points of maser operation (see Sec.2.4 and [18]). An example is shown in Fig.2 where the effective potentials $V(v)$ and the corresponding steady-state solutions $P(v, \tau)$, $\tau \rightarrow \infty$, are displayed for $\Theta = 4\pi$ and two different amounts of fluctuations in Ωt_{int}. The two equally attractive minima in $V(v)$ lead to two maxima in $P(v, \tau)$, $\tau \rightarrow \infty$. The situations shown in Fig.2 qualitatively correspond to the quantum result for *P(n)* shown in Fig.1 on the lower right. Fig.2 also visualizes the influence of noise: any kind of additional fluctuations lead to a smoother effective potential $V(v)$ and the steady-state solution $P(v, \tau)$, $\tau \rightarrow \infty$. The FP formalism is particularly suited for the calculation of the transition rates between adjacent minima of the effective potential $V(v)$ for large values of N_{ex} [10].

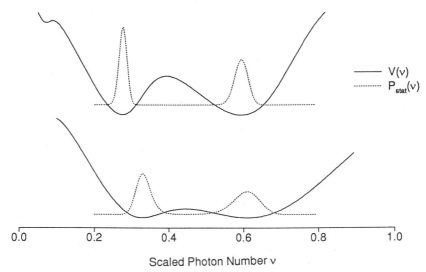

Figure 2. Effective potentials $V(v)$ (solid) and stationary solutions for $P(v, \tau)$, $\tau \rightarrow \infty$ (dotted) in the Fokker-Planck model of the micromaser for a pump parameter $\Theta = \sqrt{N_{ex}}\,\Omega t_{int} \approx 4\pi$. The upper (lower) plots correspond to a 4% (10%) relative RMS deviation of the product Ωt_{int}. In both cases the potential $V(v)$ displays two minima; the corresponding steady state functions $P(v, \tau)$, $\tau \rightarrow \infty$ are double-peaked. It can be noticed that larger fluctuations in Ωt_{int} lead to a flattening of the intermediate maximum of $V(v)$, and to broader peaks in $P(v, \tau)$, $\tau \rightarrow \infty$.

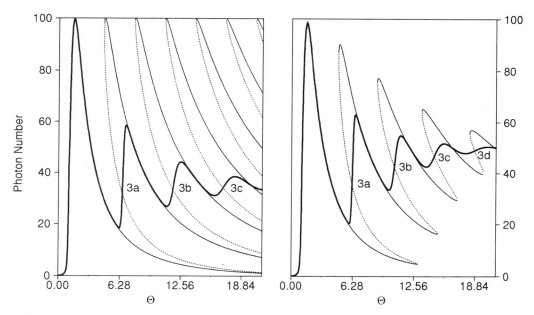

Figure 3. Stable values n_s (thin and solid), unstable gain-loss equilibrium values (dotted), and the mean photon number following from Eq.2 (thick and solid) versus the pump parameter Θ. The parameters are $N_{ex} = 100$, $\Omega = 14\ krad/s$, and $T = 0.5K$. In the left plot it is assumed that the product Ωt_{int} does not fluctuate, whereas in the right plot a Gaussian distribution with 5% relative RMS deviation is assumed. The crossing points between the lines $\Theta = const$ and the thin and solid curves correspond to the values where minima in the Fokker-Planck potential $V(v)$ occur. The phase transition $3a$ is investigated in the experiments displayed in Figs.7 and 9, whereas the experiments shown in Figs.5,6 and 8 are related to $3b$. By comparing the two diagrams it can be noticed that the widths of the intervals in which the semiclassically stable branches exist are considerably reduced by additional fluctuations. The noise also shifts the quantum result (thick and solid) to higher photon numbers, and the phase transition points to lower values of Θ.

The inverse transition rates equal the hopping time constants τ_{field} of spontaneous field jumps mentioned in Sec. 2.1. Additional fluctuations flatten the maximum in between adjacent minima of $V(v)$, and thus enhance the transition rates between the minima.

The mechanism at the first-order phase transitions mentioned in Sec.2.2 is always the same: one minimum of $V(v)$ loses its global character when Θ is increased, and is replaced in this role by the next one. This reasoning is a variation of the Landau theory of first-order phase transitions, with $\sqrt{v}$ being the order parameter. This analogy actually leads to the notion that in the limit $N_{ex} \to \infty$ the changes of the micromaser field around integer multiples of $\Theta = 2\pi$ can be interpreted as first-order phase transitions.

2.4 Semiclassical Maser Model

The physical content of the different minima in $V(v)$ becomes clear when considering a semiclassical model of the one-atom maser [18]. We assume that the field is described by a fixed, non-integer "photon number" n_0. This basicly means that we assume a classical maser field with well defined, non-quantized amplitude. The stable points n_s of maser operation are identified by gain-loss equilibrium,

$$G(n_s) = N_{ex} \sin^2(\sqrt{n_s +1}\,\Omega t_{int}) + n_b - n_s = 0 \quad , \tag{3}$$

and stability with respect to small photon number variations, i.e.

$$\frac{d}{dn_0}G(n_0) < 0 \quad \text{for} \quad n_0 = n_s \quad . \tag{4}$$

97

Here, G is the photon gain per cavity decay time. An example is shown in Fig.3 where the stable branches of maser operation correspond to the thin solid lines. The result for $\langle v \rangle$ obtained with the micromaser model treated in Sec.2.2, which is also shown in Fig.3, usually coincides with one of the semiclassical solutions. Each minimum of the potential $V(v)$ discussed in Sec.2.3 corresponds to a stable semiclassical solution n_s. The deepest minimum of $V(v)$ corresponds to the most stable semiclassical solution, the latter also approximating the quantum result for $\langle n \rangle / N_{ex}$ following from Eq.(2). At the first-order phase transition points the system evolves from one particular semiclassical branch to another, neighboring one. Analogous to Fig.1, in Fig.3 the first-order phase transition points are labeled by $3a$, $3b$,....

The theoretical considerations show that for large N_{ex} it should be possible to detect spontaneous jumps of the maser field between metastable field states. The average dwell times of the field in the metastable field states equal long evolution time constants observed in the ensemble-averaged field evolution. One also expects to find bistability and hysteresis of the maser field. Experimental demonstrations of these properties are presented in Sec.4. Some of the phenomena are also predicted for the two-photon micromaser[19], for which qualitative evidence of first-order phase transitions and hysteresis is reported in[20,21].

3 EXPERIMENTAL SETUP

The experimental setup used is shown in Fig.4. It is similar to that described in [8,12]. As before, ^{85}Rb atoms were used to pump the maser. They are excited from the $5S_{1/2}, F = 3$ ground state to $63P_{3/2}, m_J = \pm 1/2$ states by linearly polarized light of a frequency-doubled c.w. ring dye laser. The polarization of the laser light is linear and parallel to the likewise linearly polarized maser field. Superconducting niobium cavities resonant with the transition to the $61D_{3/2}$ or $61D_{5/2}$ states were used; the corresponding resonance frequencies are approximately $21.5GHz$. The experiments were performed in a $^{3}He/^{4}He$-dilution refrigerator with cavity temperatures $T \approx 0.15K$. The cavity Q-values ranged from 4×10^9 to 8×10^9. The Q-values are measured by observing the reflection of weak frequency-stabilized microwave radiation at one of the cavity holes. The velocity of the Rydberg atoms and thus their interaction time t_{int} with the cavity field were preselected by exciting a particular velocity subgroup with the laser. For this purpose, the laser beam irradiated the atomic beam at an angle of approximately 82^0. As a consequence, in the frame of reference of the atoms

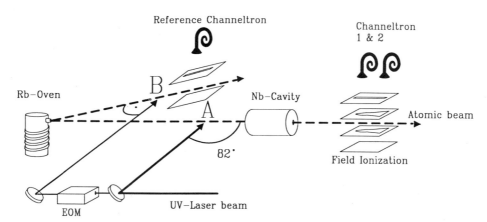

Figure 4. Sketch of the experimental setup. The rubidium atoms emerge from an atomic beam oven and are excited at an angle of 82^0 at location A. After interaction with the cavity field, they enter a state-selective field ionization region, where channeltrons 1 and 2 detect atoms in the upper and lower maser levels, respectively. A small fraction of the UV radiation passes through an electrooptic modulator (EOM), which generates sidebands of the UV radiation. The blue-shifted sideband is used to stabilize the frequency of the laser onto the Doppler-free resonance monitored with a secondary atomic beam produced by the same oven (location B). (Figure reprinted with permission from Ref. [13])

the UV laser light (linewidth $\approx 2\,MHz$) is blue-shifted by *50-200MHz* by the Doppler effect, depending on the velocity of the atoms. For an average atomic velocity of the excited Rydberg atoms of *700m/s* the obtained velocity distribution has a RMS deviation of 2% of the average velocity.

Information on the maser field and on the atom-field interaction can be solely obtained by state-selective field ionization of the atoms in the upper or lower maser level after they have passed through the cavity. The field ionization detector was recently modified, so that there is now a detection efficiency of $\eta = (40 \pm 10)\%$. For different values of t_{int} the atomic inversion has been measured as a function of the pump rate. By comparing the results with the micromaser theory [10], the coupling constants Ω are found to be $\Omega = (20 \pm 10)\,krad/s$ for the maser transition to the $61D_{3/2}$ state, and $\Omega = (40 \pm 10)\,krad/s$ for the $61D_{5/2}$ state. Another experimental method which is applied to find the detection efficiency η is the mesurement of the Mandel Q-parameters of the counting statistics of the lower state atoms after the atom-field-interaction [8,12]. With the present maser setup, experimental Q-values as low as -0.191 ± 0.004 have been obtained.

4 MEASUREMENTS OF MASER FIELD JUMPS AND RELATED PHENOMENA

4.1 Field Jumps

Depending on the parameter range, essentially three regimes of the field evolution time constant τ_{field} can be distinguished. First we discuss the results for intermediate time constants. The maser was operated on the $63P_{3/2} - 61D_{3/2}$ transition under steady-state conditions close to the second first-order phase transition (*3b* in Fig.1). The interaction time was $t_{int} = 47\,\mu s$ and the cavity decay time $\tau_{cav} = 60\,ms$. The value of N_{ex} necessary to reach the second first-order phase transition was $N_{ex} \approx 180$. Due to the large cavity field decay time, the average number of atoms in the cavity was still as low as 0.17. For these parameters, the two maxima in *P(n)* manifest themselves in spontaneous jumps of the maser field between the two maxima with a time constant of approximately *10s*. This fact and the relatively large pump rate lead to the clearly observable field jumps shown in Fig.5. The average number of photons in the cavity field approximately equals the counting rate in the lower state detector (CT2) times the cavity decay time, divided by the detection efficiency. The two discrete values for the counting rates correspond to two metastable operating points of the maser with ≈ 70 and ≈ 130 photons, respectively. In the FP description, the two values correspond to two equally attractive minima in the FP potential $V(\nu)$. If one considers, for instance, the counting rate of lower-state atoms (CT2 in Fig.5), the lower (higher) plateaus correspond to time intervals during which the field is in the low (high) metastable operating point.

Based on recordings like those shown in Fig.5 it is possible to find the fractions of time, f_h and $f_l = 1 - f_h$ during which the maser operates in the high-field and the low-field mode, respectively. The fractions f_h and f_l are very sensitive to the pump parameter $\Theta = \sqrt{N_{ex}} \Omega t_{int}$, as following from the large slope of the curve $\langle \nu \rangle (\Theta)$ in Fig.1 in the vicinity of the locations *3a,3b*, etc.. Thus, there should be significant changes of the "attractivity-ratio" f_l / f_h when N_{ex} is slightly varied about the value chosen in Fig.5. An experimental result is shown in Fig.6. The plot in the middle of Fig.6 is identical to the lower recording of Fig.5, whereas the upper (lower) plot shows the counts of lower-state atoms with reduced (increased) value of N_{ex}. The changes in N_{ex} amount to only 4, this corresponding to a relative change of the pump parameter Θ of about 1%. Fig.6 clearly demonstrates that the attractivity ratio of the two metastable operating points is, in fact, very sensitive to changes of N_{ex}. In the upper plot of Fig.6, for instance, the escape rate from the high to the low field metastable state is strongly enhanced, leading to a reduced value of the long-time average of the photon number. On the right hand side of Fig.6 the interpretation in the FP picture is visualized: the variation of N_{ex} leads to a change of the relative depths of the two competing minima in $V(\nu)$. This gives rise to the observed preference of the system for a particular minimum.

4.2 Ensemble-Averaged Evolution

In a parameter range where the sontaneous switching occurs much faster than in the cases shown in Figs.5 and 6, the individual jumps cannot be resolved any more owing to the reduced N_{ex}

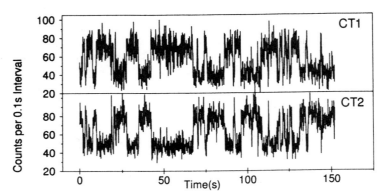

Figure 5. Quantum jumps between two equally stable operation points of the maser field. The channeltron counts are plotted versus time (CT1 = upper state, CT2 = lower state). (Figure reprinted with permission from Ref. [13])

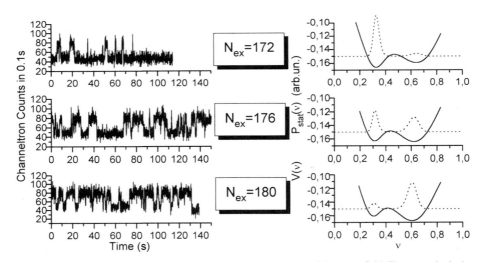

Figure 6. Left: Spontaneous jumps between two metastable operation points of the maser field. The counts in the lower state detector (CT2) are plotted versus time. The pump rate is slightly varied, as indicated in the figure by the values of N_{ex}. As a result, the upper and higher field states change their relative attractivity.

Right: Interpretation of the experimental results shown on the left in terms of the Fokker-Planck model. The solid lines show the effective potentials $V(v)$, and the dotted lines the long-time limit $P(v, \tau)$, $\tau \to \infty$

necessary in this case. Therefore, a different procedure has to be chosen for the investigation. The maser pump rate N_{ex} is periodically switched between two values differing by about 30% of their mean value. The individual values correspond to pump parameters Θ slightly below and above a first-order phase transition. The corresponding FP potential $V(v)$ has two minima at both values of Θ, the switching of Θ causing the first or the second to be the globally more stable one. In Fig.7, N_{ex} was switched between ≈ 35 and ≈ 25 with a period of $2s\ (= 33\,\tau_{cav}$, corresponding to the full time scale of Fig.7). $N_{ex} = 35(25)$ corresponds to 0.027 (0.020) atoms on the average in the cavity. The experiment runs over many switching cycles. The time is measured modulo the switching period, i.e. the time is reset to zero whenever N_{ex} is switched from the lower to the larger value. In the experiment the signal transients are averaged over many measuring periods, and hence in this case the *ensemble average* is obtained. After the switching instants, the evolution of the average signal is dominated by an exponential approach to the new steady-state values, which in this case correspond to mean photon numbers of roughly 21 and 6. Fig.7 clearly displays the long evolution time constants τ_{field}. It is expected from theory that τ_{field} rapidly increases as N_{ex} is increased. This has been confirmed by a series of other measurements. In the limit of small N_{ex} time constants of the order of $\tau_{field} \approx 5\,\tau_{cav}$ were obtained with $N_{ex} \approx 19$. In those cases, the metastable operating points of the maser differ by only 7 photons.

4.3 Bistability and Hysteresis

The third situation arises when the switching time constant becomes much longer than in the measurements shown in Figs.5 and 6. By using fast atoms with an interaction time of $35\mu s$ and operating the maser on the $63P_{3/2} - 61D_{3/2}$ transition at the second first-order phase transition (*3b* in Fig.1), it was possible to increase τ_{field} up to very long values (of the order of *15min*). In this case, a bistable cavity QED system with both low photon and atom numbers is realized.

The bistability is shown in Fig.8, where the actually detected rate of pump atoms was linearly varied between $1000s^{-1}$ and $2000s^{-1}$ with a cycle period $T = 10.2s \approx 170\,\tau_{cav}$. Fig.8 exhibits a wide hysteresis cycle, the width of the hysteresis region corresponding to a change in the rate of detected pump atoms of $\approx 800s^{-1}$. In the bistable region, the maser can be permanently prepared in either the upper or the lower branch of the hysteresis loop. The number of photons in the cavity is $\approx 80\,(160)$ in the lower (upper) field state, and the average number of pump atoms in the cavity 0.15.

According to the theoretical expectation, the area of the hysteresis cycle can be increased by scanning the pump rate more rapidly. This experiment was performed on the first first-order phase transition, corresponding to location *3a* in Fig.1. Owing to the comparatively small spontaneous hopping time constant of $\approx 1s$, the hysteresis cycle had to be scanned quite rapidly. The result is shown in Fig.9. It can be clearly recognized that with increasing scan velocity larger parts of the metastable but globally unstable branches of maser operation can be mapped out.

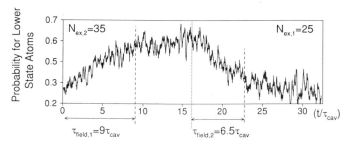

Figure 7. Measurements at low atomic flux. By switching the *UV* laser intensity the pump rate was periodically changed back and forth across phase transition *3a* (see Fig.1). The counting events in the channeltrons were recorded as a function of time modulo the cycle period. The resulting probability for lower-state atoms is plotted versus time. The experiment was performed with an average value of $N_{ex} \approx 30$, $t_{int} = 47\mu s$, and $\tau_{cav} = 60ms$. The long evolution time constants $\tau_{field,1} \approx 9\,\tau_{cav}$ and $\tau_{field,2} \approx 6.5\,\tau_{cav}$ can be clearly identified. (Figure reprinted with permission from Ref. [13])

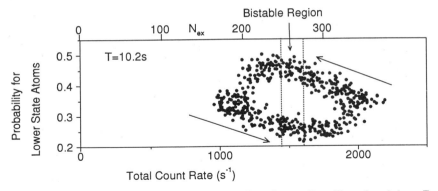

Figure 8. Hysteresis cycle of the micromaser observed with $t_{int} = 35\mu s$ and $\tau_{cav} = 60ms$. The cycle period was $T = 10.2s$. In the central part of the cycle the system is actually bistable, i.e. no spontaneous jumps between the two branches could be observed. (Figure reprinted with permission from Ref. [13])

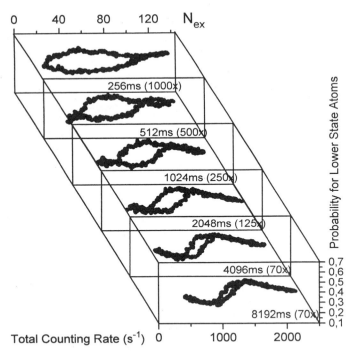

Figure 9. Time dependence of the hysteresis cycle observed with $t_{int} = 35\mu s$ and $\tau_{cav} = 23ms$. The cycle periods and the numbers of recorded cycles (in parentheses) are indicated.

5 EXPLANATION OF MASER FIELD JUMPS AND RELATED PHENOMENA

5.1 Methods

The aim of quantitative calculations is the reproduction of the experimentally observed hopping times and evolution time constants. The experiments can be compared to quantum Monte-Carlo simulations of the micromaser (see Sec.2.1). The simulations show that even after a very long time of maser operation there can be significant deviations between the actual field state and the temporal or ensemble average. This is most striking on the first-order phase transitions: The temporal average of the photon number distribution exhibits the discussed pair of maxima, whereas the time-dependent photon number distribution of the quantum Monte-Carlo calculation performs statistical jumps between two situations where one of the two peaks dominates. The characteristic hopping time between the two metastable field states equals the evolution time constant τ_{field} of the ensemble-averaged evolution. In the Monte-Carlo simulations, the experimental broadening of the velocity distribution and the spontaneous decay of the Rydberg atoms (lifetime $\approx 350 \mu s$) can be taken into account.

Alternatively, the experimental results are compared to theoretical investigations performed on the basis of the FP approach outlined in Sec.2.3. By analogy with the experiment, we consider situations in which the effective potential $V(v)$ exhibits two equally attractive minima. The hopping time τ_{field} between these minima is calculated by Kramers' analysis [10]. In all cases investigated, the hopping times obtained by the FP calculations and Monte Carlo simulations agree. If in addition to the uncertainty of t_{int} no other fluctuations are assumed, the theoretical values of the hopping times are significantly larger than the experimental values.

5.2 Fluctuations

Any kind of additional randomness included in the theoretical model flattens the effective potential $V(v)$, as can be recognized from Fig.2 where two different amounts of fluctuations in the product Ωt_{int} are assumed. The flattening of $V(v)$ reduces the hopping times between the metastable field states. An example visualizing the influence of additional noise on the hopping times is given in Fig.10. In a heuristic way, we account for all unknown fluctuations by assuming a Gaussian distribution for Ωt_{int} with relative RMS deviation $\delta = RMS(\Omega t_{\text{int}})/\langle \Omega t_{\text{int}}\rangle$. The contour plot in Fig.10 shows the hopping time as a function of the atomic lifetime and δ. The parameters of the calculation correspond to the experiment shown in Fig.5. Ideal maser performance would result in a time of stability of the metastable states of $10^8 \tau_{\text{cav}}$, which in our experiment would be about 100 days (!). The hatched area shows the parameter regime which would be compatible with the experimental estimate on the radiative lifetime of the employed Rydberg states. From Fig.10 it is concluded that the decay of the Rydberg atoms reduces the hopping rate by less than 2 orders of magnitude. The maximum time of stability feasible with the employed short-living low angular momentum Rydberg states would be about one day. This value is still 4 orders of magnitude larger than the actually observed value. In order to reproduce the experimental result we have to assume an additional 8% fluctuation of Ωt_{int}. Since the fluctuation of t_{int} is known to be only 2%, the fluctuation of Ωt_{int} must mainly result from a fluctuation of the atom-field coupling Ω.

In general it is found that fluctuations corresponding to $\delta = (15 \pm 1)\%$ have to be assumed at the first first-order phase transition point (*3a* in Fig.1), and $\delta = (8 \pm 1)\%$ at the second (*3b* in Fig. 1). Other experimental investigations, not discussed in detail here, using a cavity resonant with the $63 P_{3/2} - 61 D_{5/2}$ transition, operated at the second first-order phase transition, can be reproduced by assuming $\delta = (7 \pm 2)\%$. The question arises what are the sources of additional fluctuations. It can be excluded that fluctuations of N_{ex} reduce the field evolution time constants. The laser intensity was stabilized, and the flux of Rydberg atoms displays ideal Poissonian statistics. It is, however, important to note that the experiment deviates from the ideal situation treated in [10]. Stray electric fields in the holes through which the atoms enter the cavity could lead to adiabatic transitions between the magnetic sublevels. The maser would then be pumped by a mixture of magnetic sublevels, which have different atom-field couplings Ω. Another source of perturbations may be stray electric fields in the interior of the cavity. An estimate on this phenomenon we can get from the $(25 \pm 5)kHz$ FWHM width of the $63P_{3/2} - 61D_{3/2}$ maser resonance at low atomic flux, being larger than the expected value of *13kHz*. If the broadening is totally attributed to inhomogeneous electric stray fields in the cavity interior, an upper limit of $\approx 5mV/cm$ is obtained. As a consequence, different Rydberg atoms experience slightly different electric stray fields, leading to a spread in the Rabi flopping frequencies and amplitudes. The Rabi flopping is less affected by intracavity stray electric fields when a large number of photons is present. This may explain why the above mentioned values of δ are smaller for higher photon numbers. Finally, fluctuations in the effective Rabi frequency also occur when two atoms simultaneously interact with the cavity field; there is a small probability that this may happen in some of the described experiments. To date it is not possible to determine the contributions of the mentioned noise sources to the fluctuations of Ω.

6 EXPERIMENTAL EVIDENCE OF PERIODIC STRUCTURES IN MICROMASER RESONANCE LINES

In this Section, we report on the observation of the maser resonance under conditions that atomic interference phenoma in the cavity get observable. Since a nonclassical field is generated in the maser cavity we are able to investigate for the first time atomic interference phenomena under the influence of non-classical radiation. Interferences occur since a coherent superposition of dressed states is produced by mixing the dressed states at the entrance and exit hole of the cavity. Inside the cavity the dressed states develop differently in time giving rise to Ramsey-type interferences when the maser cavity is tuned through resonance. The new experimental findings are obtained on the $63P_{3/2} - 61D_{5/2}$ transition with a value of N_{ex} larger than about *60*. The maser resonance lines are measured by recording the counting rates of upper and lower state atoms as a function of the cavity frequency. The cavity frequency is scanned using the piezoelectric elements attached to the cavity walls. From the recordings the probability of finding lower state atoms is obtained as a function of the cavity frequency.

Fig.11 shows that the lower limit of the maser line shifts towards lower frequency with increasing N_{ex}, the shift reaching *200kHz* for $N_{ex} = 190$. Under those conditions, on the average there are roughly 100 photons in the cavity. The large red-shift cannot be explained by AC Stark effect which for 100 photons would amount to less than one *kHz* for the used transitions. It is, however, known from the previous maser experiments that there are small static electric fields in the entrance and exit holes of the cavity. Those fields can in fact cause considerable shifts of the micromaser resonance

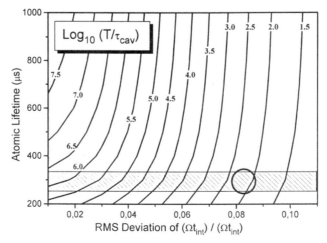

Figure 10. Theoretical values of the hopping time on the second first-order phase transition for parameters corresponding to Fig.5 as a function of the radiative atomic lifetime and the relative fluctuations of Ωt_{int}. The hatched area is compatible with experimental estimates on the atomic lifetime. The circle shows the actual experimental situation.

line (see Sec. 7.2). Fig.11 also shows that for large values of N_{ex} ($N_{ex} \geq 89$) sharp, periodic structures appear. Those typically consist of a smooth red wing and a vertical step on the blue side. The clarity of the pattern rapidly reduces when N_{ex} increases to 190 or beyond. We will see later that those structures have to be interpreted as interferences.

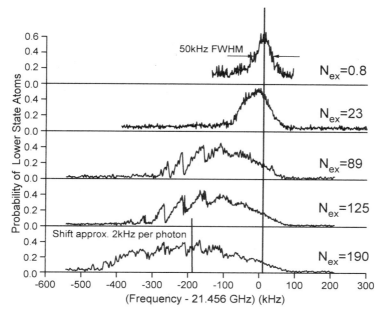

Figure 11. Shift of the maser resonance $63P_{3/2} - 61D_{5/2}$ for fast atoms ($t_{int} = 35\mu s$). The upper plot shows the maser line for low pump rate ($N_{ex} < 1$). The FWHM linewidth ($50kHz$) sets an upper limit of $\approx 5mV / cm$ to residual stray electric fields in the center of the cavity. The lower resonance lines are taken for the indicated large values of N_{ex}. The plots show that the center of the maser line shifts by about $2kHz$ per photon. In addition, there is a considerable field-induced line-broadening which is approximately proportional to $\sqrt{N_{ex}}$. For $N_{ex} \geq 89$ the lines display periodic structures which are discussed in the text. (Figure reprinted with permission from Ref. [25])

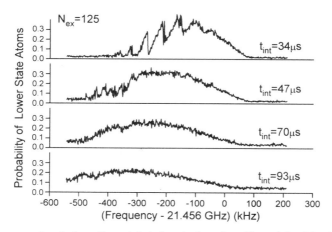

Figure12. Maser resonance lines for large N_{ex} and the indicated values of t_{int}. The period and the clarity of the steps reduce, if t_{int} is increased. The center of the resonance shifts to lower frequency with increasing t_{int}. (Figure reprinted with permission from Ref. [25])

Fig.12 shows the variation of the structure when the interaction time t_{int} between the atoms and the cavity field is changed. The uppermost recording is the same as the second from below in Fig.11. For large t_{int} no clear substructures can be observed; the substructure disappears for $t_{int} > 47 \mu s$. In the second plot from the top some substructure is still present on the left side, however, it is less pronounced than in the uppermost recording. The upper two plots in Fig.12 show that the period of the substructures reduces with increasing interaction time. Furthermore, an increasing shift of the whole resonance structure to low frequencies is observed when t_{int} is increased.

7 EXTENDED MICROMASER MODEL

7.1 The Ramsey-Type Atom-Field-Interaction

In order to understand the observed structures we first have to analyze the Jaynes-Cummings dynamics of the atoms in the cavity. The dynamics is more complicated than in previous experiments[6,7,8], since the higher maser field requires a detailed consideration of the atom-field dynamics in the periphery of the cavity volume where weak stray electric fields shift the atomic resonace frequency to lower values. The further analysis is based on the micromaser theory [10,11].

The usual formalism for the description of the coupling of an atom to the radiation field is the dressed atom approach [24] leading to a splitting of the coupled atom-field states depending on the vacuum Rabi-flopping frequency Ω, the photon number n, and the atom-field detuning $\delta = \omega - \omega_0$. We face a special situation at the entrance and the exit hole of the cavity. There we have a position dependent variation of the field of the resonant cavity mode. As a consequence, Ω is position dependent as well. In the present case of the TE_{121}-mode, the atom-field coupling Ω depends on the spatial coordinate z in atomic beam direction according to $\Omega(z) = \Omega_0 \cos(z\pi / L)$. Hereby, the cavity interior ranges from $-L/2$ to $L/2$; outside this range the coupling is zero. The fact that the mode function actually decays exponentially inside the cavity entrance and exit holes is not particularly important. An additional variation results from the fact that there are weak DC eletric stray fields in the holes which may be caused by contact potentials of rubidium deposits. The stray fields could also be induced by the microstructure of the surface of the superconducting material, since the work functions of different crystal orientations differ by values of the order of $0.5V$ [22,23]. The stray electric fields lead to a position dependent atom-field detuning $\delta(z)$. The time-dependence of the coupling Ω and the detuning δ entering in the Jaynes-Cummings Hamiltonian (Eq.1) follows from the velocity of the pump atoms and the position-dependence of $\Omega(z)$ and $\delta(z)$.

The Jaynes-Cummings-Hamiltonian only couples pairs of atom-field states $\{|e,n\rangle, |g,n+1\rangle\}$ ($n \geq 0$). Therefore, it is sufficient to consider the dynamics within such a pair of states. At each location along the atomic beam, we have well defined $\Omega(z)$ and $\delta(z)$, leading to eigenstates (dressed states) of the atom-field-system which depend on the position z of the atom. Thus, we have position-dependent dressed state energies and eigenvectors. Outside the cavity field the dressed states coincide with the states $\{|e,n\rangle, |g,n+1\rangle\}$. For parameters corresponding to the periodic substructures in Figs.11 and 12 the dressed state energies cross each other only at the beginning of the atom-field interaction, and at the end (see Fig.13). The reason for the crossings is the differential DC Stark effect of the atomic levels $|e\rangle$ and $|g\rangle$. Due to the fact that at the locations 1 and 2 in Fig.13 the maser field already has an appreciable strength the crossings at 1 and 2 are actually narrow anticrossings. When an atom passes through the cavity, the components of the atom-field wave-function written in the dressed state representation are mixed at the locations 1 and 2 in Fig.13. At the cavity entrance a coherent superposition of the dressed states is created (location 1 in Fig.13). Then the system evolves adiabatically, whereby the two dressed state components accumulate a differential dynamical phase Φ which strongly depends on the cavity frequency. At the cavity exit, the components of the atom-field wave-function in the dressed state representation are mixed a second time (location 2 in Fig.13). The mixing of the dressed states at the entrance and exit holes of the cavity, in combination with the intermediate adiabatic evolution, generates a situation similar to a Ramsey-type two-field interaction.

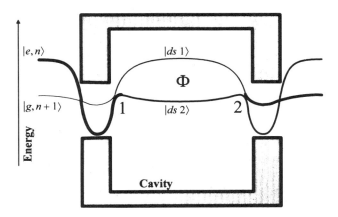

Figure 13. Energies of a pair of coupled dressed states as a function of position (schematic). The cavity volume is also indicated, whereby the diameter of the holes is exaggerated. Outside the cavity, the dressed states coincide with the indicated products of atom and field states. In the holes there are weak DC fields, which lead to a quadratic DC Stark shift being different for the atomic levels. The Stark effect causes a considerable reduction of the energies inside the holes. As a consequence, the dressed states cross each other at four points. At the inner crossing points 1 and 2, the non-zero atom-field interaction leads to some repulsion, i.e. the inner crossings are actually narrow anti-crossings. Therefore, an atom coming in via the state $|e,n\rangle$ is split at location 1 in two dressed state components coresponding to $|ds1\rangle$ and $|ds2\rangle$. The splitting is visualized by the thickness of the lines representing the energy levels: the thickness of the lines is proportional to the probability to find the atom in the respective dressed state. The two components accumulate a differential dynamical phase Φ between the locations 1 and 2, which strongly influences the outcome after the mixing point 2. The phase Φ corresponds to the dotted area.

7.2 Theoretical Micromaser Resonance Lines

The quantitative calculation leading to theoretical micromaser resonance lines can be performed in the following way. First, the variation of the static electric field in the cavity has to be estimated. This is done by numerically solving the Laplace equation with the boundaries of the cavity and assuming a particular field strength in the cavity holes. Then for different interaction times, photon numbers, and cavity frequencies the dynamics of the atom-field wave-function is calculated by numerical integration using the time-dependent Jaynes-Cummings Hamiltonian. This integration has to be performed including the local variation of $\Omega(z)$ inside the cavity owing to the mode structure of the microwave field. Furthermore, the variation of the detuning $\delta(z)$ resulting from the static electric field has to be considered. For the further analysis, which is based on the micromaser model, we extract the values $\beta(n)$ which denote the probabilities that an atom emits a photon in a field of $n-1$ photons prior to the interaction. The $\beta(n)$ have to be averaged over the probability distributions describing the properties of the pump atoms (for instance t_{int}).

In the second step of the calculation, with the averaged values of the $\beta(n)$ and using Eq.2 the photon number distribution $P(n)$ of the cavity field under steady state conditions is obtained. With $P(n)$ the average photon number $\langle n \rangle$ and $\langle n \rangle / N_{ex}$ are calculated. The latter quantity corresponds to the probability of finding an atom in the lower state, as do the experimental results displayed in Figs.11 and 12.

A theoretial result of $\langle n \rangle / N_{ex}$ obtained in this way is shown in Fig.14. The uppermost plot shows the maser resonance line expected without any static electric field. With increasing DC field strength in the cavity holes the stucture changes, the curve for *309mV/cm* coming very close to the ones displayed in Fig.11 for $N_{ex} = 89$ and 125, and on the top of Fig.12. We have to emphasize that the field values indicated in Fig.14 correspond to the maximum field strength in the cavity holes. The field value in the central part of the cavity is roughly 100 times smaller, and therefore without significance in low-flux maser experiments. Fig.14 also shows that the qualitative structure of the maser line is the same for all fields larger than about *200 mV/cm*. Thus, the conditions required to find the periodic substructures experimentally are not very stringent. The model also reproduces the

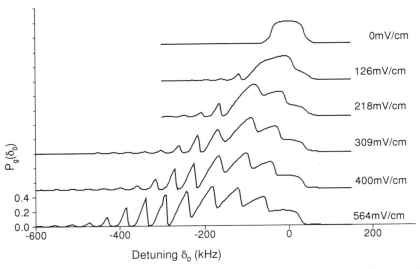

Detuning δ₀ (kHz)

Figure 14. Theoretical maser resonance lines for the indicated values of the static electric field strength in the cavity holes. The theoretical model is explained in the text. The polarizability of the transition frequency is $-1\,GHz\,/\,(V\,/\,cm)^2$. In the calculation we use $N_{ex} = 100$ and $\Omega = 45\,krad\,/\,s$. The interaction time is $t_{int} = 35\mu s$, and the RMS deviation of the interaction time $1\mu s$. In order to account for the fluctuations of the dynamical phases induced by the inhomogeneity of the stray electric fields a Gaussian distribution of the atom-field detuning with a RMS width of $5kHz$ is assumed. (Figure reprinted with permission from Ref. [25])

experimental finding that the maser line shifts to lower frequency when N_{ex} or t_{int} is increased (see Ref. [25]).

8 IMPLICATIONS OF THE EXTENDED MICROMASER MODEL

The calculations reveal that for frequencies where the maser resonance lines display sharp edges the photon number distribution has two well separated maxima. This situation is comparable to the conditions discussed in Sec.4 and Ref.[13], where maser field hysteresis and metastability is observed. In the present context, hysteresis and metastability should show up on the sharp edges in the

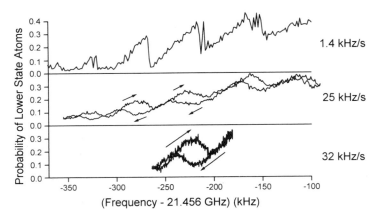

(Frequency - 21.456 GHz) (kHz)

Figure 15. Hystersis cycle of the maser. The atomic inversion is plotted versus the cavity frequency. The upper plot shows a slow scan on the red wing of a maser resonance line. The lower plots show that the vertical steps in the upper, quasi-stationary recording turn into hysteresis loops, when the cavity frequency is scanned up and down more rapidly. The values on the right indicate the scan rates. (Figure reprinted with permission from Ref. [25])

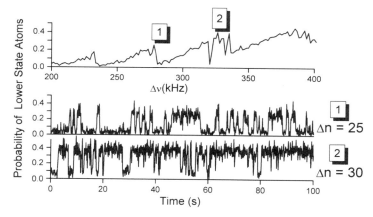

Figure 16. Spontaneous jumps of the maser field. When the maser is operated at locations 1 or 2 of the upper recording, the atomic inversion as a function of time statistically hops between two metastable situations, as shown in the lower plots. The values Δn denote the differences of the mean photon numbers in the two metastable field states. (Figure reprinted with permission from Ref. [25])

micromaser lines when the cavity frequency is varied. The experimental result displayed in Fig.15 demonstrates that each step in the steady-state maser resonance line is, indeed, related to a hysteresis loop.

Under steady-state operation on the steep edges of the quasi-stationary maser resonance line we observed spontaneous jumps of the maser field between metastable field states as well. The results displayed in Fig.16 show that the field basicly jumps between a "field on" and a "field off" state, whereby the difference in the photon numbers can be as low as 25.

The calculations also show that on the smooth wings of the more pronounced triangular-shaped structures in Fig.14 the photon number distribution $P(n)$ of the maser field is strongly sub-Poissonian. The sub-Poissonian character results from the fact that under those conditions the photon gain reduces when the photon number is increased. This feedback mechanism stabilizes the photon number.

The results can be used in order to estimate the stray electric field amplitudes in the *periphery* of the resonant cavity mode. In the present case, the maximum extension of the line towards lower frequencies is about *600kHz*, this corresponding to a stray electric field of $\approx 25mV/cm$. This estimate corresponds to the static electric field strength in a distance of a few *mm* from the cavity holes.

9 CONCLUSION

In this contribution experimental evidences of bistability, hyteresis and spontaneous field jumps in a micromaser have been presented. The observed phenomena are in principle in excellent agreement with the micromaser theory. However, the extreme sensitivity of the Rydberg atoms to perturbing influences such as stray fields and other imperfections of the experimental setup prevent us from observing extremely long switching times which could in principle be in principle realized.

The presented extended micromaser model explains line shifts and interference structures observed in the experimental micromaser resonance lines. The periodic structures in the maser lines are interpreted as Ramsey-type interferences. The model is consistent with former results in the low-pump regime (this aspect is not worked out in this contribution). Owing to the fact that the DC fields in the cavity holes are not accurately known, a detailed multi-level calculation taking into account the magnetic substructure of the involved fine-structure levels does not make sense at this point. If there was more detailed information on the DC fields, the consideration of the magnetic substructure could provide, however, a useful extension of the model.

REFERENCES

1. P. Goy, J. M. Raimond, M. Gross and S. Haroche, Observation of cavity-enhanced single-atom spontaneous emission, *Phys. Rev. Lett. 50*: 1903 (1983).
2. R. G. Hulet, E. S. Hilfer and D. Kleppner, Inhibited spontaneous emission by a Rydberg atom, *Phys. Rev. Lett. 55*: 2137 (1985).
3. D. J. Heinzen and M. S. Feld, Vacuum radiative level shift and spontaneous-emission linewidth of an atom in an optical resonator, *Phys. Rev. Lett. 59*: 2623 (1987).
4. G. Barton, Quantum-electrodynamic level shifts between parallel mirrors: analysis, *Proc. Roy. Soc. Lond. A 410*: 141 (1987).
5. G. Barton, Quantum-electrodynamic level shifts between parallel mirrors: applications, mainly to Rydberg states, *Proc. Roy. Soc. Lond. A 410*: 175 (1987).
6. D. Meschede, H. Walther and G. Müller, One-atom maser, *Phys. Rev. Lett 54*: 551 (1985).
7. G. Rempe, H. Walther and N. Klein, Observation of quantum collapse and revival in a one-atom maser, *Phys. Rev. Lett. 58*: 353 (1987).
8. G. Rempe, F. Schmidt-Kaler, and H. Walther, Observation of sub-Poissonian photon statistics in a micromaser, *Phys. Rev. Lett. 64*: 2783 (1990).
9. J. H. Eberly, N. B. Narozhny and J. J. Sanchez-Mondragon, Periodic spontaneous collapse and revival in a simple quantum model, *Phys. Rev. Lett.44*: 1323 (1980).
10. P. Filipowicz, J. Javanainen and P. Meystre, Theory of a microscopic maser, *Phys. Rev. A 34*: 3077 (1986).
11. L. A. Lugiato, M. O. Scully and H. Walther, Connection between microscopic and macroscopic maser theory, *Phys. Rev. A 36*: 740 (1987).
12. G. Rempe and H. Walther, Sub-Poissonian atomic statistics in a micromaser, *Phys. Rev. A 42:* 1650 (1990).
13. O. Benson, G. Raithel and H. Walther, Quantum jumps of the micromaser field: dynamic behavior close to phase transition points, *Phys. Rev. Lett. 72*: 3506 (1994).
14. A. M. Guzman, P. Meystre and E. M.Wright, Semiclassical theory of a micromaser, *Phys. Rev. A 40*: 2471 (1989).
15. P. Meystre and E. M. Wright, Measurements-induced dynamics of a micromaser, *Phys. Rev. A 37*: 2524 (1988).
16 G. Raithel, C. Wagner, H. Walther, L. M. Narducci and M. O. Scully, The micromaser: a proving ground for quantum physics, in: *Cavity Quantum Electrodynamics*, P. R. Berman, ed., Academic Press, Inc., San Diego (1994).
17. J. Krause, M. O. Scully and H. Walther, Quantum theory of a micromaser: symmetry breaking via off-diagonal atomic injection, *Phys. Rev. A 34*: 2032 (1986).
18. P. Meystre, Cavity quantum optics and the quantum measurement process, in: *Progress in Optics* XXX, E. Wolf, ed., Elsevier Science Publishers, Amsterdam (1992)
19. L. Davidovich, J. M. Raimond, M. Brune and S. Haroche, Quantum theory of a two-photon micromaser, *Phys. Rev. A 36*: 3771 (1987), M. Brune, J.M. Raimond, P. Goy, L. Davidovich and S. Haroche, Realization of a two-photon maser oscillator, *Phys. Rev. Lett. 59*: 1899 (1987).
20. J. M. Raimond, M. Brune, P. Goy and S. Haroche, J. de Physique Coll. 15: 17 (1990).
21. J. M. Raimond, M. Brune, L. Davidovich, P. Goy and S. Haroche, The two-photon Rydberg atom micromaser, in: *Atomic Physics* 11, S. Haroche, J. C. Gay, and G. Grynberg, eds., World Scientific, Singapore (1989)
22. *CRC Handbook of Chemistry and Physics*, R. C. Weast ed., 68th editon, CRC Press, Inc., Boca Raton (1987).
23. *American Institute of Physics Handbook*,. D. E. Gray ed., 3rd edition, McGraw-Hill Book Company, New York (1982).
24. C. Cohen-Tannoudji, J. Dupont-Roc and G. Grynberg, Atom-Photon Interactions, John Wiley and Sons, Inc., N.Y. (1992).
25. G. Raithel, O. Benson, H. Walther, Atomic Interferometry with the micromaser, *Phys. Rev. Lett. 75*: 3446 (1995).

PARADIGM FOR TWO-PHOTON CASCADE INTENSITY CORRELATION EXPERIMENTS

Georg M. Meyer,[1,3] Girish S. Agarwal,[2]
Hu Huang,[1] and Marlan O. Scully [1,3]

[1] Department of Physics
   Texas A&M University
   College Station, TX 77843
[2] School of Physics
   University of Hyderabad
   Hyderabad 500134, India
[3] Max-Planck-Institut für Quantenoptik
   Hans-Kopfermann-Str. 1
   D-85748 Garching, Germany

INTRODUCTION

Quantum electrodynamics is a spectacularly successful theory, nowhere are its successes more numerous than in modern quantum optics. However, fully quantum field theoretical calculations are frequently quite involved, and it is therefore important to have and perfect paradigms which provide physical insight, thus helping us to decide which calculations to do. A case in point is the Welton treatment of the Lamb shift via heuristic "vacuum-fluctuations" arguments.[1]

Another example of a useful paradigm concerns the intensity correlation interferometry of light arising from the two-photon cascade emission of a single three-level atom. The atom decays from the upper state a to an intermediate level b with a rate γ_a and from b to the ground state c with a rate γ_b, the atom emits two photons ψ and ϕ, respectively. Two different experimental approaches were carried out.

In the setup considered by Franson,[2] see Figure 1(a), the radiation travels from the atom through interferometers of Mach-Zehnder type to the detectors D_1 and D_2. Each photon can take either a long or a short path. An interesting new photon-photon correlation effect emerged after an ansatz for the second-order correlation function. Several groups[3-6] carried out the experiment with light produced by spontaneous parametric down-conversion in nonlinear crystals and found the results to be in good agreement

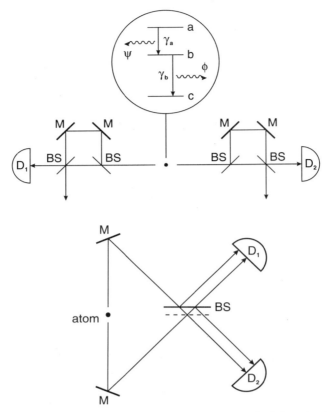

Figure 1. Two-photon interferometer with a single excited three-level atom as the light source. Decaying from the upper state a to an intermediate level b with a rate γ_a and from b to the ground state c with a rate γ_b, the atom emits two photons ψ and ϕ, respectively. (a) Scheme I: The radiation travels from the atom to the detectors D_1 and D_2 via short and long optical paths determined by beam splitters BS and mirrors M. (b) Scheme II: The displacement of the beam splitter produces a path difference between the two reflected paths. The two photons can be emitted in the same direction or in opposite directions.

with the heuristic analysis.

The other approach adopted by Hong, Ou, and Mandel[7] employed a beam splitter (BS) to create asymmetric path differences as in Figure 1(b). With a pair of frequency filters placed in front of the detectors, a beat signal at the difference frequency between the two transitions has been observed.

In order to provide a thorough discussion of two-photon coincidence interferometry, several theoretical models have been investigated.[8,9] Here we give a quantized-field analysis based on the quantum mechanical treatment of the three-level casade problem for the whole range of atomic parameters and different experimental configurations. Using diagram representations, we are able to identify the origin of each interference term in our calculation. We find that for the cascade model the Franson configuration and the Hong-Ou-Mandel configuration (henceforth referred to as Scheme I and II, respectively) give identical results in spite of their apparent differences. In the following we first present the full quantum treatment of the problem and then discuss two limiting cases.

QUANTUM TREATMENT OF INTENSITY CORRELATIONS

Within the framework of the Weisskopf-Wigner theory of spontaneous emission, the two-photon state produced by cascade emission from a single excited three-level atom at the origin is found to be the entangled state[10,11]

$$|\Psi\rangle = \sum_{\mathbf{k},\mathbf{q}} \frac{-g_{a\mathbf{k}}\,g_{b\mathbf{q}}}{[i(\nu_k + \nu_q - \omega_{ac}) - \gamma_a/2][i(\nu_q - \omega_{bc}) - \gamma_b/2]} |1_{\mathbf{k}}, 1_{\mathbf{q}}\rangle , \tag{1}$$

where $g_{a,\mathbf{k}}$ $(g_{b,\mathbf{q}})$ is the atom-field coupling constant for the $a \to b$ $(b \to c)$ transition involving wave vector $\mathbf{k}$ $(\mathbf{q})$; the atomic energy difference is given by ω_{ab} (ω_{bc}).

The state $|\Psi\rangle$ allows us to calculate the joint count probability that both detectors are excited (within an infinite coincidence window), which is proportional to the time integral

$$P(1,2) = \int_0^\infty dt_1 \int_0^\infty dt_2\, G^{(2)}(1,2) \tag{2}$$

over the Glauber second-order correlation function

$$G^{(2)}(1,2) = \langle\Psi|E^{(-)}(\mathbf{r}_1,t_1)E^{(-)}(\mathbf{r}_2,t_2)E^{(+)}(\mathbf{r}_2,t_2)E^{(+)}(\mathbf{r}_1,t_1)|\Psi\rangle . \tag{3}$$

Here $E^{(+)}(\mathbf{r}_i, t_i)$ is the usual annihilation operator part of the field at the site $\mathbf{r}_i$ of detector D_i at time t_i

$$E^{(+)}(\mathbf{r}_i, t_i) = \sum_{\mathbf{k}} \mathcal{E}_k a_{\mathbf{k}} e^{i(\mathbf{k}\mathbf{r}_i - \nu_k t_i)}, \tag{4}$$

where $\mathcal{E}_k = (\hbar\nu_k/2\varepsilon_0 V)^{1/2}$ with the free space permittivity ε_0 and the quantization volume V, $E^{(-)}$ is the corresponding creation operator part.

The action of the delay loops (i.e., the beam splitters and mirrors) may be modeled via a Heisenberg treatment of the operators

$$E^{(+)}(\mathbf{r}_i, t_i) = \frac{1}{2}\left[E^{(+)}(\mathbf{S}_i, t_i) + E^{(+)}(\mathbf{L}_i, t_i)\right] , \tag{5}$$

where $\mathbf{S}_i = S_i \hat{\mathbf{r}}_i$, $\mathbf{L}_i = L_i \hat{\mathbf{r}}_i$, and $\hat{\mathbf{r}}_i = \mathbf{r}_i/|\mathbf{r}_i|$. Since the state $|\Psi\rangle$ is a two-photon state, we obtain from Eqs. (2) and (3)

$$G^{(2)}(1,2) = \Psi^*(1,2)\,\Psi(1,2) \tag{6}$$

with the "probability amplitude for a photo-electron pair"

$$\Psi(1,2) = \frac{1}{4} \langle 0|\left[E^{(+)}(\mathbf{S}_2, t_2) + E^{(+)}(\mathbf{L}_2, t_2)\right]\left[E^{(+)}(\mathbf{S}_1, t_1) + E^{(+)}(\mathbf{L}_1, t_1)\right]|\psi\rangle . \tag{7}$$

For simplicity, we focus on the situation when the detectors are opposite of each other, i.e., $\hat{\mathbf{r}}_1 = -\hat{\mathbf{r}}_2$. Then we have

$$\Psi(1,2) = \Psi(S_1, S_2) + \Psi(S_1, L_2) + \Psi(L_1, S_2) + \Psi(L_1, L_2) \tag{8}$$

with

$$\begin{aligned}
\Psi(r_1, r_2) &= \frac{\kappa_a \kappa_b}{4 r_1 r_2}\, e^{-(i\omega_{ac}+\gamma_a/2)(t_1 - r_1/c)}\, \Theta(t_1 - r_1/c) \\
&\quad \times e^{-(i\omega_{bc}+\gamma_b/2)[(t_2 - r_2/c)-(t_1-r_1/c)]}\, \Theta[(t_2 - r_2/c) - (t_1 - r_1/c)] \\
&\quad + (1 \leftrightarrow 2) ,
\end{aligned} \tag{9}$$

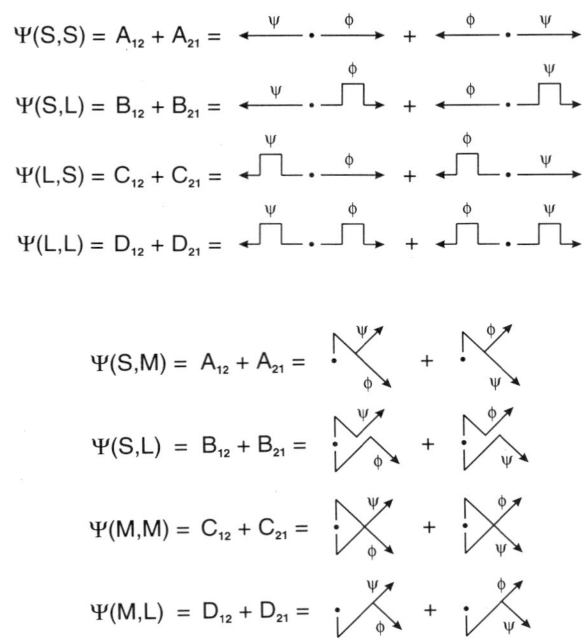

$\Psi(S,S) = A_{12} + A_{21} =$

$\Psi(S,L) = B_{12} + B_{21} =$

$\Psi(L,S) = C_{12} + C_{21} =$

$\Psi(L,L) = D_{12} + D_{21} =$

$\Psi(S,M) = A_{12} + A_{21} =$

$\Psi(S,L) = B_{12} + B_{21} =$

$\Psi(M,M) = C_{12} + C_{21} =$

$\Psi(M,L) = D_{12} + D_{21} =$

Figure 2. Illustration of the various terms in Eq. (8) for (a) Scheme I and (b) Scheme II, e.g., B_{12} can be interpreted as the contribution arising from photon ψ taking the short path S to detector D_1 and photon ϕ taking the long path L to D_2.

where $(1 \leftrightarrow 2)$ represents the same terms with subscripts 1 and 2 interchanged, $\Theta(x)$ is the usual step function, $r_i = S_i$ or L_i $(i = 1, 2)$, $\kappa_a = \pi g_{ak_0} \mathcal{E}_{k_0} D(\omega_{ab})/k_0$ and $\kappa_b = \pi g_{bq_0} \mathcal{E}_{q_0} D(\omega_{bc})/q_0$, where $D(\nu) = V\nu^2/2\pi^2 c^3$ is the mode density of free space, $k_0 = \omega_{ab}/c$, $q_0 = \omega_{bc}/c$, and c is the velocity of light.

It is useful to represent the various terms in Eq. (8) by the diagrams in Figure 2. While generally the lengths S_i and L_i can be arbitrary, we consider here the configurations with symmetric delays (Scheme I), as in Figure 2(a), and with antisymmetric delays (Scheme II), as in Figure 2(b). For example in the case of $\Psi(S, L)$, the first diagram associated with B_{12} denotes photon ψ taking the short path S to detector D_1 and photon ϕ the long path L to detector D_2. In Scheme I we set $S_1 = S_2 \equiv S$, $L_1 = L_2 \equiv L$, and $T \equiv (L - S)/c$, while in Scheme II we have $S_1 \equiv S$, $S_2 = L_1 \equiv M$, $L_2 \equiv L$, and $L - M = M - S \equiv cT$. In Scheme II the photon arriving at D_2 has traveled an additional time T compared to Scheme I.

Using Eqs. (6)–(9) and the notation of Figure 2, we find after a somewhat lengthy calculation for $S \approx L$ and $S/c \gg \gamma_a^{-1} + \gamma_b^{-1}$

$$
\begin{aligned}
P(1,2) = & \int_0^\infty dt_1 \int_0^\infty dt_2 \left\{ \left[(|A_{12}|^2 + |B_{12}|^2 + |C_{12}|^2 + |D_{12}|^2) + (1 \leftrightarrow 2) \right]_1 \right. \\
& + \left[(A_{12}^* D_{12} + \text{c.c.}) + (1 \leftrightarrow 2) \right]_2 \\
& + \left[(A_{12}^* B_{12} + C_{12}^* D_{12} + \text{c.c.}) + (1 \leftrightarrow 2) \right]_3 \\
& + \left[(A_{12}^* C_{12} + B_{12}^* D_{12} + \text{c.c.}) + (1 \leftrightarrow 2) \right]_4 \\
& + \left[(B_{12}^* C_{12} + \text{c.c.}) + (1 \leftrightarrow 2) \right]_5
\end{aligned}
$$

$$+ \left[A_{12}^* B_{21} + C_{12}^* A_{21} + C_{12}^* B_{21} + C_{12}^* D_{21} + D_{12}^* B_{21} + \text{c.c.} \right]_6 \Bigg\} .$$

(10)

After the integration over t_1 and t_2, both schemes give the same result

$$
\begin{aligned}
P(1,2) &= \left[\frac{2\kappa}{\gamma_a \gamma_b} \right]_1 + \left[\frac{\kappa e^{-\gamma_a T/2}}{\gamma_a \gamma_b} \cos(\omega_{ac} T) \right]_2 + \left[\frac{2\kappa e^{-\gamma_b T/2}}{\gamma_a \gamma_b} \cos(\omega_{bc} T) \right]_3 \\
&+ \left[\frac{2\kappa e^{-(\gamma_a + \gamma_b) T/2}}{\gamma_a \gamma_b} \cos(\omega_{ab} T) \right]_4 + \left[\frac{\kappa e^{-(\gamma_a/2 + \gamma_b) T}}{\gamma_a \gamma_b} \cos(\Delta T) \right]_5 \\
&+ \left[\frac{4\kappa e^{-\gamma_b T/2}}{4\Delta^2 + \gamma_a^2} \left\{ \cos(\omega_{bc} T) - \frac{2\Delta}{\gamma_a} \sin(\omega_{bc} T) + \frac{e^{-\gamma_b T/2}}{2} \right. \right. \\
&\qquad - e^{-\gamma_a T/2} \left(\cos(\omega_{ab} T) - \frac{2\Delta}{\gamma_a} \sin(\omega_{ab} T) \right) \\
&\qquad \left. \left. - \frac{e^{-(\gamma_a + \gamma_b) T/2}}{2} \left(\cos(\Delta T) - \frac{2\Delta}{\gamma_a} \sin(\Delta T) \right) \right\} \right]_6 ,
\end{aligned}
$$

(11)

where $\kappa = (\kappa_a \kappa_b / 2SL)^2$, $\omega_{ac} = \omega_{ab} + \omega_{bc}$, $\Delta = \omega_{ab} - \omega_{bc}$. The square bracket expressions in Eq. (10), enumerated by subscripts, are the origin of the corresponding terms in Eq. (11).

DISCUSSION OF SPECIAL CASES

We next consider two limits of the atomic parameters. For $\gamma_b \gg \gamma_a$ as in Figure 3(a) and $\gamma_b T \gg 1$, we obtain from Eq. (11)

$$P(1,2) = \frac{2\kappa}{\gamma_a \gamma_b} \left[1 + \frac{e^{-\gamma_a T/2}}{2} \cos(\omega_{ac} T) \right] ,$$

(12)

the sum frequency oscillation comes from the $A_{12}^* D_{12}$ term. This is the limit discussed by Franson[2] for $\gamma_a T \ll 1$. In his derivation, the two-photon state of the radiation field is assumed to have the following two properties

$$\langle 0 | E^{(+)}(\mathbf{r}_1, t) E^{(+)}(\mathbf{r}_2, t \pm T) | \Psi \rangle = 0 ,$$

(13)

$$\langle 0 | E^{(+)}(\mathbf{r}_1, t - T) E^{(+)}(\mathbf{r}_2, t - T) | \Psi \rangle = e^{i\omega_{ac} T} \langle 0 | E^{(+)}(\mathbf{r}_1, t) E^{(+)}(\mathbf{r}_2, t) | \Psi \rangle .$$

(14)

Under these conditions the second-order correlation function is

$$G^{(2)}(1,2) = 2 \left| \langle 0 | E^{(+)}(\mathbf{r}_1, t) E^{(+)}(\mathbf{r}_2, t) | \Psi \rangle \right|^2 \left[1 + \cos(\omega_{ac} T) \right].$$

(15)

The conditions (13) and (14) are not only fulfilled for two-photon cascade emission from single atoms with $\gamma_b \gg T \gg \gamma_a$, but also for the light produced by spontaneous parametric down-conversion in nonlinear crystals.

In the case when $\gamma_a \gg \gamma_b$ as in Figure 3(b) and $\gamma_a T \gg 1$, we obtain

$$
\begin{aligned}
P(1,2) &= \frac{2\kappa}{\gamma_a \gamma_b} \left[1 + e^{-\gamma_b T/2} \cos(\omega_{bc} T) \right] \\
&+ \frac{4\kappa e^{-\gamma_b T/2}}{4\Delta^2 + \gamma_a^2} \left[\frac{e^{-\gamma_b T/2}}{2} + \cos(\omega_{bc} T) - \frac{2\Delta}{\gamma_a} \sin(\omega_{bc} T) \right] .
\end{aligned}
$$

(16)

115

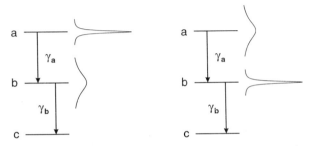

Figure 3. The radiative broadening of the upper and intermediate level of a three-level atom in the cascade configuration is depicted schematically for two cases: (a) $\gamma_b \gg \gamma_a$ and (b) $\gamma_a \gg \gamma_b$.

For comparison, we now consider the radiation field to be in the product state

$$|\Psi\rangle = |\psi\rangle \otimes |\phi\rangle \tag{17}$$

with

$$|\psi\rangle = \sum_k \frac{g_{ak}}{(\nu_k - \omega_{ab}) + i\gamma_a/2}|1_k\rangle \,,$$

$$|\phi\rangle = \sum_q \frac{g_{aq}}{(\nu_q - \omega_{bc}) + i\gamma_b/2}|1_q\rangle \,. \tag{18}$$

This two-photon state is produced by interrupted cascade emission, which is depicted in Figure 4 and has been studied by Scully and Drühl[12] in the context of quantum erasers. We obtain from Eq. (2)

$$
\begin{aligned}
P(1,2) &= \frac{2\kappa}{\gamma_a\gamma_b}\left[1 + e^{-\gamma_a T/2}\cos(\omega_{ab}T)\right]\left[1 + e^{-\gamma_b T/2}\cos(\omega_{bc}T)\right] \\
&+ \frac{8\kappa}{4\Delta^2 + (\gamma_a + \gamma_b)^2}\left[1 + \frac{e^{-\gamma_a T} + e^{-\gamma_b T}}{4} + e^{-\gamma_a T/2}\cos(\omega_{ab}T)\right. \\
&\left. + e^{-\gamma_b T/2}\cos(\omega_{bc}T) + \frac{e^{-(\gamma_a + \gamma_b)T/2}}{2}\cos(\omega_{ac}T)\right] \,.
\end{aligned}
\tag{19}
$$

In the limit $\gamma_a \gg \gamma_b$ and $\gamma_a T \gg 1$, the joint count probabilities for the two-photon state of Eq. (1) and the product state of Eq. (17) have the same dominant terms.

The eight diagrams in Figures 2(a) and 2(b) represent all the possible paths the two cascade photons can take. The advantage of organizing the various terms in Eqs. (10) and (11) according to the corresponding paths becomes clear when we consider the following variations.

The $X_{12}^* Y_{21}$ terms in the last group $[\ldots]_6$ of Eq. (10) are due to the inability to determine which photon arrives at each detector. They are removed when frequency filters are placed before the detectors.

In addition, in the original experiments by Hong, Ou, and Mandel, the photons always enter the interferometer from different ports. In our calculation this corresponds

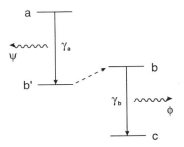

Figure 4. The scheme of interrupted two-photon cascade emission. After decaying from a to b', the atom is excited to level b, from where it decays to the ground state c.

to the absence of all terms containing A or D, and only the B and C terms remain, which lead to difference frequency oscillations,

$$P(1,2) = \frac{\kappa}{2\gamma_a \gamma_b} \left[1 + e^{-(\gamma_a/2 + \gamma_b)T} \cos(\Delta T) \right] , \tag{20}$$

in accord with the experimental observation.

If one considers a setup similar to the one used by Larchuk *et al.*,[13] the correlated photons can also enter the interferometer from the same port, so that all the terms in Eqs. (10) and (11) occur.

On the other hand, if we select the A and D terms, we obtain only the sum frequency oscillations, like in the Franson limit, but here reproduced without the need of approximations concerning the decay rates.

CONCLUSIONS

We have presented a rigorous quantized-field analysis of two-photon coincidence interferometry with cascade emission from a single three-level atom. Our discussion uses diagram representations and provides physical interpretations of the various interference terms for the whole range of atomic parameters and different intensity correlation experiments. The two well-known interferometric configurations used by Franson and by Hong, Ou, and Mandel are shown to give similar results. The limiting cases of broad and narrow middle level have been discussed. We have regained the results obtained by Franson and by Mandel and co-workers and have shown the connection to the Scully-Drühl situation of interrupted cascade emission. In addition, we found interference terms that were absent in the heuristic analysis. Finally we note that the interference effects may be studied by confining atoms in traps. In this way, a Young-type interference pattern was produced by Eichmann *et al.*[14] in a recent beautiful experiment.

ACKNOWLEDGMENTS

This work has been supported by the Office of Naval Research, the Welch Foundation, and the Texas Advanced Research Program.

REFERENCES

1. T.A. Welton, *Phys. Rev.* 74:1157 (1948).
2. J.D. Franson, *Phys. Rev. Lett.* 62:2205 (1989); 67:290 (1991).
3. J.D. Franson, *Phys. Rev. A* 44:4552 (1991).
4. P.G. Kwiat, W.A. Vareka, C.K. Hong, H. Nathel, and R.Y. Chiao, *Phys. Rev. A* 41:2910 (1990).
5. Z.Y. Ou, X.Y. Zou, L.J. Wang, and L. Mandel, *Phys. Rev. Lett.* 65:321 (1990).
6. J.G. Rarity and P.R. Tapster, *Phys. Rev. A* 45:2052 (1992).
7. C.K. Hong, Z.Y. Ou, and L. Mandel, *Phys. Rev. Lett.* 59:2044 (1987); Z.Y. Ou and L. Mandel, *ibid.* 61:54 (1988); 62:2941 (1989).
8. R.A. Campos, B.E.A. Saleh, and M.C. Teich, *Phys. Rev. A* 42:4127 (1990).
9. M.H. Rubin and Y.H. Shih, *Phys. Rev. A* 45:8138 (1992).
10. M.O. Scully and M.S. Zubairy. *Quantum Optics*, Cambridge University Press (to be published).
11. H. Huang and J.H. Eberly, *J. Mod. Optics* 40:915 (1993).
12. M.O. Scully and K. Drühl, *Phys. Rev. A* 25:2208 (1982).
13. T.S. Larchuk, R.A. Campos, J.G. Rarity, P.R. Tapster, E. Jakeman, B.E.A. Saleh, and M.C. Teich, *Phys. Rev. Lett.* 70:1603 (1993).
14. U. Eichmann, J.C. Bergquist, J.J. Bollinger, J.M. Gilligan, W.M. Itano, D.J. Wineland, and M.G. Raizen, *Phys. Rev. Lett.* 70:2359 (1993).

ENTANGLEMENT IN TWO–PHOTON CASCADE RADIATION AND THE VISIBILITY OF SECOND–ORDER INTERFERENCE

Susanne F. Yelin,[1,2,3] Ulrich W. Rathe,[1,2,3]
and Marlan O. Scully[1,3]

[1] Department of Physics
Texas A&M University
College Station, TX 77843
[2] Sektion Physik
Ludwig-Maximilians–Universität München
Theresienstr. 37
D-80333 München, Germany
[3] Max-Planck-Institut für Quantenoptik
Hans-Kopfermann-Str. 1
D-85748 Garching, Germany

INTRODUCTION

Two–photon coincidence interferometry has recently been the subject of growing experimental and theoretical interest.[1–7] From an experimental point of view, the visibility of the second–order interference is an important issue in such interferometers. For example, if the violation of a Bell inequality is tested in a given setup, then the visibility of the interference as predicted by quantum mechanics has to be high enough to contradict the existence of hidden variables. In the present article, we show that the entanglement present in spontaneous cascade radiation emitted by an excited three–level atom potentially enhances the visibility of second–order interference in comparison to the level produced by radiation from other sources.

The analyzed atomic level scheme and the interferometer are depicted in Figures 1 and 2, respectively. In the existing literature on this interferometric setup[8] and on related problems, a heuristically motivated product state $|\psi_f\rangle = |\gamma\rangle \otimes |\phi\rangle$ of two one–photon states $|\gamma\rangle$ and $|\phi\rangle$ is frequently used for the radiation field produced by a cascading atom (for notation, see Figure 1). Such a state leads to a joint count probability of the form (see Eq. (14))

$$P_{\text{factorized}} \propto 1 + e^{-\gamma_a |\Delta r|/c - \gamma_b |\Delta \rho|/c} \cos(k_0 \Delta r + q_0 \Delta \rho) , \qquad (1)$$

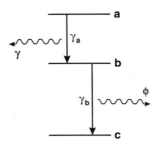

Figure 1. Atomic level scheme. Upper level a decays to intermediate level b with rate γ_a, emitting γ–radiation. b decays to ground state c with rate γ_b, emitting ϕ–radiation.

where ck_0 and γ_a (cq_0 and γ_b) are the resonance frequency and decay rate of the upper (lower) transition, respectively. $\Delta r = r_1 - r_2$ and $\Delta\rho = \rho_1 - \rho_2$ are the characteristic pathlength differences of the interferometer as shown in Figure 2.

The full quantum–field theoretical result for the radiation field state produced by a cascading atom, however, does not factorize (for details, see Eq. (6)). In the present paper, we show that such an entangled two–photon state leads to a joint count probability (see Eq. (15))

$$P_{\text{entangled}} \propto 1 + e^{-\gamma_a|\Delta r|/c - \gamma_b|\Delta\rho - \Delta r|/c} \cos(k_0\Delta r + q_0\Delta\rho) . \tag{2}$$

If the decay rates of the atoms in question fulfill $\gamma_a \gg \gamma_b$ (see Figure 3(a)), then the two results in Eqs. (1) and (2) coincide. In this sense we confirm the validity of the earlier treatment with the factorized field state. However, the more general result Eq. (2) allows for a higher visibility of the interference term than the one in Eq. (1) because it is possible to cancel the influence of γ_b in the exponential damping term $e^{-\gamma_b|\Delta\rho - \Delta r|}$ by using $\Delta r = \Delta\rho$ in the experiment. This is particularly advantageous for atoms with decay rates fulfilling $\gamma_a \ll \gamma_b$ (see Figure 3(b)).

Before we proceed with the detailed derivation of the results outlined so far, we like to clarify that the entanglement mentioned in the previous paragraphs is not the "Bell-type" entanglement due to the superposition of two photon pairs, each coming from a different source atom. Instead, the enhancement of the visibility is due to the entanglement of the two photons in each one of the two superposed two–photon states.

VISIBILITY OF SECOND–ORDER INTERFERENCE

Two atoms of the kind introduced in Figure 1 are placed in an interferometer as depicted in Figure 2. The field is initially in the vacuum state and the atoms are prepared in a linear superposition of the two states where either the atom at site $\mathbf{R}_1$ is excited and the atom at site $\mathbf{R}_2$ is in the ground state or vice versa,

$$|\Psi(t = 0)\rangle = \frac{1}{\sqrt{2}} (|a_1, c_2\rangle + |c_1, a_2\rangle) \otimes |0\rangle . \tag{3}$$

Such an atomic state can be generated, for example, by pumping the system weakly on the $|c\rangle \rightarrow |a\rangle$ transition such that the probability of simultaneous excitation of both

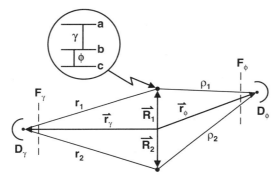

Figure 2. Two–photon coincidence interferometer. Atoms at site $\mathbf{R}_1$ and $\mathbf{R}_2$ emit γ– and ϕ–radiation picked up by detectors D_γ and D_ϕ after selection by filters F_γ and F_ϕ, respectively.

atoms is negligible. Each of the two detectors is shielded by a filter susceptible only for radiation from one of the transitions, such that the detector D_γ detects only γ–radiation ($|a\rangle \to |b\rangle$) and D_ϕ only ϕ–radiation ($|b\rangle \to |c\rangle$). The second–order correlation term we are interested in is then obtained by measuring the coincidence count rate of the two detectors.

The initial state Eq. (3) decays via spontaneous emission, cascading through the two involved transitions, and evolves to

$$|\Psi(t \to \infty)\rangle = |c_1, c_2\rangle \otimes |\psi_f\rangle , \tag{4}$$

where the state of the radiation field only is

$$|\psi_f\rangle = \frac{1}{\sqrt{2}} \left[|\psi_f(\mathbf{R}_1)\rangle + |\psi_f(\mathbf{R}_2)\rangle \right] . \tag{5}$$

Here $|\psi_f(\mathbf{R}_i)\rangle$ ($i = 1, 2$) is the field state generated by the spontaneous cascade of one excited atom.

The state of the radiation field generated by one excited atom at $\mathbf{R}$ reads within the Wigner–Weisskopf approximation

$$|\psi_f(\mathbf{R})\rangle = \sum_{\mathbf{k},\mathbf{q}} \frac{g_{a,\mathbf{k}} g_{b,\mathbf{q}} e^{-i(\mathbf{k}+\mathbf{q}) \cdot \mathbf{R}}}{[c(q - q_0) + i\gamma_b][c(k - k_0 + q - q_0) + i\gamma_a]} |1_\mathbf{q}, 1_\mathbf{k}\rangle , \tag{6}$$

as derived, for example, by Scully. [9] Here, $\mathbf{k}$ ($\mathbf{q}$) is the wavevector of the upper (lower) transition with the subscript 0 denoting the resonance, $g_{a,\mathbf{k}}$ and $g_{b,\mathbf{q}}$ are the respective coupling constants between the atom and the radiation field, and γ_a (γ_b) is the upper (lower) decay rate. Note that this state is not simply the product of two one–photon states $|\gamma\rangle \otimes |\phi\rangle$, but instead involves entanglement between the "γ–radiation" emitted by the upper transition $|a\rangle \to |b\rangle$ and "ϕ–radiation" emitted by the lower transition $|b\rangle \to |c\rangle$. As shown elsewhere [10] such an entanglement can lead, among other interesting effects, to features of subnatural width in the second–order coherence function, thus providing an in–principle means for subnatural atomic spectroscopy.

In addition to using the radiation field state of Eq. (5) in Eq. (4), we also consider a field state with factorized two–photon states in the spirit of the earlier treatment [8]

$$\left|\psi_f^{\text{fact}}\right\rangle = \frac{1}{\sqrt{2}} \left(|\gamma_1\rangle |\phi_1\rangle + |\gamma_2\rangle |\phi_2\rangle \right) , \tag{7}$$

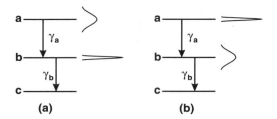

Figure 3. a) Three–level scheme of Figure 1 with broad uppermost level and sharp intermediate level, producing a two–photon field state with weak entanglement. (b) Three–level scheme with narrow uppermost level and broad intermediate level, producing a two–photon field state with strong entanglement.

where the single–photon states are given by

$$|\gamma_i\rangle = \sum_k \frac{g_{a,k} e^{-i\mathbf{k}\cdot\mathbf{R}_i}}{c(k-k_0)+i\gamma_a} |1_k\rangle, \tag{8}$$

$$|\phi_i\rangle = \sum_q \frac{g_{b,q} e^{-i\mathbf{q}\cdot\mathbf{R}_i}}{c(q-q_0)+i\gamma_b} |1_q\rangle. \tag{9}$$

Indeed, since a long lifetime of the intermediate state $|b\rangle$ practically decouples the two transitions $|a\rangle \to |b\rangle$ and $|b\rangle \to |c\rangle$, the state given by Eq. (5) can be replaced by the state of Eq. (7) in the case of a short lifetime of the upper state $|a\rangle$ (large decay rate γ_a) and a narrow intermediate state $|b\rangle$ (small γ_b, see Fig. 3(a)).

Glauber's second–order correlation function $G^{(2)}$ gives a quantity proportional to the coincidence count rate of the two detectors and reads for the field state in Eq. (5)

$$
\begin{aligned}
G^{(2)}(1,2) &= \left\langle \psi_f \left| \hat{E}_\gamma^{(-)} \hat{E}_\phi^{(-)} \hat{E}_\phi^{(+)} \hat{E}_\gamma^{(+)} \right| \psi_f \right\rangle \\
&= \left\langle \psi_f \left| \hat{E}_\gamma^{(-)} \hat{E}_\phi^{(-)} \right| 0 \right\rangle \left\langle 0 \left| \hat{E}_\phi^{(+)} \hat{E}_\gamma^{(+)} \right| \psi_f \right\rangle \\
&= \frac{1}{2} \left| \left\langle 0 \left| \hat{E}_\phi^{(+)} \hat{E}_\gamma^{(+)} \right| \psi_f(\mathbf{R}_1) \right\rangle + \left\langle 0 \left| \hat{E}_\phi^{(+)} \hat{E}_\gamma^{(+)} \right| \psi_f(\mathbf{R}_2) \right\rangle \right|^2.
\end{aligned}
\tag{10}
$$

The two detectors are described by the electric field operators

$$\hat{E}_\xi^{(+)} = \sum_k \mathcal{E}_k \hat{a}_k e^{i\mathbf{k}\cdot\mathbf{r}_\xi - ickt_\xi} = \hat{E}_\xi^{(-)\dagger}, \tag{11}$$

where $\xi = \gamma$ or ϕ, indicating that each detector picks up radiation from only one transition due to the filters. We obtain with the photon arrival times t_γ and t_ϕ at the detector D_γ (at $\mathbf{r}_\gamma$) and D_ϕ (at $\mathbf{r}_\phi$), respectively

$$
\begin{aligned}
G^{(2)}(1,2) = \; &\frac{1}{2} \left(\frac{V}{2\pi c}\right)^4 |\mathcal{E}_{k_0} \mathcal{E}_{q_0} g_{a,k_0} g_{b,q_0} k_0 q_0|^2 \times \\
&\times \left\{ \frac{e^{-2\gamma_a(t_\gamma - r_1/c) - 2\gamma_b[(t_\phi - \rho_1/c)-(t_\gamma - r_1/c)]}}{r_1^2 \rho_1^2} \Theta\left(t_\gamma - \frac{r_1}{c}\right) \Theta\left[\left(t_\phi - \frac{\rho_1}{c}\right) - \left(t_\gamma - \frac{r_1}{c}\right)\right] \right. \\
&\left. + \frac{e^{-2\gamma_a(t_\gamma - r_2/c) - 2\gamma_b[(t_\phi - \rho_2/c)-(t_\gamma - r_2/c)]}}{r_2^2 \rho_2^2} \Theta\left(t_\gamma - \frac{r_2}{c}\right) \Theta\left[\left(t_\phi - \frac{\rho_2}{c}\right) - \left(t_\gamma - \frac{r_2}{c}\right)\right] \right.
\end{aligned}
$$

$$+2\frac{e^{-\gamma_a(2t_\gamma-(r_1+r_2)/c)-\gamma_b[(2t_\phi-(\rho_1+\rho_2)/c)-(2t_\gamma-(r_1+r_2)/c)]}}{r_1r_2\rho_1\rho_2}\Theta\left(t_\gamma-\frac{r_1}{c}\right)\Theta\left(t_\gamma-\frac{r_2}{c}\right)\times$$

$$\times\Theta\left[\left(t_\phi-\frac{\rho_1}{c}\right)-\left(t_\gamma-\frac{r_1}{c}\right)\right]\Theta\left[\left(t_\phi-\frac{\rho_2}{c}\right)-\left(t_\gamma-\frac{r_2}{c}\right)\right]\cos(k_0\Delta r+q_0\Delta\rho)\Bigg\},$$

$$(12)$$

where Θ denotes the step function and V the quantization volume; $r_i=|\mathbf{R}_i-\mathbf{r}_\gamma|$ and $\rho_i=|\mathbf{R}_i-\mathbf{r}_\phi|$ are the distances between atoms and detectors.

In the case of factorized two–photon states, Eq. (7), $G^{(2)}$ calculated along the same lines takes the form

$$G^{(2)}_{\text{fact}}(1,2) = \frac{1}{2}\left(\frac{V}{2\pi c}\right)^4|\mathcal{E}_{k_0}\mathcal{E}_{q_0}g_{a,k_0}g_{b,q_0}k_0q_0|^2\times$$

$$\times\Bigg\{\frac{e^{-2\gamma_a(t_\gamma-r_1/c)-2\gamma_b(t_\phi-\rho_1/c)}}{r_1^2\rho_1^2}\Theta\left(t_\gamma-\frac{r_1}{c}\right)\Theta\left(t_\phi-\frac{\rho_1}{c}\right)$$

$$+\frac{e^{-2\gamma_a(t_\gamma-r_2/c)-2\gamma_b(t_\phi-\rho_2/c)}}{r_2^2\rho_2^2}\Theta\left(t_\gamma-\frac{r_2}{c}\right)\Theta\left(t_\phi-\frac{\rho_2}{c}\right)$$

$$+2\frac{e^{-\gamma_a(2t_\gamma-(r_1+r_2)/c)-\gamma_b(2t_\phi-(\rho_1+\rho_2)/c)}}{r_1r_2\rho_1\rho_2}\Theta\left(t_\gamma-\frac{r_1}{c}\right)\Theta\left(t_\gamma-\frac{r_2}{c}\right)\times$$

$$\times\Theta\left(t_\phi-\frac{\rho_1}{c}\right)\Theta\left(t_\phi-\frac{\rho_2}{c}\right)\cos(k_0\Delta r+q_0\Delta\rho)\Bigg\}.$$

$$(13)$$

After integrating over the photon arrival times we obtain the joint count probability of the coincidence for this factorized case, depending on the pathlength differences $\Delta r=r_1-r_2$ and $\Delta\rho=\rho_1-\rho_2$, respectively,

$$P^{(2)}_{\text{fact}}(1,2) = \frac{\kappa}{4\gamma_a\gamma_b}\Bigg\{\frac{1}{r_1^2\rho_1^2}+\frac{1}{r_2^2\rho_2^2}+2\frac{e^{-\gamma_a|\Delta r|/c-\gamma_b|\Delta\rho|/c}}{r_1r_2\rho_1\rho_2}\cos\left(k_0\Delta r+q_0\Delta\rho\right)\Bigg\},\quad (14)$$

where κ is a scaling factor. For the state Eq. (5) with entangled two–photon states this joint count probability reads

$$P^{(2)}_{\text{ent}}(1,2) = \frac{\kappa}{4\gamma_a\gamma_b}\Bigg\{\frac{1}{r_1^2\rho_1^2}+\frac{1}{r_2^2\rho_2^2}+2\frac{e^{-\gamma_a|\Delta r|/c-\gamma_b|\Delta\rho-\Delta r|/c}}{r_1r_2\rho_1\rho_2}\cos\left(k_0\Delta r+q_0\Delta\rho\right)\Bigg\}.\quad (15)$$

Eqs. (14) and (15) are the main result of the present paper. Choosing r_1, r_2, ρ_1, and ρ_2 of the same order of magnitude, they reduce to the results given in the Introduction, Eqs. (1) and (2). Note that the interference term in the case of factorizing two–photon states is exponentially suppressed by $\exp(-\gamma_a|\Delta r|/c-\gamma_b|\Delta\rho|/c)$, see Eq. (14). In the case with entangled two–photon states the corresponding term is $\exp(-\gamma_a|\Delta r|/c-\gamma_b|\Delta\rho-\Delta r|/c)$, see Eq. (15). Thus, experimentally mapping out the interference term $\cos(k_0\Delta r+q_0\Delta\rho)$ by varying $\delta\equiv\Delta\rho=\Delta r$ takes advantage of the entanglement — the influence of γ_b is simply cancelled and does not influence the visibility as it does if one assumes a field state without entanglement.

This result is quite general and does not depend on the ratio of γ_a/γ_b. However, the advantage of cancelling the influence of γ_b by keeping $\Delta\rho=\Delta r$ during a measurement is negligible if $\gamma_a\gg\gamma_b$ as mentioned in the Introduction. In other words, the atomic level scheme depicted in Figure 3(a) produces a field state that can for all practical purposes be replaced by the factorizing two–photon field state. This confirms the validity of the common choice for a factorizing two–photon field state in the literature on EPR-type correlations and Bell inequalities.

The most advantageous case in our setup, however, is the one depicted in Figure 3(b), where γ_b is large compared to γ_a. There the visibility can indeed reach 100% for r_1, r_2, ρ_1, $\rho_2 \approx r$ and $\Delta\rho = \Delta r$.

ACKNOWLEDGMENTS

This work was supported by the Office of Naval Research, the Texas Advanced Research Program, and the Welch Foundation. We thank J. D. Franson and R. Y. Chiao for stimulating discussions. UR and SY would like to thank the "Studienstiftung des Deutschen Volkes" and SY the "Deutscher Akademischer Austauschdienst" for support.

REFERENCES

1. R. Gosh and L. Mandel, *Phys. Rev. Lett.* 59:1903 (1987); C. K. Hong, Z. Y. Ou, and L. Mandel, *Phys. Rev. Lett.* 59:2044 (1987); Z. Y. Ou and L. Mandel, *Phys. Rev. Lett.* 61:50 (1988); Z. Y. Ou and L. Mandel, *Phys. Rev. Lett.* 61:54 (1988); Z. Y. Ou, X. Y. Zou, L. J. Wang, and L. Mandel, Phys. Rev. Lett. 65:321 (1990).
2. J. D. Franson, *Phys. Rev. Lett.* 62:2205 (1989); J. D. Franson, *Phys. Rev. Lett.* 67:290 (1991); J. D. Franson, *Phys. Rev. A* 44:4552 (1991).
3. M. O. Scully and K. Drühl, *Phys. Rev. A* 25:2208 (1982).
4. M. A. Horne, A. Shimony, and A. Zeilinger, *Phys. Rev. Lett.* 62:2209 (1989).
5. P. G. Kwiat, W. A. Vareka, C. K. Hong, H. Nathel, and R. Y. Chiao, *Phys. Rev. A* 41:2910 (1990); P. G. Kwiat, A. M. Steinberg, and R. Y. Chiao, *Phys. Rev. A* 45:7729 (1992); P. G. Kwiat, A. M. Steinberg, and R. Y. Chiao, *Phys. Rev. A* 47:R2472 (1993); P. G. Kwiat, A. M. Steinberg, and R. Y. Chiao, *Phys. Rev. A* 49:61 (1994).
6. J. G. Rarity and P. R. Tapster, *Phys. Rev. Lett.* 64:2495 (1990); J. G. Rarity and P. R. Tapster, *Phys. Rev. A* 45:2052 (1992); T. S. Larchuk, R. A. Campos, J. G. Rarity, P. R. Tapster, E. Jakeman, B. E. A. Saleh, and M. C. Teich, *Phys. Rev. Lett.* 70:1603 (1993).
7. U. Eichmann, J. C. Bergquist, J. J. Bollinger, J. M. Gilligan, W. M. Itano, D. J. Wineland, M. G. Raizen, *Phys. Rev. Lett.* 70:2359 (1993).
8. M. O. Scully, in: *Laser Spectroscopy V*, Springer Series in Optical Sciences, Vol. 30, A. R. W. McKellar, T. Oka, and B. P. Stoicheff, eds., Springer-Verlag, Berlin (1981).
9. M. O. Scully, Photon–photon correlations from single atoms, in: *Proceedings of "Advances in Quantum Phenomena," Erice, Italy, 1994* (in press).
10. U. W. Rathe, M. O. Scully, and S. F. Yelin, Second–order photon–photon correlations and atomic spectroscopy, *Ann. NY Acad. of Sciences* 755:28 (1995)

QUANTUM BEATS DESCRIBED
BY DE BROGLIAN PROBABILITIES

Mirjana Božić, Zvonko Marić, and Dušan Arsenović

Institute of Physics, P.O.Box 57, Belgrade, Yugoslavia

ABSTRACT

Compatible statistical interpretation is used to give objective interpretation of quantum beats in atomic fluorescence.

I. INTRODUCTION

In the first lecture of this school, Asim Barut summarized the development of the theory of electron, and draw on the blackboard the list which is shown on the left hand side of the Table 1, given below. By analyzing Barut's list one may conclude that it contains that part of the development of the theory of electron in which the emphasis is on phenomena of the interaction of electron with radiation. The study of phenomena which reveal the properties of electron itself lead to a new field - Electron interferometry[1,2], which evolved later into Quantum Interferometry[3,4]. This development is shortly described on the right hand side of the Table 1.

One may say, by looking at the Table 1, that it summarizes the development of the Quantum theory, as a whole, nonrelativistic as well as relativistic. This is because the development of Quantum theory is tightly bound to the study of phenomena associated with electron. That is why Barut considers[5] that the most fundamental problem of contemporary physics has been to answer to the question: "What is an electron?" But, any attempt to answer to this question would meet with the basic problems in the understanding of physical reality in quantum domain, which has been the subject of the debate since the birth of Quantum mechanics[6]. Quantum interferometry has grown from this debate, which has developed mainly between followers of Copenhagen (standard) interpretation of Quantum mechanics[7] and followers of Wave mechanics[8,9,10].

Table 1. The summary of the development of the theory of electron.

1894, G. Stoney
Electron
Atom of electricity
1897. J.J. Thompson, Wiechert, Zeeman
The most fundamental particle,
absolutely stable.

Electron according to			
	H. A. Lorentz		
— ‖ —	Poincare-Einstein		
— ‖ —	De Broglie-Schrödinger (wave property)	$\Rightarrow$	Electron interferometry
— ‖ —	Born-Heisenberg (probability wave)		$\Downarrow$
			Interferometry with quantons (electrons, photons, neutrons, protons)
— ‖ —	Dirac (spin)		$\Downarrow$
— ‖ —	Pauli (spin)		
— ‖ —	QED + Renormalization	$\Rightarrow$	Ingenious quantum interference experiments:
— ‖ —	"Self-energy"		- Conference on Matter Wave Interferometry[3] (1987. - hundred years of Schrödinger's birth) - Conference on Quantum Interferometry[4] (1993)

II. INTERFERENCE PATTERN AS A RESULT OF ACCUMULATION OF INDIVIDUAL EVENTS

Characteristic features of the phenomena studied in quantum interferometry are determined by relative phases of different components present in the quantum state

$$\psi(\mathbf{r}, t) = \sum_n c_n(t)\varphi_n(\mathbf{r}) \tag{1}$$

Standard (Copenhagen) interpretation of QM avoids to attribute[11] physical meaning to phases of wave functions as well as to relative phases of components in an arbitrary superposition. This has been, the source of numerous difficulties, controversies and paradoxes which have appeared in the endeavor to understand quantum interference phenomena.

Recent experiments, performed with beams of one per one electron[2], neutron[12], atom[13], in double-slit and Mach-Zehnder interferometer clearly show that the interference pattern emerges through the process of accumulation of many dots (events) on the screen (in the detector).

The standard quantum mechanical formalism does not explain neither the appearance of single dots (events) on the screen (in the detector) nor the process of accumulation of the dots. It describes only the regularities of the interference pattern when enough dots (events) are collected. Barut used this fact in his argumentation[14-18] about the incompleteness of the standard description of double-slit interference pattern, as well as about the incompleteness of Quantum mechanics: "Probabilistic quantum mechanics is incomplete because it cannot make statements about individual events."[18]

The essence of the disagreement in the understanding of interference phenomena is related to the question "what is the cause of quantum interference phenomena?" According to the Quantum theory of measurement[7], interference (addition of probability amplitudes) is due to our lack of knowledge of a particular transition chosen by the system.

According to the realistic wave interpretation (de Broglie[8], Vigier[9], Selleri and Tarozzi[19]), the addition of probability amplitudes (wave functions) corresponds to the objectively existing superposition of real component waves, which emerge from two slits, when the particle and its accompanying wave hit the slits. Each particle has a trajectory inside the interferometer. The trajectory is determined by the superposed wave and mechanical quantities of the particle. Recently proposed compatible statistical interpretation of double-slit experiment[20,21] uses de Broglian probabilities in order to express statistically the concepts and ideas of the realistic wave interpretation of quantum interference.

Compatible statistical interpretation (CSI) may be applied to an arbitrary superposition (1) of quantum states. In addition to usual probabilities and probability densities: $|\psi(\mathbf{r}, t)|^2$, $|c_n(t)|^2$, $|\varphi_n(\mathbf{r})|^2$, CSI uses de Broglian probabilities, defined by[22]:

$$P_n(\mathbf{r}, t) = |c_n(t)|^2 |\varphi_n(\mathbf{r})|^2 \left[1 + \frac{\sum'_{mk} c_m^*(t) c_k(t) \varphi_m^*(\mathbf{r}) \varphi_k(\mathbf{r})}{\sum_m |c_m(t)|^2 |\varphi_m(\mathbf{r})|^2} \right] \tag{2}$$

$P_n(\mathbf{r}, t)$ is the probability density that quanton's quantum number is n and that particle is situated around $\mathbf{r}$ at the moment t when the quanton wave function is $\psi(\mathbf{r}, t)$.

In addition to the double-slit experiment, compatible statistical interpretation was applied to the interpretation of various experiments in neutron interferometry[23], wave packets[24] and Rabi oscillations[25]. Here, we shall present compatible statistical interpretation of quantum beats.

III. QUANTUM BEATS

Quantum beat phenomena is a very good example of a quantum interference. Quantum beats may be observed in the fluorescence from atoms excited initially, impulsively, into a coherent superposition $\Psi(\mathbf{R}, \mathbf{r_e}, 0)$ of excited states $\Phi_i(\mathbf{R}, \mathbf{r_e})$, having slightly different energies[26,27,28] ($\mathbf{R}$ is the radius vector of the atomic center of mass and $\mathbf{r_e}$ is the relative radius vector of the electron). A typical quantum beat experiment involving atoms is symbolized in Fig. 1. The atom, located at point $\mathbf{R}$ and initially in its ground state $\Phi_g(\mathbf{R}, \mathbf{r_e})$, is suddenly excited at time $t = 0$ by a short pulse of broad band light irradiation into a coherent superposition of excited states $\Phi_i(\mathbf{R}, \mathbf{r_e})$, $i = 1, 2$; having slightly different internal energies E_i.

$$\Psi(\mathbf{R}, \mathbf{r_e}, 0) = \sum_i c_i(0) \Phi_i(\mathbf{R}, \mathbf{r_e}) \qquad i = 1, 2 \tag{3}$$

After the excitation, the atom starts emitting light by spontaneous emission (wavy lines on Fig. 1) and decays back to the ground state $\Phi_g(\mathbf{R}, \mathbf{r_e})$.

$$\Psi(\mathbf{R}, \mathbf{r_e}, t) = \sum_i c_i(0)e^{-iE_it/\hbar}e^{-\Gamma t/2}\Phi_i(\mathbf{R}, \mathbf{r_e}) + c_g(t)\Phi_g(\mathbf{R}, \mathbf{r_e}) \qquad (4)$$

Here Γ is the rate of spontaneous emission ($\tau = 1/\Gamma$ is the common radiative lifetime of the Φ_i substates).

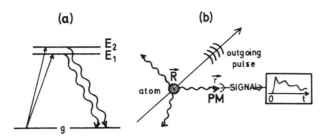

Fig. 1 a and b. Single-atom quantum beats. (a) Energy diagram show-ing the transitions involved in the preparation of the excited state coherences (straight line arrows) and in the subsequent fluorescence (wavy lines). (b) Sym-bolic sketch of the experiment showing the emitting atom located at point $\mathbf{R}$ and the detector at point $\mathbf{r}$.

The emitted light is detected by a photomultiplier P.M. located at point $\mathbf{r}$. Zee-man quantum beats observed in the fluorescence of the 6^3P_1 level of mercury after excitation by a shuttered spectral lamp are shown in Fig.2.

According to the interpretation based on the quantum theory of measurement[29], quantum beat signal is due to the impossibility to tell whether the photon has been emmitted from the level Φ_1 or from the level Φ_2. Therefore, quantum theory of mea-surement attributes an objectively existing phenomena to observer's lack of knowl-edge.

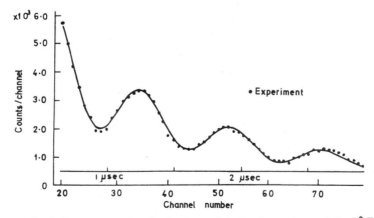

Fig. 2. Zeeman quantum beats observed in the fluorescence of the 6^3P_1 level of mercury after excitation by a shuttered spectral lamp (from [26]).

We propose here an explanation of quantum beat phenomena which is objective and therefore independent of the observer knowledge. In the explanation based on the compatible statistical interpretation, quantum beat phenomena is due to the coherence in the atomic wave function (3), which is a real superposition of two real waves.

If there would be no spontaneous emission the initial state (3) would evolve in time as

$$\Psi(\mathbf{R}, \mathbf{r_e}, t) = \sum_n c_n(0) \exp(-iE_n t/\hbar) \Phi_n(\mathbf{R}, \mathbf{r_e}) \tag{5}$$

Therefore, the initial coherent state remains coherent during the evolution.

Supposing that the atomic function is a product of the function of external (atomic) variable ($\mathbf{R}$) by the function of the internal (electronic) variable ($\mathbf{r_e}$):

$$\Phi_i(\mathbf{R}, \mathbf{r_e}) = A e^{i\mathbf{K}\cdot\mathbf{R}} \varphi_i(\mathbf{r_e}) \tag{6}$$

(A is the normalization constant), it is straightforward to find that the space distribution of the electron inside the atom which is in the state

$$\psi(\mathbf{r_e}, t) = c_1(0)\varphi_1(\mathbf{r_e})e^{-iE_1 t/\hbar} + c_2(0)\varphi_2(\mathbf{r_e})e^{-iE_2 t/\hbar} \tag{7}$$

oscillates in time with frequency $\omega_b = |E_1 - E_2|/\hbar$

$$|\psi(\mathbf{r_e}, t)|^2 = \int |\Psi(\mathbf{R}, \mathbf{r_e}, t)|^2 d^3\mathbf{R} = |c_1(0)|^2|\varphi_1(\mathbf{r_e})|^2 + |c_2(0)|^2|\varphi_2(\mathbf{r_e})|^2 +$$
$$+ 2\mathrm{Re}\, c_1(0)c_2^*(0)\varphi_1(\mathbf{r_e})\varphi_2^*(\mathbf{r_e})e^{-i(E_1-E_2)t/\hbar} \tag{8}$$

The ω_b is identical to the frequency of beats in the fluorescence signal.

According to the compatible statistical interpretation, quantum beats reflect the oscillations of the electronic distribution. Electronic distribution in the state $\psi(\mathbf{r_e}, t)$ oscillates because this state is not a stationary state but a solution of the time dependent Schrödinger equation. $\psi(\mathbf{r_e}, t)$ is the wave function of an electron with energy E_1 as well as of an electron with energy E_2. But, according to CSI electronic space distribution in the state $\psi(\mathbf{r_e}, t)$ differ from the space distributions in the states $\varphi_1(\mathbf{r_e})$ and $\varphi_2(\mathbf{r_e})$. They are determined by de Broglian probabilities (2) which for the state (7) read

$$P_i(\mathbf{r_e}, t) = |c_i(0)|^2|\varphi_i(\mathbf{r_e})|^2 \left[1 + \frac{2\mathrm{Re}\, c_1(0)c_2^*(0)\varphi_1(\mathbf{r_e})\varphi_2^*(\mathbf{r_e})e^{-i(E_1-E_2)t/\hbar}}{\sum_{n=1}^{2} |c_n(0)|^2|\varphi_n(\mathbf{r_e})|^2}\right] \tag{9}$$

Their sum is equal to $|\psi(\mathbf{r_e}, t)|^2$;

$$|\psi(\mathbf{r_e}, t)|^2 = P_1(\mathbf{r_e}, t) + P_2(\mathbf{r_e}, t), \tag{10}$$

which is the probability density that any electron (either with energy E_1 or with energy E_2) is situated around $\mathbf{r_e}$ at time t.

This interpretation is similar to the CSI interpretation of Rabi oscillations, proposed previously[25]. The similarity is understandable since the time dependence in the state (7) is a special case of the time-dependent state associated with Rabi oscillations[30−32].

The main point of the above argumentation is that any electron in the atom, either with energy E_1 or with energy E_2 is accompanied by the same real wave $\psi(\mathbf{r_e}, t)$. This wave is different from waves $\varphi_1(\mathbf{r_e})e^{-iE_1 t/\hbar}$ and $\varphi_2(\mathbf{r_e})e^{-iE_2 t/\hbar}$ in the stationary states. Electron energy in the state $\psi(\mathbf{r_e}, t)$ may be E_1 or E_2. Probability density of the coordinate of the electron with energy E_1 is neither $|\psi(\mathbf{r_e}, t)|^2$, nor $|c_1|^2|\varphi_1(\mathbf{r_e})|^2$, but it is $P_1(\mathbf{r_e}, t)$. Similarly, the probability density of the coordinate of the electron with energy E_2 is neither $|\psi(\mathbf{r_e}, t)|^2$, nor $|c_2(0)|^2|\varphi_2(\mathbf{r_e})|^2$ but it is $P_2(\mathbf{r_e}, t)$. Probability densities $P_1(\mathbf{r_e}, t)$ and $P_2(\mathbf{r_e}, t)$ are determined by the coherent state $\psi(\mathbf{r_e}, t)$ as a whole and they are oscillating functions in time. Consequently, the total probability density $|\psi(\mathbf{r_e}, t)|^2$ oscillates in time too. Oscillations of photon number in time at a given position are the consequence of oscillations of the electronic distribution.

In the presence of spontaneous emission the time evolution (4) of the initial state (3) is more complicated. In addition to the oscillating factors it contains also exponential factors determined by the rate of spontaneous emission. Consequently, intensity of fluorescence is an exponentially oscillating function of time, in agreement with the experimental results.

REFERENCES

1. G. Mollenstedt and C. Jönsson, Some remarks on the quantum mechanics of the electron, *Z. Phys.* 155:472 (1959); G. Möllenstedt, *Physica* B151:201 (1988).
2. A. Tonomura, "Electron Holography," Springer-Verlag, Berlin (1993); A. Tonomura, J. Endo, T. Matsuda, T. Kawasaki, H. Ezawa, *Am. J. Phys.* 57:? (1989).
3. "Matter-Wave Interferometry," G. Badurek, H. Rauch and A. Zeilinger eds. *Physica* B151 North-Holland, Amsterdam (1988).
4. "Quantum Interferometry", F. de Martini, G. Denardo and A. Zeilinger eds., World Scientific, Singapore (1994).
5. A.O. Barut, Brief history and recent developments in electron theory and quantum electrodynamics, *in*: "The Electron," D. Hestenes and A. Weingartshofen, eds., Kluwer Acad. Pub., Dordrecht (1991).
6. E. Schrödinger, Are there quantum jumps?, *The British Journal for the Philosophy of Science*, 3:109 (1952); What is an elementary particle, *Endeavour* 9:109 (1950).
7. "Quantum Measurement Theory," J.A. Wheeler and W.H. Zurek, eds., Princeton Univ. Press, Princeton-New Jersey (1983).
8. L. de Broglie, "Etude Critique des Bases de l'Interpretation Actuelle da la Mécanique Ondulatoire", Gauthier-Villars, Paris (1963).
9. J.P. Vigier, New theoretical implications of neutron interferometric experiments, *Physica* B151:386 (1988).
10. F. Selleri, "Quantum Paradoxes and Physical Reality", Kluwer Acad. Publ., Dordrecht (1990).
11. M. Bozić, Neutron optics, *in* "New Frontiers of Quantum Electrodynamics and Quantum Optics", A.O. Barut, ed., Plenum, New York, (1990).
12. "Neutron Interferometry," U. Bonse and H. Rauch, Clarendon Press, Oxford (1979).
13. O. Carnel, J. Mlynek, Young's double-slit experiment with atoms: a simple atom interferometer, *Phys. Rev. Letters*, 66:2689 (1991).
14. A.O. Barut, De Broglie wavelets, Schrödinger waves and classical trajectories, *Found. Phys.* 20:1233 (1990); How to avoid quantum paradoxes, *Found. Phys.* 22:137 (1992).
15. A.O. Barut, $E = \hbar\omega$, *Phys. Lett.* 143:349 (1990).
16. A.O. Barut, Quantum theory of single events, *in*: "Symposium on the Foundations of Modern Physics", P. Lahti et al. eds. World Scientific, Singapore (1991) p. 31.

17. A.O. Barut and S. Basri, Path integrals and quantum interference, *Amer. J. Phys.* 60:896 (1992).
18. A.O. Barut, The deterministic wave mechanics. A bridge between classical mechanics and probabilistic quantum theory, *in*: "The Interpretation of quantum Theory: Where we stand," L. Acardi, ed., Acta Encyclopedia, (1993).
19. F. Selleri and G. Tarozzi, Is nondistributivity for microsystems empirically founded?, *Nuovo Cimento* 43:31 (1978).
20. M. Bozić, Z. Marić and J.P. Vigier, De-Broglian probabilities in the double-slit experiment, *Found. Phys.* 22:1325 (1992).
21. M. Bozić, Compatible setatistical interpretation of interference in double-slit interferometer, *in*: "Waves and Paricles in Light and Matter", A. van der Merwe and A. Garuccio, eds., Plenum, New Yrok (1994).
22. M. Bozić and Z. Marić, Probability and interference, *in*: "Courants, amers et écueils en microphysique," G. Lochac and P. Lochac, eds., Fondation Louis de Broglie, Paris (1994).
23. M. Bozić and Z. Marić, Probabilities of de Broglie's trajectories in Mach-Zehnder and in neutron interferometers, *Phys. Lett.* A158:33 (1991).
24. M. Bozić and Z. Marić, Compatible statistical interpretation of a wave packet, *Found. Phys.* 25:159 (1995).
25. M. Božić and D. Arsenović, *in*: "Frontiers of Fundamental Physics," M. Barone and F. Selleri, eds., Plenum, New York (1994).
26. J.N. Dodd, W.J. Sandle, D. Zissermann, Study of resonance fluorescence in cadmium: Modulation effects and lifetime measure, *Proc. Phys. Soc.* London 92:497 (1967).
27. S. Haroche, J.A. Paisner, and A.L. Schawlaw, Hyperfine quantum beaats observed in Cs vapor under pulsed dye laser excitation, *Phys. Rev. Lett.* 30:948 (1973).
28. I.A. Sellin, J.R. Mowal, R.S. Peterson, P.M. Griffin, R. Lanbert, and H.H. Haselton, Observation of coherent electron-density-distribution oscillations in collision-averaged foil excitation of the $n = 2$ hydrogen levels, *Phys. Rev. Lett.* 31:1335 (1973).
29. W.W. Chow, M.O. Scully, J.O. Stoner, Jr; Quantum beat phenomena described by quantum electrodynamics and neoclassical theory, *Phys. Rev. A* 11:1380 (1975).
30. P.L. Knight and P.W. Milonni, The Rabi frequency in optical spectra, *Phys. Reports* 66:21 (1980)
31. L. Allen and J.H. Eberly, "Optical Resonance and Two-level Atoms", Dover, New York (1987)
32. G. Raithel, O. Benson and H. Walther, Dynamics of the micromaser field, this Proceedings.

QUANTUM OSCILLATOR WITH KRONIG-PENNEY
EXCITATION IN DIFFERENT REGIMES OF DAMPING

Olga Man'ko

Lebedev Physical Institute

Leninsky prospekt 53, Moscow 117924, Russia

fax: (095) 938 22 51, e-mail: manko@sci.fian.msk.su

INTRODUCTION

The behaviour of an oscillator may be controlled by the frequency time–dependence. For example, one can kick the oscillator frequency by short pulses and this kicking produces an excitation of the parametric oscillator. The amplitude of the oscillator vibrations and its energy may increase due to the external influence expressed as the frequency time–dependence. Also the statistical properties of the oscillator state may be changed due to the action of external forces. The aim of the talk is to discuss the exact solution of the time–dependent Schrödinger equation for a damped quantum oscillator subject to a periodical frequency delta–kicks describing squeezed states which are expressed in terms of Chebyshev polynomials. The cases of strong and weak damping are investigated in the frame of Caldirola–Kanai model [1], [2].

The problem of quantum oscillator with a time–dependent frequency was solved in Refs. [3]–[11]. It was shown that the wave function and, consequently, all physical characteristics of the oscillator can be expressed in terms of the solution of the classical equation of motion

$$\ddot{\varepsilon}(t) + 2\gamma\dot{\varepsilon}(t) + \omega^2(t)\varepsilon(t) = 0, \tag{1}$$

with initial conditions

$$\varepsilon(0) = 1,$$

$$\dot{\varepsilon}(0) = i\Omega(0). \tag{2}$$

where $\Omega(0) = \Omega$ will be defined below. The remaining problem is to find explicit expression for the function ε.

DIFFERENT REGIMES OF DAMPING

Here we consider the case of a periodically kicked oscillator, where the frequency depends on time as follows

$$\omega^2(t) = \omega_0^2 - 2\kappa \sum_{k=1}^{N-1} \delta(t - k\tau),$$

where ω_0 is constant part of frequency, δ is Dirac delta–function, γ is the damping coefficient, and κ is the force of delta–kicks. We consider the damping in the frame of Caldirola–Kanai model, and take into account three cases:

(i) undamped case ($\gamma = 0$);

(ii) the case of weak damping ($\omega_0 > \gamma$);

(iii) the case of strong damping ($\omega_0 < \gamma$).

The undamped case was considered in [10]; following [10] we have the equation for function $\varepsilon(t)$

$$\ddot{\varepsilon}(t) + 2\gamma\dot{\varepsilon}(t) + \omega_0^2\varepsilon(t) - 2\kappa \sum_{k=1}^{N-1} \delta(t - \kappa\tau) = 0. \tag{3}$$

It is obvious, due to substitutions t by x, ε by ψ, and $\omega_0^2/2$ by E, that if the damping is absent this equation coincides with the equation for the wave function of a quantum particle of unit mass in a Kronig–Penney potential (the sequence of δ–potentials). For every interval of time $(k - 1)\tau < t < k\tau$ the solution for the classical equation of motion is given by

$$\varepsilon_k(t) = A_k e^{\mu_1 t} + B_k e^{\mu_2 t}, \quad k = 0, 1, \ldots, N, \tag{4}$$

μ_1 and μ_2 are complex numbers. Due to continuity conditions we have

$$\varepsilon_{k-1}(k\tau) = \varepsilon_k(k\tau),$$

$$\dot{\varepsilon}_k(k\tau) - \dot{\varepsilon}_{k-1}(k\tau) = 2\kappa\varepsilon_{k-1}(k\tau). \tag{5}$$

134

Formulae (5) are obtained by integrating Eq. (3) over the infinitely small time interval $n\tau - 0 < t < n\tau + 0$. The coefficients A_k and B_k must satisfy the relations which can be expressed in the matrix form

$$\begin{pmatrix} A_k \\ B_k \end{pmatrix} = \begin{pmatrix} 1 - \frac{2\kappa}{D} & -\frac{2\kappa}{D}e^{D\tau k} \\ \frac{2\kappa}{D}e^{-D\tau k} & 1 + \frac{2\kappa}{D} \end{pmatrix} \begin{pmatrix} A_{k-1} \\ B_{k-1} \end{pmatrix},$$ (6)

where $D = \mu_2 - \mu_1$. After the sequence of δ-kicks the coefficients A_n, B_n are connected with the initial ones A_0, B_0 through the equation

$$\begin{pmatrix} A_n \\ B_n \end{pmatrix} = S^{(n)} \begin{pmatrix} A_0 \\ B_0 \end{pmatrix}, \qquad S^{(n)} = T^{-(N-1)}(MT)^n,$$ (7)

with matrices T and M given by

$$T = \begin{pmatrix} e^{-D\tau/2} & 0 \\ 0 & e^{D\tau/2} \end{pmatrix},$$

$$M = \begin{pmatrix} 1 - \frac{2\kappa}{D} & -\frac{2\kappa}{D} \\ \frac{2\kappa}{D} & 1 + \frac{2\kappa}{D} \end{pmatrix}.$$

Thus the elements of the matrix $S^{(n)}$ are of the form

$$S_{11}^{(n)} = (1 - \frac{2\kappa}{D})U_{n-1}(\chi/2)e^{D\tau(n-2)/2} - U_{n-2}(\chi/2)e^{D\tau(n-1)/2},$$

$$S_{12}^{(n)} = -\frac{2\kappa}{D}U_{n-1}(\chi/2)e^{Dn\tau/2},$$

$$S_{21}^{(n)} = \frac{2\kappa}{D}U_{n-1}(\chi/2)e^{-Dn\tau/2},$$

$$S_{22}^{(n)} = (1 + \frac{2\kappa}{D})U_{n-1}(\chi/2)e^{-D(n-2)\tau/2} - U_{n-2}(\chi/2)e^{-D(n-1)\tau/2}.$$ (8)

where U_{n-1}, U_{n-2} are Chebyshev polynomials of the second kind defined by the expression:

$$U_n(\cos\varphi) = \frac{\sin(n+1)\varphi}{\sin\varphi};$$

with argument $\chi/2 = \frac{1}{2}\mathrm{Tr}\, MT$.

If at the initial moment of time the quantum oscillator was in a coherent state the parametric excitation will transform it into a squeezed correlated state with coordinate variances $\sigma_x(t) = \frac{\hbar}{2m\Omega} \mid \varepsilon \mid^2$, and squeezing coefficient $K = \frac{\sigma_x(t)}{\sigma_x(0)} = \mid \varepsilon \mid^2$. Thus after the sequence of δ–kicks one has

$$\sigma_x(t) = \mid A_n \mid^2 \exp(\mu_1 + \mu_1^*)t + \mid B_n \mid^2 \exp(\mu_2 + \mu_2^*)t$$
$$+ \quad B_n A_n^* \exp(\mu_2 + \mu_1^*)t + A_n B_n^* \exp(\mu_1 + \mu_2^*)t. \tag{9}$$

In the case of zero damping ($\gamma = 0$)

$$\mu_1 = i\omega_0, \quad \mu_2 = -i\omega_0,$$

$$\cos\varphi = \frac{\chi}{2} = \cos\omega_0\tau + \frac{\kappa}{\omega_0}\sin\Omega_0\tau, \quad \Omega = \omega_0,$$

and from initial conditions (2) one has $A_0 = 1$, $B_0 = 0$. The explicit expression for squeezing coefficient is

$$K = U_{n-1}^2 + U_{n-2}^2 + \frac{2\kappa}{\omega_0}U_{n-1}^2\sin 2\omega_0[t - (n-1)\tau] - \chi U_{n-1}U_{n-2}$$

$$+ \quad \frac{4\kappa^2}{\omega_0^2}U_{n-1}^2(\sin\omega_0[t - (n-1)\tau])^2 - \frac{2\kappa}{\omega_0}U_{n-1}U_{n-2}\sin 2\omega_0[t - (n - 1/2)\tau]. \tag{10}$$

In the case of weak damping the squeezing coefficient is determined by Eq. (9) with following parameters

$$A_0 = 1 - i\gamma/2\Omega,$$
$$B_0 = \frac{i\gamma}{2\Omega},$$
$$\Omega = (\omega_0^2 - \gamma^2)^{1/2},$$
$$\frac{\chi}{2} = \cos\Omega\tau + \frac{\kappa}{\Omega}\sin\Omega\tau,$$
$$\mu_1 = -\gamma + i(\omega_0^2 - \gamma^2)^{1/2},$$
$$\mu_2 = -\gamma - i(\omega_0^2 - \gamma^2)^{1/2}. \tag{11}$$

One has the squeezing coefficient

$$K = e^{-2\gamma t}\{K(\gamma = 0) + \frac{\gamma}{\Omega}[\frac{2\kappa}{\Omega}U_{n-1}^2\cos 2\Omega\tau + \frac{2\kappa^2}{\Omega^2}U_{n-1}^2\sin 2\Omega\tau - \frac{2\kappa}{\Omega}U_{n-1}U_{n-2}\cos\Omega\tau$$

$$+ \quad (1 - \frac{\kappa^2}{\Omega^2})U_{n-1}^2\sin 2\Omega(t - \tau(n-2)) + U_{n-2}^2\sin 2\Omega(t - \tau(n-1))$$
$$- \quad 2\frac{\kappa}{\Omega}U_{n-1}^2\cos 2\Omega(t - (n-2)\tau) - 2U_{n-1}U_{n-2}\sin 2\Omega(t - (n - 3/2)\tau)$$

$$+ \quad \frac{2\kappa}{\Omega}U_{n-1}U_{n-2}\cos 2\Omega(t - (n - 3/2)\tau) + \frac{\kappa^2}{\Omega^2}U_{n-1}^2\sin 2\Omega(t - n\tau)]$$

$$+ \frac{\gamma^2}{2\Omega^2}[(1 + \frac{2\kappa^2}{\Omega^2})U_{n-1}^2 + U_{n-2}^2 - \chi U_{n-1}U_{n-2} + \frac{2\kappa}{\Omega}U_{n-1}^2(\sin 2\Omega\tau - \frac{\kappa}{\Omega}\cos 2\Omega\tau)$$

$$- 2\frac{\kappa}{\Omega}U_{n-1}U_{n-2}\sin\Omega\tau + \frac{2\kappa}{\Omega}U_{n-1}^2\sin 2\Omega(t - (n-1)\tau)$$

$$- \frac{2\kappa^2}{\Omega^2}U_{n-1}^2\cos 2\Omega(t - (n-1)\tau) - \frac{2\kappa}{\Omega}U_{n-1}U_{n-2}\sin 2\Omega(t - (n-1/2)\tau)$$

$$+ \frac{\kappa^2}{\Omega^2}U_{n-1}^2\cos 2\Omega(t - n\tau) - (1 - \frac{\kappa^2}{\Omega^2})U_{n-1}^2\cos 2\Omega(t - (n-2)\tau)$$

$$- \frac{2\kappa}{\Omega}U_{n-1}^2\sin 2\Omega(t - (n-2)\tau) - U_{n-2}^2\cos 2\Omega(t - (n-1)\tau)$$

$$+ 2U_{n-1}U_{n-2}\cos 2\Omega(t - (n-3/2)\tau) + \frac{2\kappa}{\Omega}U_{n-1}U_{n-2}\sin 2\Omega(t - (n-3/2)\tau))]\}(12)$$

The squeezing phenomenon appears when the squeezing coefficient starts to be less then 1. The force of delta–kicks κ plays the main role in appearing of the squeezing phenomenon at initial moments of time as can be seen from the previous formula, with time increasing the damping begins to play the main role through the exponential function. Let us mention for simplicity the expression for squeezing coefficient in the case of one delta–kick of frequency at the moment of time $t = 0$

$$K = e^{-2\gamma t}[K(\gamma = 0) + \frac{\gamma}{\Omega}(\sin 2\Omega t + \frac{4}{\Omega}(\kappa + \frac{\gamma}{4})\sin^2\Omega t].$$

In the case of strong damping one has the following expressions for the parameters

$$
\begin{aligned}
A_0 &= 1/2 + i/2 + \gamma/2\Omega, \\
B_0 &= 1/2 - i/2 - \gamma/2\Omega, \\
\Omega &= (\gamma^2 - \omega_0^2)^{1/2}, \\
\mu_1 &= -\gamma + (\gamma^2 - \omega_0^2)^{1/2}, \\
\mu_2 &= -\gamma - (\gamma^2 - \omega_0^2)^{1/2}, \\
\frac{\chi}{2} &= \cosh\Omega\tau + \frac{\kappa}{\Omega}\sinh\Omega\tau.
\end{aligned}
\tag{13}
$$

Thus we have considered the parametric excitation of damped oscillator in the frame of Caldirola–Kanai model and discussed the influence of different regimes of damping on the squeezing coefficient which describes squeezing phenomenon in the system. The parametric excitation is chosen in the form of periodical δ–kicks of frequency and the formulae for squeezing coefficient are expressed through the Chebyshev polynomials. It is necessary to add that different aspects of the damped oscillator problem in the frame of Caldirola–Kanai model were considered in Refs. [12]–[17].

ACKNOWLEDGMENTS

The research was supported by the Russian Science Foundation under grant 94-02-04715. The author wishes to thank the organizers of NATO Advanced Study Institute "Electron Theory and Quantum Electrodynamics" for the support of the participation in this conference.

References

[1] P. Caldirola, Nuovo Cim. **18**, 393 (1941).

[2] E. Kanai, Progr. Theor. Phys. **3**, 440 (1948).

[3] K. Husimi, Progr. Theor. Phys. **9**, 381 (1953).

[4] H. R. Lewis and W. B. Reisenfeld, J. Math. Phys. **10**, 381 (1969).

[5] I. A. Malkin and V. I. Man'ko, Phys. Lett. A **31**, 243 (1970).

[6] R. J. Glauber, in: *Quantum Measurements in Optics*, ed. by P. Tombesi and D. F. Walls (Plenum Press, N. Y., 1992) p. 3.

[7] G. S. Agarwal and S. A. Kumar, Phys. Rev. Lett. **67**, 3665 (1991).

[8] L. S. Brown, Phys. Lett. **66**, 527 (1991).

[9] V. V. Dodonov and V. I. Man'ko, *Invariants and Evolution of Nonstationary Quantum Systems*, Proc. of Lebedev Physical Institute, **183** (Nova Science, N. Y., 1989).

[10] V. V. Dodonov, O. V. Man'ko, and V. I. Man'ko, Phys. Lett. A **175**, 1 (1993).

[11] O. Man'ko and Leehwa Yeh, Phys. Lett. A **189**, 268 (1994).

[12] P. Caldirola and L. Lugiato, Physica A **116**, 133 (1982).

[13] L. H. Buch and H. H. Denman, Amer. J. Phys. **42**, 304 (1974).

[14] R. W. Hasse, J. Math. Phys. **16**, 2005 (1975).

[15] V. W. Myers, Amer. J. Phys. **27**, 507 (1959).

[16] H. Dekker, Phys. Repts. **80**, 1 (1981).

[17] S. Baskoutas, A. Jannussis, and R. Miznani, Il Nuovo Cim. B **108**, 953 (1993).

PHASE DISTRIBUTIONS AND QUASIDISTRIBUTIONS OF AMPLITUDE–SQUARED SQUEEZED STATES

Alexei Chizhov and Bolat Murzakhmetov

Bogolubov Laboratory of Theoretical Physics
Joint Institute for Nuclear Research
141980 Dubna, Moscow Region, Russia

INTRODUCTION

In recent years, special attention in quantum optics has been paid to a class of optical field states that are called squeezed states (for a recent review see, for example, special issues of two optical journals [1] devoted to this subject). These states show reduced fluctuations in one quadrature component of the electromagnetic field and enhanced fluctuations in the other.

By considering higher–order correlation functions of the field amplitude it is possible to define higher–order squeezing effects. Hong and Mandel defined a state to be squeezed to $2N$th order if the expectation value of the $2N$th power of the difference between a field quadrature component and its average value is less than it would be in a coherent state [2]. They found this type of squeezing in a number of nonlinear optical processes. Another approach has been explored by Braunstein and McLachlan [3]. They considered higher–order analogs of the squeezed operator in order to define what they called generalized squeezed states. As is shown by their Q representations, these states have highly unusual noise properties.

Another form of higher–order squeezing in terms of the real and imaginary parts of the square (or higher powers) of the field amplitude has been proposed by Hillery [4, 5]. The amplitude–squared squeezing was shown to arise in a natural way in second harmonic generation. In particular, in the paper [6] general solutions for amplitude–squared squeezed minimum uncertainty states and their photon statistics were derived and investigated.

The purpose of this paper is to study the phase properties of the amplitude–squared squeezed states. We use the Pegg–Barnett Hermitian phase formalism to find the phase distribution function for such states. The Pegg–Barnett phase distribution is compared to the phase quasiprobability distributions obtained from the Wigner function and Q function integrating them over the radial variable.

Electron Theory and Quantum Electrodynamics: 100 Years Later
Edited by Dowling, Plenum Press, New York, 1997

MINIMUM UNCERTAINTY STATES FOR AMPLITUDE–SQUARED SQUEEZING

Amplitude–squared squeezing is described in terms of the real and imaginary parts of the square of the field amplitude. These quantities correspond to the operators

$$Y_1 = (a^{\dagger 2} + a^2)/2, \tag{1}$$
$$Y_2 = i(a^{\dagger 2} - a^2)/2, \tag{2}$$

respectively. These operators obey the uncertainty relation [4]

$$\Delta Y_1 \Delta Y_2 \geq \left\langle N + \frac{1}{2} \right\rangle, \tag{3}$$

where $N = a^{\dagger} a$ is the photon number. A state is said to be amplitude–squared squeezed in the Y_1 direction if

$$(\Delta Y_1)^2 < \left\langle N + \frac{1}{2} \right\rangle. \tag{4}$$

Amplitude–squared squeezed minimum uncertainty states minimize the inequality (3) and their number state decomposition can be written as

$$|\lambda, \beta\rangle = \sum_n b_n |n\rangle \tag{5}$$

where the coefficients b_n are defined by the following recurrent relations [6]

$$b_{n+4} = b_{n+2} \frac{2\beta}{(\lambda+1)\sqrt{(n+4)(n+3)}} + b_n \frac{\lambda-1}{\lambda+1} \sqrt{\frac{(n+2)(n+1)}{(n+4)(n+3)}},$$

$$b_2 = \sqrt{2} \frac{\beta}{\lambda+1} b_0, \quad b_3 = \sqrt{\frac{2}{3}} \frac{\beta}{\lambda+1} b_1, \quad 0 \leq \lambda < \infty, \quad 0 \leq |\beta| < \infty, \tag{6}$$

and normalization condition

$$\sum_{n=0}^{\infty} |b_n|^2 = 1. \tag{7}$$

From eq. (6) we see that the coefficients b_n are split into two series with even and odd indices. Thus we can determine two kinds of the states depending on the choice of the decomposition coefficients — even and odd amplitude–squared squeezed states.

Let us consider the quasiprobability functions for the states under consideration. The Q–function was investigated in [6]. Using the number state decomposition (5) it can be written in the following form

$$Q(\alpha) = \frac{1}{\pi} |\langle \alpha | \psi \rangle|^2 = \frac{e^{-|\alpha|^2}}{\pi} \left| \sum_{n=0}^{\infty} b_n \frac{\alpha^n}{\sqrt{n!}} \right|^2. \tag{8}$$

The Wigner function can be defined as [19]

$$W(\alpha) = \frac{2}{\pi} \text{Tr}\{\hat{\rho} \hat{D}(2\alpha) \exp(i\pi \hat{a}^{\dagger} \hat{a})\} \tag{9}$$

142

and in terms of the decomposition coefficients b_n we represent it as

$$
\begin{aligned}
W(\alpha) \; = \; & \frac{2}{\pi} e^{-2|\alpha|^2} \left\{ \sum_{n=0}^{\infty} (-1)^n L_n(|2\alpha|^2)(b_n)^2 \right. \\
& \left. + \; 2 \sum_{n>m=0}^{\infty} \left(\frac{m!}{n!} \right)^{1/2} (-1)^m (2\alpha)^{n-m} L_m^{n-m}(|2\alpha|^2) b_n b_m \right\}. \qquad (10)
\end{aligned}
$$

The Q function and the Wigner function for the amplitude–squared squeezed states have a dramatic difference in their behaviors for β being equal and nonequal to zero. In the first case the functions posses a fourfold symmetry while in the second case the symmetry reduces to a twofold one and the functions are appeared to be stretched along $\arg(\beta)$–direction. Here it is to note that the Q function, due to its definition (8), is positive defined in the phase space but the Wigner function, as it is seen in Fig. 1, might have negative regions. As will be shown in next section, this fact can lead to the problems in determination of the Wigner phase quasidistribution.

PHASE DISTRIBUTIONS

The problem of the quantum description of the optical field phase has been the subject of considerable study for many years [7]. This is connected with the difficulty in constructing a linear Hermitian phase operator. Within the past few years the notion of phase variables in quantum systems has been greatly clarified. Pegg and Barnett [8]–[10] have shown how such an operator can be defined for quantized electromagnetic fields. This new formalism makes it possible to describe the quantum properties of optical phase in a direct way within quantum mechanics on the basis of the Hermitian phase operator and its eigenstates.

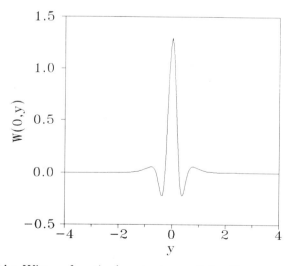

Fig. 1. Plot of the Wigner function's cut at $x = 0$ for the even amplitude–squared squeezed states with $\lambda = 3$, $\beta = 5$.

A quite different approach to the concepts of the phase variable has also been widely used in quantum optics [11]–[14] and which involves quantum quasiprobability distributions such as the Glauber–Sudarshan P function, the Q function and the Wigner function rather than Hermitian operators and their eigenstates. These quasiprobability distributions depend upon the complex eigenvalue α of the non-Hermitian annihilation operator, which can be expressed in terms of a radial variable $|\alpha|$ and a "phase" θ both of which are real. If we integrate over the radius, the resulting distributions are periodic in the phase angle and, for most states they satisfy all properties required by a proper phase distribution [12, 13]. In recent papers, we have compared the Pegg–Barnett phase distribution with those distributions obtained from the Wigner and Q functions by integrating them over the radius for the displaced number states [16], squeezed number and squeezed thermal states [17], two–mode squeezed number states [18].

Having the number state decomposition (5) of the amplitude–squared squeezed states, we can employ the Pegg–Barnett [8]–[10] Hermitian phase formalism to find the phase distribution function for such states

$$P^{(\mathrm{PB})}(\theta) = \frac{1}{2\pi}\left\{1 + 2\sum_{n>k}^{\infty} b_n b_k \cos[(n-k)\theta]\right\}, \tag{11}$$

where b_n are given by Eq. (6), and the phase window is from $-\pi$ to π. This form of the phase distribution is common for the partial phase states [9, 10]. However, due to the particular choice of b_n this phase distribution shows some interesting features that characterize the amplitude–squared squeezed states.

Phase quasiprobability distributions $P^{(Q)}$ and $P^{(W)}$ can be obtained by integrating the $Q(\alpha)$ and $W(\alpha)$ functions, respectively, over the radial variable $|\alpha|$ [12, 13]. As we have previously shown [16], all three phase distributions can be unified into one analytical formula which has the form

$$P^{(s)}(\theta) = \frac{1}{2\pi}\left\{1 + 2\sum_{n>k} b_n b_k \cos[(n-k)\theta]G^{(s)}(n,k)\right\}, \tag{12}$$

where the coefficients $G^{(s)}(n,k)$ distinguish between three distributions, and they are:

(i) for the Pegg–Barnett phase distribution

$$G^{(\mathrm{PB})}(n,k) \equiv 1, \tag{13}$$

(ii) for the distribution $P^{(Q)}(\theta) = \int_0^\infty Q(\alpha)|\alpha|d|\alpha|$ obtained by integrating the Q function over the radius [15]

$$G^{(Q)}(n,k) = \frac{\Gamma[(n+k)/2+1]}{\sqrt{n!k!}}, \tag{14}$$

(iii) for the distribution $P^{(W)}(\theta) = \int_0^\infty W(\alpha)|\alpha|d|\alpha|$ obtained by integrating the Wigner function over the radius [16]

$$G^{(W)}(n,k) = \sum_{m=0}^{p}(-1)^{p-m}2^{(|n-k|+2m)/2}$$
$$\times \sqrt{\binom{p}{m}\binom{q}{p-m}}G^{(Q)}(m,|n-k|+m), \tag{15}$$

where $p = \min(n, k)$, $q = \max(n, k)$. All the coefficients $G^{(s)}(n, k)$ are symmetrical, $G^{(s)}(n, k) = G^{(s)}(k, n)$, and $G^{(s)}(n, n) = 1$. Relation (12) is quite general and can be applied to any states with known amplitudes b_n. Here, we apply it to the amplitude–squared squeezed states.

In Fig. 2, we show the plots of the three phase distributions calculated according to formula (12) with the coefficients (13) and (14) for the even amplitude–squared squeezed states with different values of λ and β. We see four–lobe structure of phase distributions for the case of zero value of parameter β. In the case of nonzero value of β all three distributions have two–lobe structure. It agrees with the symmetry of the quasiprobability distributions. We can also see that the Wigner phase quasidistribution is appeared to be the most sharp among the three distributions. As to the odd states, it can be shown that the Pegg–Barnett phase distribution is the sharpest one.

In Fig. 3 we show the Wigner phase quasidistribution for the even amplitude–squared states for different values of β. It is seen that the Wigner phase quasidistribution can obtain negative values at some phase angles. Similar properties of this distribution were found by Garraway and Knight [20] for simple superpositions of coherent and number states.

The region of positive definiteness of the Wigner phase quasidistribution and the region where it has negative values is plotted in Fig. 4.

In the region of the state parameters where the quasidistribution can become negative, the quasidistribution doesn't satisfy a requirement for a genuine phase distribution that has to be positive for all phase angles. So, we can define the Wigner phase quasidistribution for the even amplitude–squared squeezed states only in restricted area of the state parameters. It is worth to note, however, that in the case of the odd states the Wigner phase quasidistribution is always positive.

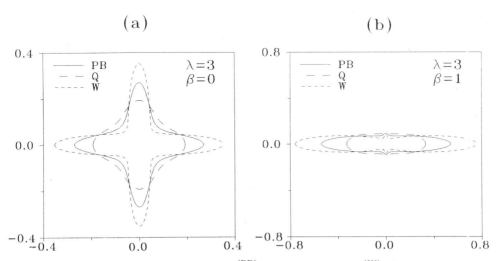

Fig. 2. Plots of the phase distributions $P^{(\mathrm{PB})}(\theta)$ (full curves), $P^{(W)}(\theta)$ (dotted curves), and $P^{(Q)}(\theta)$ (dashed curves) for the even amplitude–squared squeezed states with $\lambda = 3$ and (a) $\beta = 0$, (b) $\beta = 1$.

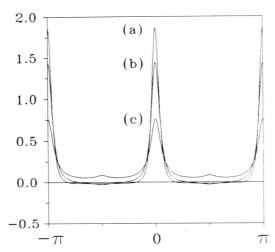

Fig. 3. Plots of the phase distributions $P^{(W)}(\theta)$ for the even amplitude–squared squeezed states with $\lambda = 3$ and (a) $\beta = 1$, (b) $\beta = 3$, and (c) $\beta = 5$.

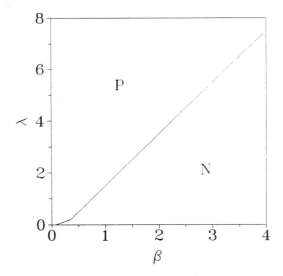

Fig. 4. The regions of the parameters λ and β for the even amplitude–squared squeezed states where the Wigner phase quasiprobability distribution $P^{(W)}(\theta)$ is always positive (P) and can take on negative values (N).

CONCLUSION

In this paper we have studied the phase properties of the amplitude–squared squeezed states applying the Pegg–Barnett Hermitian phase formalism. We have compared the Pegg–Barnett phase distribution with the phase quasiprobability distributions $P^{(W)}(\theta)$ and $P^{(Q)}(\theta)$ obtained by integrating the Wigner function and the Q function, respectively, over the radial variable. We have shown that all three distributions have a two–lobe structure in the case of nonzero value of parameter β. When β is equal to zero these distributions reveal a four–lobe structure. This fact is in agreement with the shapes of the Wigner function and the Q function for these states. The Wigner phase quasidistribution for the even amplitude–squared squeezed states is positive definite only in restricted region of the state parameters. Outside this region it gets negative values. It is not the case for the odd states for which the Wigner phase distribution is appeared to be positive for all values of the state parameters.

ACKNOWLEDGEMENT

The research described in this publication was made possible in part by Grant $N^{\underline{\circ}}$ RFD000 from the International Science Foundation.

REFERENCES

[1] Special issues: J. Opt. Soc. Am. **B4**, No. 10 (1987); J. Mod. Opt. **34**, No. 6/7 (1987).

[2] C. K. Hong and L. Mandel, Phys. Rev. Lett. **54**, 323 (1985); Phys. Rev. A **32**, 974 (1985).

[3] S. L. Braunstein and R. I. McLachlan, Phys. Rev. A **35**, 1659 (1987).

[4] M. Hillery, Opt. Commun. **62**, 135 (1987).

[5] M. Hillery, Phys. Rev. A **36**, 3796 (1987).

[6] D. Yu and M. Hillery, Quantum Optics **6**, 37 (1994).

[7] P. Carruthers and M. M. Nieto, Rev. Mod. Phys. **40**, 411 (1968);
S. M. Barnett and D. T. Pegg, J. Phys. A **19**, 3849 (1986).

[8] D. T. Pegg and S. M. Barnett, Europhys. Lett. **6**, 483 (1988).

[9] S. M. Barnett and D. T. Pegg, J. Mod. Opt. **36**, 7 (1989).

[10] D. T. Pegg and S. M. Barnett, Phys. Rev. A **39**, 1665 (1989).

[11] A. Bandilla and H. Paul, Ann. Phys. (Leipzig) **23**, 323 (1969).

[12] S. L. Braunstein and C. M. Caves, Phys. Rev. A **42**, 4115 (1990).

[13] W. Schleich, R. J. Horowicz and S. Varro, Phys. Rev. A **40**, 7405 (1989).

[14] W. Schleich, A. Bandilla and H. Paul, Phys. Rev. A **45**, 6652 (1992).

[15] R. Tanaś and Ts. Gantsog, Phys. Rev. A **45**, 5031 (1992).

[16] R. Tanaś, B. K. Murzakhmetov, Ts. Gantsog and A. V. Chizhov, Quantum Optics **4**, 1 (1992).

[17] A. V. Chizhov, Ts. Gantsog and B. K. Murzakhmetov, Quantum Optics **5**, 85 (1993).

[18] A. V. Chizhov and B. K. Murzakhmetov, Physics Letters A **176**, 33 (1993).

[19] K. E. Cahill and R. J. Glauber, Phys. Rev. **177**, 1857; 1883 (1969).

[20] B. M. Garraway and P. L. Knight, Phys. Rev. A **46**, R5346 (1992); Physica Scripta **T48**, 66 (1993).

HUNTING THE ELECTRON ELECTRIC DIPOLE MOMENT

Antoine Weis

Max-Planck-Institut für Quantenoptik
H. Kopfermannstr. 1, 85748 Garching, Germany

INTRODUCTION

The CPT theorem, the fact that the "laws of physics"[1] are invariant with respect to the combined discrete symmetry operations **C** (charge conjugation = matter-antimatter exchange), **P** (parity = mirror symmetry) and **T** (time reversal), is one of the uppermost credos of modern physics (Jost, 1960; Streater and Wightman, 1964). On the other hand there is no fundamental reason for each symmetry operation to be conserved individually. The experimental discovery of parity violation in β-decay in 1957 therefore profoundly changed our view of the discrete symmetry laws governing atomic and subatomic processes. Parity violation has since been investigated in a great variety of systems including elementary particles, nuclei and atoms and is well understood today in the frame of the so-called standard model of electroweak interactions developed in the sixties.

The related question as to whether the laws of nature in microscopic systems are invariant with respect to a reversal of the arrow of time or with the respect to the exchange of matter and antimatter exchange is less well answered. A "macroscopic" manifestation of a violation of the latter symmetry is the asymmetry in the matter/antimatter ratio found in the universe (cosmological baryon asymmetry) which could be traced back to CP violating interactions in the early universe. In 1964 it was found experimentally that the combined symmetry CP is violated in the decay of the long-lived component of the K^0 meson. Together with our belief in CPT , this violation is thus equivalent to a violation of T-invariance. The exact nature of CP violation in this decay of the neutral kaon is not understood as of today. In particular the question whether other microscopic systems do possess similar CP- or T-symmetry violating properties remains open. The puzzling fact that despite enormous experimental efforts in investigating nucleons, nuclei, β-decay, atoms and molecules, the K_L decay is the only physical process in which CP / T is violated remains one of the biggest mysteries of modern physics and presents a great challenge to experimenters as well as theoreticians for probably many more years to come.

---

[1] More precisely: a broad class of Poincaré invariant quantum field theories.

One out of several possible manifestations of T-violation is the coexistence in an elementary particle of permanent magnetic and electric dipole moments. As we shall show the coexistence of such moments also violates P. The scope of this paper is the discussion of various theoretical and experimental aspects related to the search for a permanent electric dipole moment (EDM) of the electron, the physics involved being certainly beyond QED, and probably - or as we shall see - hopefully beyond the standard model of electroweak interactions.

The paper is organized as follows: An introductory part sets the historical context and reviews the relation of EDM's and discrete symmetries. The main body of the paper is then split into a theoretical and an experimental section. In the theoretical part we discuss the electron EDM in terms of a form factor and present theoretical predictions for finite EDM values. The discussion of the feasibility of EDM experiments on the electron then naturally leads to a detailed discussion of atomic EDM's and their possible origins. The experimental section starts out by giving a general discussion of the principles and common features of all EDM experiments. A short excursion to the world of past, present and future neutron EDM-experiments is meant to illustrate the "mother"-role played by these experiments for all EDM-experiment. We then discuss in some details experiments performed on the Tl and Cs atoms which have provided the present lowest upper bounds of the electron EDM. The section concludes by a brief outlook on alternative experimental approaches to the electron EDM search using novel samples, such as paramagnetic polar molecules and paramagnetic atoms trapped in cryogenic matrices.

The paper tries to be didactic throughout by giving more or less detailed introductions into the various aspects of the topic, and is meant to serve as a first lecture to novices in the field. As this approach, and the fact that the author is an experimentalist, may imply at some points a certain loss of depth, a detailed bibliography may help the interested reader to get deeper insights into individual aspects. In this respect we want to point out the monography *"Parity Nonconservation in Atomic Phenomena"* by I.B. Khriplovich (1991), which presents an exhaustive discussion of P- and T- violating processes in atoms and which has served as a source of inspiration for many aspects of the present lecture. Another extensive bibliographical guide presenting more than 200 references to problems related to the violation of discrete symmetries is a Resource Letter by E.D. Commins (1993). Further review papers have been given by L.R. Hunter (1991), E. Hinds (1988) and N.F. Ramsey (1994).

Historical Overview

The hint by Purcell and Ramsey (1950) in 1950 that there was no direct experimental evidence for the invariance of the laws of nature with respect to each of the symmetry operations C,P, or T taken separately, and the speculation in 1956 by Lee and Yang (1956) that in particular in processes governed by the weak interaction, parity may be violated triggered a wealth of experimental searches for violations of discrete symmetries. In the year following the Nobel prize winning paper by Lee and Yang, three papers reported indeed the experimental observation of parity violating processes (Friedman and Telegdi, 1957; Garwin et al., 1957; Wu et al., 1957). These processes were various manifestations of β-decay, which is governed, in modern terms, by *charged weak currents*, in which the interaction between the involved leptonic and hadronic (or leptonic and leptonic) currents is mediated by the exchange of a charged intermediate vector boson (W^+, W^-). These particles appear as gauge bosons in the standard model of electroweak interactions developed in the sixties by Weinberg, Glashow, and Salam (Weinberg, 1974), which presents a unified description of the electromagnetic and weak interactions. The electroweak theory may be viewed as a modern continuation of the work of J. C. Maxwell who gave a unified description of the

Table 1. Historical milestones in the search for violations of discrete symmetries

Year	Milestone
early 50's	CPT theorem
1950	no experimental evidence for P-conservation
	$\rightarrow$ existence of permanent electric dipole moments
1956	P may be violated in weak interactions
1957	P is violated in β-decay
1957/58	no experimental evidence for T-conservation
60´s	standard model of electroweak interactions
1964	observation of CP-violation in K_L decay
1974	observation of neutral currents
1974	P violation might be observable in atoms
1978	P is violated in e-nucleon interaction
1983	detection of $W^{\pm}/Z^{\circ}$ bosons
70´s, 80´s	observation of P violation in atoms
60´s - 90´s	no evidence for T-violation in atoms, molecules nucleons, and nuclei

phenomena of electricity and magnetism in terms of an electromagnetic theory. Besides the charged bosons, the electro-weak theory predicts the existence of a neutral boson, the Z^0. The interactions mediated by this particle are known as *weak neutral currents* and were first observed 1973 in neutrino-electron scattering experiments (Aubert and al., 1974; Benvenuti and al., 1974; Hasert and al., 1973)

The direct observation of the $W^{\pm}$ and Z^0 bosons followed only a decade later (Arnison and al., 1983; Arnison and al., 1983; Banner and al., 1983). The standard model predicts a violation of parity in neutral current interactions. The early experiments however were not sensitive to parity violation, which could only be observed in the famous 1978 SLAC experiments on deep inelastic scattering of highly relativistic spin polarized electrons on protons and deuterons (Prescott and al., 1978). As in atoms the binding of electrons to the nucleus is also mediated by a *neutral* current interaction (via the exchange of photons) it was quite natural to expect the electronic structure of atoms to also to be affected by neutral currents mediated by the Z^0 boson, implying that atoms should possess parity violating properties. First estimations of the magnitudes of these effects provided however little hope for their experimental observation (Michel, 1965; Zeldovich, 1959). The experimental search for left-right asymmetries in atoms was triggered by the landmark papers of M.A. and C.C. Bouchiat in 1974 (Bouchiat and Bouchiat, 1974; Bouchiat and Bouchiat, 1974) which first pointed out the Z^3 scaling law[2] for parity violating effects in heavy atoms and proposed specific experiments. The first experimental results giving order of magnitude agreement with the predictions of the standard model were published at the end of the 70's and have been confirmed since on half a dozen of different atomic systems. The most precise measurements to date reach an accuracy in the percent range and are fully compatible with

---

[2] Z is the charge of the nucleus

the predictions from the standard model (Noecker et al., 1988). A recent review of the status of experiments on parity violation in atoms is given e.g. by Commins (1993). As a didactically very valuable introduction to this most fascinating subject I recommend the 1982 Les Houches lecture by M.-A. Bouchiat (1982) (see also (Fortson and Lewis, 1984)).

In their 1950 paper, Ramsey and Purcell (1950) proposed to search for a permanent electric dipole moment (EDM) of elementary particles, and in particular of the neutron as a sensitive test of P- violation. In 1957, after the fall of parity, Landau pointed out (Landau, 1957) that the existence of EDM's is not only forbidden by P-invariance, but as well by CP-invariance[3], i.e. by T-invariance, to which Jackson et al. (1957) and independently Ramsey (1958) replied that there was no experimental evidence that T was a good symmetry and that the search for a neutron EDM was hence a sensitive test of P and T violation. This gave birth to an enormous amount of experimental and theoretical work aimed at demonstrating T-violating effects in atomic and subatomic systems. In 1964 another Nobel prize winning high energy physics experiment showed that in the decay of K_L, the long lived component of the K^0 meson, CP is violated (Christenson et al., 1964). With our belief in CPT, a violation of CP is equivalent to a violation of T, although some direct evidence for T-violation may be inferred from the decay of the long lived component of the K^0 meson (Schubert and al, 1970). No other experiment up today has revealed any indication for a violation of T invariance, be it in searching for permanent electric dipole moments of elementary particles or atoms, or for a violation of T-reversal invariance in fundamental interactions such as nuclear or neutron β-decay. The status of T-violation in the nuclear sector has recently been reviewed at the XXXth Rencontre de Moriond (Moriond, 1995). As we shall see below spectacularly low upper bounds have been set by experiments, in particular in the atomic physics sector, on possible T-violating processes, making these experiments among the most precise experiments known in physics.

Permanent Electric Dipole Moments

If a particle at rest with spin $\vec{s}$ has an electric ($\vec{d}$) or a magnetic ($\vec{\mu}$) dipole moment, $\vec{d}$ and $\vec{\mu}$ being vector quantities, have to be oriented parallel or antiparallel to $\vec{s}$, the only vector quantity which characterizes the orientation of the particle, when at rest. The parallelism or antiparallelism of $\vec{s}$ and $\vec{\mu}$ in elementary particles is a well known fact and the proportionality factor is known as the gyromagnetic ratio. It is quite natural to assume that a similar scalar relationship holds between spin $\vec{s}$ and electric dipole moment $\vec{d}$. The possibility of an independant orientation of $\vec{d}$ and $\vec{s}$ can be ruled out on the basis of the Pauli principle, as can be seen as follows: the complete description of the quantum state of the particle would require, besides the specification of the magnetic quantum number s_z the specification of the projection of $\vec{d}$ along the direction of $\vec{s}$. In the case of bound electrons in atoms this additional degree of freedom would thus allow to violate the Pauli princple. Note that in this case an arbitrarily small, but finite electric dipole moment would lead to a full violation of the principle. It is therefore legitimate to assume that $\vec{d} = d\dfrac{\vec{s}}{\hbar}$ the same way as $\vec{\mu} = \mu\dfrac{\vec{s}}{\hbar}$.

---

[3] CP is the combined symmetry operation of charge-conjugation and parity

Consider the coexistence of $\vec{s}$ and $\vec{d}$: the application of either T or P reverses the relative orientation of $\vec{s}$ and $\vec{d}$ (Fig.1). If P and T are to be good symmetries in a spin polarized system, i.e. in a system with a well defined value of $<\vec{s}>$, the left and right hand sides of Fig.1 have to occur with equal probabilities. This is only possible if $<\vec{d}> = 0$. Conversely, $<\vec{d}> \neq 0$ in a spin polarized system implies that P and T have to be violated. For the demonstration we have made use of rotational invariance by applying a 180° rotation about the y-axis to give the same orientation to $\vec{s}$ on the left and right hand sides. One can verify in the same way that the coexistence of $\vec{s}$ and $\vec{\mu}$ conserves P and T. In this case the P and T operations lead to mirrored systems with the same physical properties.

Another way to state the symmetry properties formulated above is to characterize the system by a rotationally invariant quantity, as $<\vec{s}\cdot\vec{d}>$ in Fig.1, or equivalently $<\vec{s}\cdot\vec{\mu}>$, where $< >$ indicates an ensemble average. These invariants are pure numbers, which may however change sign under the considered symmetry operation, in which case one calls the invariant a *pseudoscalar* in contrast to a *true scalar*. A violation of a symmetry is then equivalent to the system having a non-vanishing ensemble averaged expectation value of a pseudoscalar. In this sense $<\vec{s}\cdot\vec{d}>$ or $<\vec{\mu}\cdot\vec{d}>$ are P and T pseudoscalars, while $<\vec{s}\cdot\vec{\mu}>$ is a true P and T scalar.

As we shall see below, in actual experiments one investigates dynamic effects induced by the coupling of $\vec{d}$ and $\vec{\mu}$ to static electric and magnetic fields $\vec{E}$ and $\vec{B}$, the typical non-relativistic interaction Hamiltonian being

$$H = -\vec{\mu}\cdot\vec{B} - \vec{d}\cdot\vec{E}. \tag{1}$$

It is clear that the symmetry operations could be applied either to the system under investigation (characterized by $\vec{d}$ and $\vec{\mu}$) *or* to the laboratory quantities $\vec{E}$ and $\vec{B}$. Therefore $<\vec{\mu}\cdot\vec{d}>$ and $<\vec{B}\cdot\vec{E}>$ are equivalent P,T-odd invariants.

It is worthwhile to note that the above arguments do not hold for oscillating dipole moments. As discussed e.g. by Bouchiat and Bouchiat (1974) or Fortson (1984) the coexistence of an oscillating magnetic dipole moment $\vec{M}(t)$ and an oscillating electric dipole moment $\vec{D}(t)$ violates P, but conserves T if $\vec{M}(t)$ and $\vec{D}(t)$ are dephased by 90 degrees. Interference effects between such amplitudes are the basis of experiments investigating the parity violation processes in atoms discussed above.

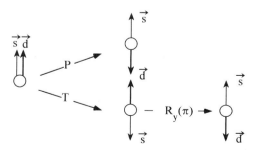

Figure 1. Coexistence of $\vec{s}$ and $\vec{d}$ violates P and T

THEORY

The Electric Dipole Moment of The Electron

The experimental search for permanent EDM's always involves the coupling of the EDM to an external static electric field; for this reason EDM experiments on charged particles, such as the electron or the proton are hindered by the charge of these particles; the strong electric field needed to induce a precession of the dipole moment will accelerate the particle to the field electrode, thus critically reducing the useful time for the experiment. Sensitive experiments on charged particles therefore have to be carried out on neutral, orientable systems which contain these particles as constituents, such as atoms or molecules. One is thus led quite naturally to a discussion of permanent electric dipole moments of such systems. Here we will focus our main attention on the problem of the EDM of *atoms*. In this context two questions immediately arise:

- if the electron (nucleon) has an EDM, does it transfer it to the atom?

- is the electron EDM the only possible source of an atomic EDM?

Definition. Before discussing the different effects that contribute to atomic electric dipole moments, we start by giving a definition of the electric dipole moment of the electron. Formally the permanent electric dipole moment d of a neutral particle may be defined via

$$d \equiv - \lim_{E \to 0} \frac{\partial W(E)}{\partial E},$$

where W(E) is the energy of the particle in an external static electric field E. The limit E→0 excludes dipole moments induced by a finite electric polarizability.

In a classical picture the EDM of a finite sized particle, as the neutron, is defined in a straightforward way via $\vec{d} \equiv \int \vec{r} \, \rho(\vec{r}) \, d^3 r$, where $\rho(\vec{r})$ is the charge density of the particle. The electric dipole moment of a pointlike particle as the electron is defined similarly to its anomalous magnetic moment in terms of a form factor.

Following the discussion given by Bernreuther and Suzuki (1991) the electron EDM may be introduced by writing down all possible Lorentz invariant couplings of the relativistic electron current

$$j_\mu = \bar{e}(p') \, \Gamma_\mu (q^2) \, e(p)$$

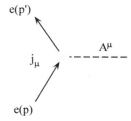

Figure 2. Coupling of electron to external electromagnetic fields

to external electromagnetic fields represented by the potential A^μ_{ext} (Fig.2), i.e.

$$L = j_\mu A^\mu.$$

The decomposition of the electron 4-current is given by

$$\Gamma_\mu(q^2) = F_1(q^2)\,\gamma_\mu + F_2(q^2)\,i\,\frac{\sigma_{\mu\nu}\,q^\nu}{2m} + F_3(q^2)\,\frac{\sigma_{\mu\nu}\,\gamma_5\,q^\nu}{2m} + F_A(q^2)\left[\gamma_\mu\gamma_5\,q^2 - 2m\,\gamma_5\,q_\mu\right]$$

where the form factors $F_i(q^2)$ depend on the square of the 4-momentum transfer $q_\mu = P'_\mu - P_\mu$ and describe the interaction at the electron-photon vertex. In the non-relativistic limit the term proportional to F_1 reduces to the well known expression $\rho\,V_{ext} - \vec{j}\cdot\vec{A}_{ext}$, whereas the second term describes the Zeeman interaction of the anomalous magnetic moment of the electron. The third term describes a coupling of the electron to external electric and magnetic fields, and the proportionality factor of the electric term may hence be identified with the electric dipole moment of the electron. This term, proportional to F_3, then reads

$$L_{EDM} = -d_e\,\gamma_5\,\vec{\gamma}\cdot\vec{E} \tag{2},$$

which becomes in the non-relativistic limit

$$L_{EDM} = -d_e\,\vec{\sigma}\cdot\vec{E}_{ext} \equiv -\vec{d}\cdot\vec{E}_{ext} \tag{3}.$$

The electric dipole moment is thus expressed in terms of the form factor by

$$d_e \equiv -\frac{F_3(o)}{2m} \tag{4}$$

Theoretical predictions

The Standard Model. CP violation appears in the standard model through two complex phase angles: δ_{KM}, a CP violating phase angle in the Kobayashi-Maskawa matrix (Kobayashi and Maskawa, 1973) describing the coupling of charged weak currents, and θ, the QCD vacuum angle. The latter is only relevant in strong interaction physics and hence of no interest in connection with the electron EDM. As we have seen above, the electron EDM is given by the form factor of the electron-photon vertex at zero momentum transfer, which includes diagrams of the type depicted in Fig.3.

In the standard model one-loop diagrams as the one shown in Fig.3.a do not produce a finite EDM, as the relevant CP-violating phases cancel. The same holds for two-loop diagrams, and a finite EDM is only expected from three-loop diagrams (Fig. 3.b) with gluonic corrections (Bernreuther and Suzuki, 1991). No reliable calculation of such a diagram exist. Its contribution to the electron EDM may only be estimated to be

$$d_e \text{ (standard model)} < 10^{-37} \text{ e cm,}$$

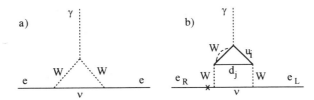

Figure 3. Electron EDM in standard model: one-loop (a) and three-loop (b) contributions to electron-photon vertex describing the coupling of the electron e to the electromagnetic field mediated by the photon γ. (W: $W^{\pm}$-boson, u,d: quarks, v: neutrino)

which is about 10 orders of magnitude below the state-of-the-art sensitivity

$$|d_e \, (exp)| < 4 \times 10^{-27} \text{ e cm}$$

of present day experiments .

If one considers that, averaged over the past quarter century, the experimental sensitivity has improved by one order of magnitude every 8 years, there is little hope for a detection of such a small EDM in a foreseeable future. If, on the other hand, experiments will detect a non-zero EDM in the near future, it will be a clear signature of physics beyond the standard model of electroweak interactions.

Beyond the standard model. The fact that the predictions of the standard model for the EDM of the electron are so desperately small is only bad news at first glance: as mentioned above the experimental search for EDM's may thus be looked at as the search for physics beyond the standard model. There are indeed several reasons - discussed e.g. by Barr (1994) - to believe that the standard model is not complete. A vast number of alternative models have been put forward in the past few years (for a review and discussion see e.g. (Barr, 1993; Bernreuther and Suzuki, 1991)), some of which predict electron and neutron EDM's much larger than the upper limit set by the standard model. Among these models the most optimistic predictions are given by

- **Multi-Higgs models**

  A model with two Higgs boson doublets of mass around 100 GeV and CP violating phases of order unity predicts

  $$|d_e| = 10^{-27} - 10^{-26} \text{ e cm}$$

- **Supersymmetry (SUSY)**

  Supersymmetric theories in their simplest form predict

  $$|d_e| = 10^{-25} \text{ e cm}$$

  assuming masses of the relevant superparticles in the 100 GeV range and CP violating phases of order unity. More sophisticated arguments predict a suppression factor of 2-3 orders of magnitude, so that
  $$|d_e| = 10^{-28} - 10^{-27} \text{ e cm}$$

  seems to be a more realistic prediction of SUSY models.

156

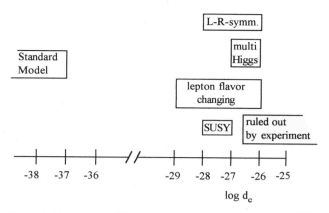

Figure 4. Comparison of theoretical predictions for the electron EDM

- **Left-right symmetric models**

A group of models with symmetric sets of left and right handed leptons and quarks based on the gauge group $SU(2)_L$ x $SU(2)_R$ x $U(1)$ predict

$$|d_e| = 10^{-28} - 10^{-26} \text{ e cm.}$$

Two remarks have to be made:

- The theoretical values quoted above are to be understood as coarse estimates based on the assumptions of "reasonable" values of the model parameters.
- From Fig.4 it becomes evident that the experimental hunt for the electron EDM is now entering a most challenging and thrilling phase. If experimenters are able to improve the current experimental sensitivity to d_e by 1 to 2 orders of magnitude, atomic physics experiments will - within the next decade - make substantial contributions to the test of a great variety of elementary particle models, by either ruling them out or setting stringent upper limits on their parameters in case of null results. In the case where finite sized EDM's are observed, a comparison of values obtained from atoms (both diamagnetic and paramagnetic), molecules and the neutron should help to identify the mechanisms responsible for the generation of CP violation.

For more details on the various models the interested reader is referred to exhaustive theoretical review articles on the electron EDM (Barr, 1993; Bernreuther and Suzuki, 1991).

The Electric Dipole Moment of Atoms

It is well known that atoms, which do not have degenerate states of opposite parity cannot have a permanent electric dipole moment, and hence do not show a linear Stark effect[4]. A net atomic dipole moment arises from interactions mixing states of opposite parity, i.e. states whose wave functions are even and odd respectively under the operation of

---

[4] Strictly speaking this holds for all atoms, although some cases of very close lying states of opposite parity are known, the most famous example being the hydrogen atom.

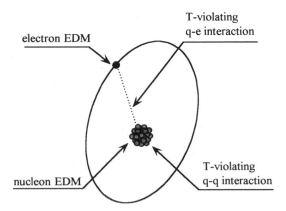

electron EDM

T-violating
q-e interaction

nucleon EDM

T-violating
q-q interaction

Figure 5. Possible contributions to the electric dipole moment (EDM) of an atom

spatial inversion. Consider an atom with 2 states (s and p) of opposite, well defined parity. Parity conservation implies that neither state can have a permanent electric dipole moment, $\langle s|d|s\rangle = \langle p|d|p\rangle \equiv 0$. However, if the states are coupled by an interaction V_{mix}, the perturbed s-state e.g. is given by

$$|s'\rangle = |s\rangle + \varepsilon|p\rangle, \text{ where } \varepsilon = \frac{\langle p|V_{mix}|s\rangle}{E_s - E_p} \tag{5}.$$

By the admixture to the symmetric (with respect to spatial inversion) s-state of a small fraction of antisymmetric p-state, the atom thus acquires an electric dipole moment

$$\langle \vec{d}_A\rangle = \langle s'|e\vec{r}|s'\rangle$$
$$\approx (\varepsilon + \varepsilon')\langle p|e\vec{r}|s\rangle$$
$$\propto \frac{\langle s|V_{mix}|p\rangle\langle p|e\vec{r}|s\rangle}{E_s - E_p} \tag{6}$$

From Eq.6 it becomes evident that only interactions with real matrix elements can give rise to an atomic dipole moment. The dipole moment of interest here should not be confused with the the atomic dipole moment induced by the Stark interaction $V_{mix} = V_{Stark} = -e\vec{r}\cdot\vec{E}_{ext}$ which is proportional to the applied electric field E_{ext}; in this case the energy of the atom in the external field $H = -\vec{d}_A\cdot\vec{E}_{ext}$ is a *quadratic* function of the field strength, whereas dipole moments arising from P- and T- violating internal atomic couplings, lead to an atomic energy *linear* in the external field.

Various P- and T- violating interactions may mix states of opposite parity and thus give rise to a net atomic electric dipole moment (Fig. 5):

- The mixing of states may be induced by the coupling of an EDM of the electrons or the nucleons to an external field, in which case

$$V_{mix} = -\vec{d}_{e,N}\cdot\vec{E}_{ext}.$$

- The state mixing interaction may be a P and T-violating interaction between the nucleons (quarks) and electrons.

- T-violating interactions between the nucleons (quarks) may be the origin of the state mixing interaction.

We start the discussion by showing under which circumstances an electron EDM can contribute to a net atomic EDM. We then address the case of mixing via a P- and T-violating electron-nucleon (quark) interaction and briefly discuss nuclear contributions to atomic EDM's. A more detailed discussion of these and other mechanisms can be found in the literature (Khriplovich, 1988; Khriplovich, 1991; Mårtensson-Pendrill, 1992).

Contributions from Electron EDM; Schiff's Theorem

A theorem discussed by L.I. Schiff (1963) states that a neutral system of electrostatically bound particles with electric dipole moments cannot possess a permanent electric dipole moment. At first glance this is bad news, as it implies that an eventual electronic dipole moment will not be transferred to the atom. However, the assumptions under which the theorem is valid, i.e. pointlike constituents, non-realtivistic motion, and no magnetic interaction between constituents are not fulfilled in most atoms, and in particular not in heavy atoms. Moreover, as was first pointed out by Sandars (1965), for certain atoms the atomic EDM may surpass the electronic one by several orders of magnitude. In order to get a better understanding of the origin of an atomic EDM via a violation of the assumptions of the Schiff theorem we first outline the proof of the latter using classical and quantum mechanical arguments.

Classical argument. Consider an atom in an external field $\vec{E}_{ext}$ which adds to the local internal field $\vec{E}_{int}(t)$. Each of the constituents then sees a field $\vec{E}_{tot}(t) = \vec{E}_{ext} + \vec{E}_{int}(t)$, whose time averaged values have to vanish; otherwise the resulting net forces exerted by the field would accelerate the atom or tear it apart. The atom therefore can have no linear Stark effect and by definition no electric dipole moment. There is a complete screening of the external field at the positions of the electrons and nucleons in the atom.

Quantum mechanical argument. The proof given here follows the one found in (Khriplovich, 1991). Consider a system of particles with charges e_k and electric dipole moments $\vec{d}_k$ at positions $\vec{r}_k$. Two contributions to the net atomic dipole moment arise, viz.

- a *bare* dipole moment given as the sum of the dipole moments of all the constituents

$$\vec{D}_{bare} = \sum_k \vec{d}_k, \tag{7}$$

and

- an *induced* dipole moment[5]

$$\vec{D}_{ind} = \sum_l \langle \tilde{s} | e_l \vec{r}_l | \tilde{s} \rangle \tag{8}$$

---

[5] without loss of generality we chose as model atomic state a perturbed ($\tilde{\ }$) s-state.

arising from the mixing of states of opposite parity by the (nonrelativistic) interaction

$$V_{mix} = -\sum_k \vec{d}_k \cdot \vec{E}_{int}(\vec{r}_k) = \sum_k \vec{d}_k \cdot \vec{\nabla}_k V(\vec{r}_k),$$

We now show that these two contributions exactly cancel. Writing the internal field in terms of the potential $V(r)$ seen by the particles and using the general commutator property of the momentum operator $[\vec{p}, f(r)] = \frac{\hbar}{i} \vec{\nabla} f(r)$ the mixing potential may be rewritten as

$$V_{mix} = \frac{i}{\hbar} \sum_k \frac{\vec{d}_k}{e_k} \cdot [\vec{p}_k, H_0], \text{ where } H_0 = \sum_k \frac{p_k^2}{2m_k} + V(r_k).$$

The first order perturbation of an S state by V_{mix} is then given by

$$|\tilde{s}\rangle = |s\rangle + \sum_p \frac{\langle p | V_{mix} | s \rangle}{E_s - E_p} |p\rangle \qquad (9),$$

which leads to

$$|\tilde{s}\rangle = \left(1 + \frac{i}{\hbar} \sum_k \frac{\vec{d}_k \cdot \vec{p}_k}{e_k}\right) |s\rangle \qquad (10)$$

and may be interpreted as the lowest order Taylor expansion of the displacement transformation $|\tilde{s}\rangle = e^{iQ} |s\rangle$, where Q is the sum over the (charge) displacement operators $Q = \frac{i}{\hbar} \sum_k \frac{\vec{d}_k \cdot \vec{p}_k}{e_k}$ induced by the dipole moments $\vec{d}_k$ (Mårtensson-Pendrill, 1992; Schiff, 1963). After some simple algebra the induced atomic dipole moment then becomes

$$\langle \vec{D}_{ind} \rangle = \sum_l \langle \tilde{s} | e_l \vec{r}_l | \tilde{s} \rangle$$

$$= \frac{i}{\hbar} \left\langle s \left| \left[\sum_l e_l \vec{r}_l , \sum_k \frac{\vec{d}_k}{e_k} \cdot \vec{p}_k \right] \right| s \right\rangle \qquad (11)$$

$$= \frac{i}{\hbar} \left\langle s \left| \sum_{k,l} \frac{e_l}{e_k} \vec{d}_k [\vec{r}_l, \vec{p}_k] \right| s \right\rangle$$

The commutator is given by

$$-\frac{\hbar}{i} \vec{\nabla}_k \cdot \vec{r}_l = -\frac{\hbar}{i} \delta_{kl}$$

and the induced dipole moment becomes

$$\langle \vec{D}_{ind} \rangle = - \langle s | \sum_k \vec{d}_k | s \rangle = - \langle \vec{D}_{bare} \rangle$$

The induced dipole moment thus cancels the bare dipole moment exactly and the atom has no net dipole moment. Is there a way out? Is there another mechanism, by which the electron dipole moment may be transferred to the atom? The answer is yes and relies on the fact that the motion of the electron - in particular in heavy atoms - is relativistic.

The atomic enhancement factor. For the correct description of the interaction between the electron electric dipole moment and an external electric field we have to replace the non relativistic Hamiltonian (Eq.3)

$$V_{mix} = -\vec{d}\cdot\vec{E} = -d_e\,\vec{\sigma}\cdot\vec{E}$$

by its relativistic counterpart (Eq.2)

$$H_{EDM} = -d_e\,\gamma_5\vec{\gamma}\cdot\vec{E},$$

which can be rewritten as

$$H_{EDM} = -d_e\,\vec{\Sigma}\cdot\vec{E} - d_e(\gamma_0-1)\vec{\Sigma}\cdot\vec{E} \tag{12},$$

where $\vec{\Sigma} = \begin{pmatrix} \vec{\sigma} & 0 \\ 0 & \vec{\sigma} \end{pmatrix}$ with the Pauli matrices $\vec{\sigma}$. In Eq. 12 we have explicitly split off the part which vanishes in the non-relativistic limit (first term), so that the atomic dipole moment induced by the electron dipole moment becomes

$$<\vec{d}_A> = d_e \sum_m \frac{<n|\vec{r}|m><m|(\gamma_0-1)\vec{\Sigma}\cdot\vec{\nabla}U(r)|n>}{E_n - E_m} \tag{13}.$$

From inspection of this expression it becomes evident that the atomic dipole moment arises from a relativistic coupling of the electron spin to the field of the nucleus, so that atomic EDM´s are expected to be particularly large in heavy atoms with unpaired spins.

For a Coulomb potential $U(r) = \dfrac{Ze^2}{r}$ Eq.13 can be evaluated analytically for hydrogenic wavefunctions and leads to the result

$$R \equiv \frac{d_A}{d_e} \approx \frac{16}{3}Z^3\alpha^2 \frac{<s|r|p>}{\gamma(4\gamma^2-1)(n_s n_p)^{3/2}} \frac{1}{E_p - E_s} \tag{14}$$

with $\gamma^2 = (j+1/2)^2 - Z^2\alpha^2$.

R is called the enhancement or antiscreening factor and may reach, due to its scaling with $Z^3\alpha^2$, values in the range of 100-1000 for heavy atoms. We are thus faced with the remarkable fact that while the Schiff theorem forbids a transfer of the electron dipole moment to the atom, the violation of the theorem leads to atomic dipole moments which may be orders of magnitude larger than the free electron dipole moment. The fact that nature provides us with this strong lever arm is, by itself, a strong motivation to push experiments looking for electric dipole moments in heavy paramagnetic atoms.

Enhancement factors have been calculated for many atoms. Table 2 shows theoretical predictions for different one-electron systems. The results can be seen to present quite some scatter, which is discussed by several authors (Hartley et al., 1990; Johnson et al., 1985; Johnson et al., 1986). Hartley et al., e.g. believe their result for Cs (R_{Cs} = 114) to be reliable at the 3% level. The same authors believe the value R_{Tl} = -600±400 to "be likely to cover the true value" for Tl (Hartley et al., 1990), whereas Liu and Kelly (1992) believe their

Table 2. Atomic enhancement factors $R = d_A/d_e$

Atom	State	$R = \dfrac{d_A}{d_e}$
Li	$2\,^2S_{1/2}$	0.0043^1, 0.0042^8
Na	$3\,^2S_{1/2}$	0.318^1, 0.314^8
K	$4\,^2S_{1/2}$	2.42^1, 2.76^8
Rb	$5\,^2S_{1/2}$	24^1, $16\text{-}25^3$, 24.6^8, 25.7^9
Cs	$6\,^2S_{1/2}$	119^1, $140\text{-}160^2$, $80-110^3$, 114^4, 138^8
Fr	$7\,^2S_{1/2}$	1150^1
Tl	$6\,^2P_{1/2}$	-716^5, -500^6, $-500..-1000^3$, -600^4, -585^7
Au	$6\,^2S_{1/2}$	$130\text{-}250^3$

[1]Sandars (1966),   [4]Hartley et al. (1990),   [7]Liu et al., (1992),
[2]Johnson et al. (1985),   [5]Sandars et al. (1975),   [8]Sternheimer, (1969),
[3]Johnson et al. (1986),   [6]Flambaum (1976),   [9]Shukla et al. (1994)

calculation of R_{Tl} = -585 to be accurate at the 5% level. When deducing values for the electron EDM from measurements in the paramagnetic atoms Tl and Cs values of R_{Tl} = -585 (Commins et al., 1994) and R_{Cs} = 120±20 (Murthy et al., 1989) were used.

Diamagnetic atoms. In diamagnetic atoms with paired spins the enhancement factor from the second order contribution (Eq. 13) vanishes as the latter is proportional to the electronic spin. The coupling of the electron electric dipole moment to the atom arises only in third order through the hyperfine interaction. The general scaling behaviour of the enhancement factor $R \propto Z^2 \alpha^2 \left(\dfrac{m_e}{m_N} \right)$ leads to an actual reduction of the atomic dipole moment as compared to the electronic one. One finds for instance R=-8 x 10^{-4} and 1.2 x 10^{-2} for the ground states of ^{129}Xe and ^{199}Hg respectively (Hartley et al., 1990).

Contributions from P-, T-violating Electron-Nucleon Interactions

Another possible source of atomic dipole moment is a P- and T-violating contribution $V_{mix} = V_{PT}$ to the coupling of electron and nucleons. In electroweak theory the Lagrangian of the electron-nucleon(quark) interaction mediated by an infinitely heavy particle is commonly written as a pointlike current-current interaction

$$L = j^{(e)} \cdot j^{(N)} \tag{15},$$

where the current 4-vectors are given by $j_\mu^{(x)} = \bar{u}(p') \Gamma_\mu u(p)$, the u(p) being the Dirac spinors of particle *(x)* with 4-momentum p. Γ_μ is one of the 5 Lorentz invariant currents involving no derivatives which transform under rotations as scalar (S), pseudoscalar (P), vector (V), axial vector (A), or tensor (T) respectively.

The Lagrangian then consists of pairwise combinations of currents with specific transformation properties, the strength of each combination being described by a coupling constant G'_x.

162

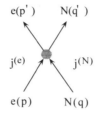

Figure 6. *e-N interaction*

Among all the possible combinations (x = S, P, V, A, T) the expressions

$$G'_V \, V^N \cdot A^e \text{ and } G'_A \, A^N \cdot V^e \tag{16}$$

are, P-odd and T-even, while the combinations

$$i G'_S \, S^N \cdot P^e, \quad i G'_P \, P^N \cdot S^e, \quad \text{and} \quad i G'_T \, T^N \cdot T^e \tag{17}$$

are P-odd and T-odd. The P-odd, T-even terms have been the focus of interest in experiments studying parity violation in atoms mediated by the Z^0 boson, while the P-odd, T-odd terms may contribute to an atomic electric dipole moment. Assuming an infinitely heavy nucleus the second term in (17) vanishes and in the non-relativistic limit and the PT-violating Lagrangian is given by

$$L_{PT} = i \, \frac{G_F}{\sqrt{2}} \, \delta(r) \left[\kappa_1 \, \gamma_0 \, \gamma_5 + \kappa_2 \, \vec{\sigma}_N \cdot \vec{\gamma} \right] \tag{18},$$

where the constants κ_1 and κ_2 measure the strengths of the nuclear spin independent and dependent contributions in units of the Fermi coupling constant G_F. They are related to the coupling constants G'_x by $G_S = \frac{G_F}{\sqrt{2}} \kappa_1$, and $G_T = \frac{G_F}{\sqrt{2}} \kappa_2$. The constants G_S, G_T are sometimes referred to as C_S and C_T in the literature.

The contribution from the scalar interaction C_S is proportional to the electron spin and is therefore most sensitively tested in experiments on paramagnetic systems. For a discussion of theoretical predictions for C_S and C_T the reader is referred to the discussions by Khriplovich (1988; 1991) and Mårtensson-Pendrill (1992).

Contributions from Nuclear T-violating Effects and Nuclear EDM's

Besides the electron EDM and P-, T-violating electron-nucleon interactions several other mechanisms related to T-violating effects in the nucleus can create a finite *atomic* EDM. However, in contrast to electron EDM's nuclear EDM's typically play a minor role in heavy paramagnetic atoms and we shall be brief in the discussion of these mechanisms. The lowest upper limits on these effects do indeed not come from EDM experiments on paramagnetic atoms, but rather from experiments on diamagnetic atoms and the polar molecule TlF.

Two sources for an EDM of the nucleus can be considered: EDM's of the constituents (protons and neutrons) and P- and T-violating nucleon-nucleon interactions. Both effects can be modeled in terms of P-, T-violating nucleon-pion interactions; while in the first case a dipole moment may arise from emission/reabsorption processes of virtual pions, the second contribution may be traced back to the virtual exchange of pions. The screening arguments put forward by the Schiff theorem apply of course also to the nucleus. In this case however the enhancement effect discussed for the electron EDM above does not apply, as the motion of the nucleus is nonrelativistic. There exist however other mechanisms, based on violations of the assumptions of Schiff's theorem, which can transfer T-violating effects from the nucleus to the atom.

Finite size effect (Schiff moment). One of the assumptions of Schiff's theorem was the pointlike nature of the atomic constituents. If one allows for a finite size of the particles, and if furthermore the spatial ditributions of charge and dipole moment in the particles differ, the atom can acquire a permanent dipole moment. This is easily seen by recalling the classical argument presented above; even if the total electric field acting on the center of the charge distribution vanishes, it may still have a finite value in regions of non-vanishing dipole moment and give rise to a linear Stark effect. Quantum mechanical proofs may be found e.g. in (Khriplovich, 1991; Schiff, 1963). The atomic EDM arising from the interaction of the electrons with the EDM d_N of a finite sized nucleus is given by the interaction

$$V_{mix} = 4 \pi e \, \vec{Q}_S \cdot \vec{\nabla} \delta(\vec{r})$$

where Q_S is the Schiff moment, a quantity proportional to the difference of the second moments of the charge and dipole moment distributions in the nucleus, and which scales like

$$\frac{d_A}{d_N} \propto Z^2 \left(\frac{r_{nucleus}}{r_{Bohr}} \right)^2.$$

Magnetic effects. A nuclear electric dipole moment d_N may couple to the magnetic field $\vec{B}$ produced by the electron cloud via the interaction

$$V_{mix} = -\frac{i}{e} \left[\vec{d}_N \cdot \vec{p} , \vec{\mu}_N \cdot \vec{B} \right]$$

where μ_N is the nuclear magnetic moment and p the electron momentum, leading to an atomic EDM of order (Hinds and Sandars, 1980)

$$\frac{d_A}{d_N} \propto Z \alpha^2 \left(\frac{m_{electron}}{m_{nucleon}} \right).$$

Similarly the electron EDM may couple to the nuclear magnetic moment. This effect however is further reduced due to the smallness of the nuclear magneton.

Nuclear magnetic quadrupole moment (MQM). The same way as a moving charge has a magnetic moment, a moving electric dipole will have a magnetic quadrupole moment. In atoms with nuclear spin I≥1 the coupling of this quadrupole moment produced by a nuclear EDM to a magnetic field gradient, produced by the valence electron cloud polarized by an external electric field will give rise to an atomic EDM of the order (Sandars, 1984)

$$\frac{d_A}{d_N} \propto Z^2 \alpha^2 \left(\frac{m_{electrons}}{m_{nucleonr}} \right).$$

EXPERIMENTS

Principle of EDM Experiments

Before entering a detailed discussion of various experiments searching for EDM's we want to point out some common features to all EDM experiments which rely on the fact that due to their common proportionality to the particle spin $\vec{s}$, the magnetic dipole moment $\vec{\mu}$ and the electric dipole moment $\vec{d}$ of the particle are "locked" to each other. Magnetic moments are commonly investigated by studying their coupling to static or oscillating magnetic fields. Similarly the presence of an electric dipole moment $\vec{d}$ is measured by observing its precession in a static electric field $\vec{E}$ under the action of the torque $\vec{d} \times \vec{E}$. In the experiments the magnetic moment $\vec{\mu}$ (or the spin angular momentum $\vec{s}$) is used in the preparation of the sample as well as in the detection of the E-field induced precession. More specifically one can distinguish 3 stages in a typical EDM experiment (Fig. 7) which may either occur simultaneously or in spatially (or temporally) separated regions:

1) The sample is spin polarized by orienting the magnetic moments $\vec{\mu}$ and hence the electric moments $\vec{d}$.

2) A transverse static electric field $\vec{E}$ induces a precession of the electric dipole moments which drag the magnetic moments along.

3) The change in orientation of the magnetic moments is detected.

If at the same time a magnetic field $\vec{B}$ (parallel to $\vec{E}$) is applied, which induces a precession of the magnetic moment, the presence of the electric dipole moment $\vec{d}$ manifests itself as an E-field dependent acceleration or deceleration of that precession.

The major difference between the various experiments lies in the techniques used to orient magnetic moments in the sample and to detect the change in their orientation (stages 1 and 3). The sample may be polarized by scattering in magnetized materials (neutron EDM experiments), by spin dependent focussing in inhomogeneous magnetic mutipole fields (early atomic experiments, molecular experiments) or by optical pumping (new generation of atomic experiments). The preparation and detection stages are often arranged in a symmetric way using the same technique. With a time interval T between the preparation and the detection process the moments acquire a total phase angle $\Phi = \omega_{prec} T$, with a precession frequency ω_{prec} given by

$$\hbar \omega_{prec} = \hbar (\omega_L \pm |\omega_E|)$$
$$= \vec{\mu} \cdot \vec{B} \pm |\vec{d} \cdot \vec{E}| \tag{19},$$

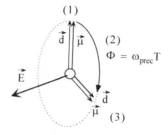

Figure 7. Precession of electric and magnetic dipole moments.

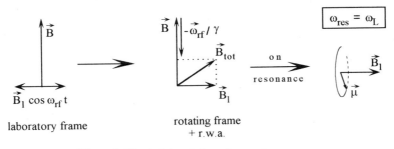

laboratory frame   rotating frame
                      + r.w.a.

Figure 8. Classical description of magnetic resonance

where ω_L is the Larmor frequency and $\omega_E = \dfrac{\vec{d} \cdot \vec{E}}{\hbar}$ its electrical equivalent. The sign refers to the relative orientation of $\vec{B}$ and $\vec{E}$.

The method of choice for the measurement of this precession frequency is magnetic resonance spectroscopy, the basic principle of which we shall briefly review. Consider a spin ½ particle with magnetic and electric dipole moments. The Zeeman degeneracy of the magnetic sublevels is lifted by the presence of static magnetic and electric fields. If the sample is polarized, i.e. if the populations in the two sublevels differ, spin flip transitions between the levels may be induced by irradiating the sample with an oscillating magnetic field $\vec{B}_{rf} = \vec{B}_1 \cos\omega_{rf} t$, and a resonance occurs at

$$\omega_{rf}^{res} = \omega_{prec} \equiv \omega_L \pm |\omega_E|.$$

The EDM thus manifests itself as a shift of the magnetic resonance frequency linear in E. In many experiments it is the magnetic field rather than the frequency which is scanned, and the resonance occuring at $B_{res} = \dfrac{\omega_{rf}}{\gamma}$ in absence of an EDM will be shifted due to the EDM d by the amount $\Delta B_E = \dfrac{d}{\mu} E$.

It is also instructive to visualize the MR process in the usual classical picture (Slichter, 1963). The basic set-up of a MR experiment (Fig.8) has perpendicular static and oscillating (radio-frequency) magnetic fields $\vec{B}$ and $\vec{B}_{rf} = \vec{B}_1 \cos\omega_{rf} t$. The linearly polarized field $\vec{B}_{rf}$ is decomposed into two components rotating clockwise and counterclockwise respectively around the static field. In a coordinate frame rotating with ω_{rf} around $\vec{B}$ one of these components is static, while the other rotates at an angular frequency $-2\omega_{rf}$. This component is neglected, an approximation known as rotating wave approximation (r.w.a.).

As a consequence of the Larmor theorem an additional fictitious magnetic field $\vec{B}_{fict} = -\dfrac{\bar{\omega}_{rf}}{\gamma}$ oriented to compensate the static field $\vec{B}$ appears in the rotating frame. The magnetic moments, initially oriented along the static field precess around the total field $\vec{B}_{tot}$.

The maximum probability for a reversal of the magnetic moment thus occurs when the fictitious field exactly compensates the static field, which leads to the resonance condition $\omega_{rf} = \omega_L$. If at the same time the particle has an electric dipole moment and an electric field is applied, an additional fictitious field $\Delta \vec{B}_E = \dfrac{d}{\mu} \vec{E}$ appears in the rotating frame and leads to a corresponding shift of the resonance frequency (Fig.9).

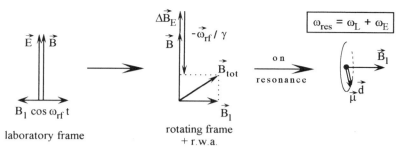

Figure 9. *Magnetic resonance in presence of EDM*

An interesting point to be noted here is the fact that in the rotating frame only static fields are present. For this reason one can perform similar MR experiments by using only static fields. We shall show later, that this is indeed possible by a technique called "zero field level crossing spectroscopy".

Equation (19) allows the interpretation of the experimental sensitivity to dipole moments in terms of magnetometric sensitivity. It is well known that magnetic resonance experiments may be used to measure magnetic moments when performed in a known magnetic field, and conversely to measure magnetic fields when using systems with known magnetic moments. In this sense the sensitivity of a given experimental arrangement to changes ΔB of the applied external field enables us to express the sensitivity Δd of that experiment to electric dipole moments as

$$\Delta d = \left(\frac{\mu}{E} \right) \Delta B \tag{20}.$$

Assuming a shot noise limited signal for which the signal noise scales with the integration time T_{int} as $T_{int}^{-1/2}$, we obtain the practical expression[6]

$$\Delta d \left[10^{-24} \, e \cdot cm \right] \approx 100 \, \Delta B[nG] \, \frac{1}{E[kV \, / \, cm]} \, \frac{1}{\sqrt{T_{int} [h]}} \tag{21},$$

which may be useful for quick estimations. In (21) ΔB is the sensitivity to magnetic field variations obtained with a 1 second integration time.

Neutron EDM Experiments

We start the detailed discussion of actual EDM experiments by an excursion into the world of neutron EDM's. The motivation for this sideline is twofold: on one hand the neutron is the simplest neutral paramagnetic particle and on the other hand the n-EDM experiments have been a source of inspiration for the early atomic EDM experiments started in the 60's. It is also worthwile to note some parallel developments of ideas and techniques in the more recent, and also in planned future neutron and atomic EDM experiments.

Early Experiments. The neutron EDM experiments - pioneered by N.F. Ramsey - have a history of almost 40 years and are the longest ongoing experimental effort to detect a T violating property of an elementary particle (Ramsey, 1990) .

---

[6] A magnetic moment of 1 Bohr magneton was assumed and the numerical factor rounded.

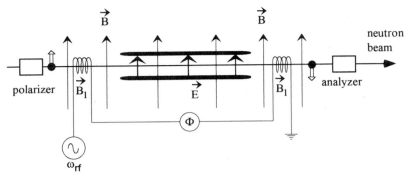

Figure 10. *Principle of early neutron and atomic EDM experiments*

The principle of these early experiments is illustrated in Fig.10. A neutron beam is spin polarized by reflection from a magnetized iron mirror and sent through a Ramsey resonance configuration (Ramsey, 1956) with separated oscillatory magnetic fields[7]. The interaction region is exposed to a homogeneous magnetic field B and a homogeneous electric field E (parallel to B) is applied in the free precession region between the r.f. zones. Resonant spin-flip transitions are driven when the frequency ω_{rf} of the oscillatory fields equals the generalized precession frequency $\omega_{prec} = \omega_L \pm |\omega_E|$ of the static fields. The magnetic resonance lineshape has basically a sinusoidal dependence on the quantity $\delta\omega\,T + \Phi$, where $\delta\omega = \omega_{rf} - \omega_{prec}$ is the detuning from resonance, T the time-of-flight between the interaction zones, and Φ the relative phase of the oscillating fields in these zones. Averaging this oscillation over the velocity distribution in the beam and the finite interaction times in the individual r.f. zones superposes an envelope onto this modulation whose width $\Delta\omega$ is determined by the interaction time with a single r.f. zone $\dfrac{<v>}{l}$ ($<v> =$ average velocity; l = length of single zone). By this averaging process only a few oscillations survive. These are usually referred to as Ramsey fringes. With a suitable choice of Φ (i.e. $\pm\pi/2$) the signal has a linear dependence on the detuning in the vicinity of the center of the resonance $S_\pm \propto (\omega_L \pm \omega_E)\,T$. The electric dipole moment d is inferred by measuring the signal change $\Delta S = S_+ - S_- \propto 2\,\dfrac{d\,E}{\hbar}\,\dfrac{L}{<v>}$ when reversing the orientation of the electric field after calibration of the proportionality constant. This technique has been used for more than 30 years to lower the upper limit on the neutron EDM (Ramsey, 1994).

New Generation of Neutron EDM Experiments.   Besides increasing neutron flux and detection efficiency, an improvement of the sensitivity to EDM's can only be obtained by narrowing the magnetic resonance line shape, increasing the time $T = \dfrac{L}{<v>}$ between the successive interactions with the oscillating fields. As there are practical limitations to the length L of the apparatus, the experimental effort was put into a lowering of the neutron velocity <v> or, more radically, by trapping the neutrons in "bottles". Starting in the 80's two independent groups at Grenoble and Leningrad have performed EDM searches on stored neutrons (Altarev et al., 1986; Pendlebury, 1992; Smith et al., 1990). This storage

---

[7] We shall call these fields radio-frequency (r.f.) fields, although in some experiments frequencies as low as 30 Hz are used.

technique relies on the fact that, similarly to optics, the passage of neutrons through matter may be described in terms of an index of refraction, n<1. For sufficiently cold neutrons the critical angle for total reflection approaches normal incidence, allowing an effective storage time of many minutes. Due to their large de Broglie wavelength the reflection of the cold neutrons by the container walls involves the coherent interaction with a large number of nuclei, so that the neutrons are not heated up by the interaction. In this way neutrons at milli-Kelvin temperatures may be stored in a room temperature container for several minutes. The oscillatory field is applied as a sequence of two pulses separated by T=70 seconds (Smith et al., 1990) yielding a Ramsey fringe width of 10^{-2} Hz. This represents reduction in the fringe width of $5 \ 10^4$ as compared to the one obtained in the early beam experiments (Smith et al., 1957). Together with better counting statistics, the value of the neutron EDM was improved over the same period by more than 6 orders of magnitude. The current limits on the neutron EDM are

$$d_n = (-3 \pm 2 \pm 4) \ 10^{-26} \text{ e cm}$$

from the ILL-Grenoble collaboration (Ramsey, 1994; Smith et al., 1990), and

$$d_n = (+2.6 \pm 4.2 \pm 1.6) \ 10^{-26} \text{ e cm}$$

from the St.Petersburg (former Leningrad) team (Altarev et al., 1992), where the first errors are statistical and the second errors reflect systematic uncertainties. The results of the experiments are compatible with a zero value of the neutron EDM with comparable uncertainties.

The numerical values of these upper limits on the neutron EDM allow a direct interpretation in terms of a separation of the centers of the positive and negative charge distributions inside the neutron. The impressively low value of the upper limit of the neutron EDM is best visualized by scaling it to a neutron blown up to the size of the earth, for which this charge separation is then known to be less than a few microns!

Future Generation of Neutron EDM Experiments.   Both neutron-EDM teams are currently improving their apparatus towards increased sensitivity by moving towards large storage volumes, larger count rates and online magnetometry. By eliminating systematic errors and improving the statistics an improvement of the EDM limit of the neutron by a factor 5 is expected (Pendlebury, 1995). The planed use of online magnetometry using atomic magnetometers (^{199}Hg vapor in the n experiments) is another common feature of the coming generation of neutron and atomic EDM experiments.

A novel, very elegant approach to n-EDM search has been proposed by Golub and Lamoreaux (1993; 1994). Ultracold neutrons are to be trapped in a container containing superfluid ^{4}He with a dilute fraction of ^{3}He, the latter serving the 3 purposes of polarizer, analyzer and magnetometer. For the first two applications one uses the enormous difference in the neutron absorption cross section when the n and ^{3}He spins are parallel (forbidden) or antiparallel (allowed). In order to keep the two precessing spins parallel to each other despite of their different gyromagnetic ratios the authors plan to use the technique of "dressed g-factors"(Golub and Lamoreaux, 1994), which allows the continuous fine-tuning of g-factors by irradiation with non-resonant r.f. radiation. As the EDM's of the two particles differ by many orders of magnitude, the effect of a n-EDM shows up as an E-field dependent angle between the two spins, and is monitored by the change in n absorption by ^{3}He.

Early Atomic EDM Experiments

Early experiments. Early experiments aimed at setting upper limits on *atomic* EDM's were inspired by the early neutron experiments and used quite similar experimental arrangements, atomic beams with Ramsey resonance zones. The major differences were the spin polarizing and analyzing techniques, which in the case of atoms used the orientation dependent force exerted on magnetic moments by inhomogeneous magnetic multipole fields (Stern-Gerlach effect). Among these early atomic experiments let us mention the one on Cs by Weisskopf et al.(1968) and the one on Tl by Gould(1970). Atoms were prepared in a well defined Zeeman sublevel IF, M_F> of the ground state hyperfine structure (Fig.11) and transitions driven to the neigbouring Zeeman level using a conventional Ramsey resonance set-up in thermal atomic beams. As in the neutron experiments, an EDM induced signal change was then sought by trying to detect a static electric field induced shift of the resonance pattern .

Motional B-fields. A major source of systematic effects in any *beam* EDM experiment is associated with the motional magnetic field $\vec{B}_{mot} = -\dfrac{\vec{v}}{c} \times \vec{E}$, seen by the atom moving in the static electric field $\vec{E}$. If the applied electric and magnetic fields are parallel, the motional field is orthogonal to the applied magnetic field and hence does not - to first order - affect the spin precession in that field. Any deviation from the perfect parallelism of $\vec{E}$ and $\vec{B}$ however gives $\vec{B}_{mot}$ a component along $\vec{B}$, which reverses sign under a reversal of $\vec{E}$, thus leading to a signal which has the same signature as the one from an EDM, an acceleration or deceleration of the Larmor precession linear in the electric field. In the early atomic beam EDM experiments the following "trick" was used to align the parallelism of $\vec{E}$ and $\vec{B}$.
An atomic beam of a lighter element was run in the same apparatus and the same experiment performed upon that beam. Due to the Z^3 scaling law of the EDM enhancement factor, any linear Stark shift signal from that beam was likely to come from an imperfect alignment of $\vec{E}$ and $\vec{B}$, and could be used to adjust the latter.

Modern Atomic EDM Experiments. A new approach to the search for atomic EDM's arose in the 80's when the technique of optical pumping, invented in the 50's by Kastler and Brossel (1950), was applied to spin polarize the atoms and to detect the spin flip magnetic resonance. The principle of this technique may be illustrated as follows (Fig. 12). When an unpolarized atom is illuminated with polarized resonant light, the light polarization is partly

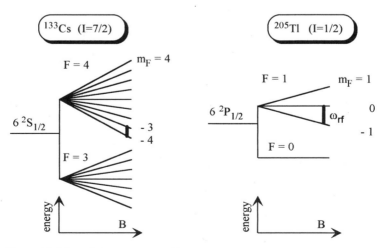

Figure 11. Magnetic resonance transitions used in early atomic EDM experiments

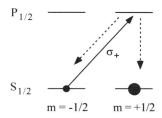

$P_{1/2}$

σ_+

$S_{1/2}$

m = -1/2   m = +1/2

Figure 12. Optical pumping with cp light

or completely transferred to the atom. Consider the case of circularly polarized light tuned to an $S_{1/2} \to P_{1/2}$ transition. The light only couples to the m = -1/2 sublevel of the ground state, and after several absorption-reemission cycles all population will be transferred to the m = +1/2 sublevel. As the m=+1/2 level does not couple to the light, it is often referred to as "dark state".

If the optical pumping rate is much larger than any relaxation rate which mixes the two ground-state components, a very large degree of spin polarization may be obtained[8]. Optical pumping can thus be detected by either monitoring a decrease of the fluorescence from the sample or by monitoring its optical susceptibility χ. In the latter case one can detect either the imaginary part of χ by monitoring a difference in the absorption coefficients for left- and right-handed circularly polarized light (circular dichroism), or one can detect the real part of χ by monitoring the corresponding difference of the indices of refraction (circular birefringence). If a r.f. field now drives spin-flip transitions in the ground state, population reappears in the m = -1/2 level and the atom can interact again with the light field, leading to a resonant change in the monitored signal.

This technique of optical pumping is at the basis of two recent EDM experiments on the Cs and Tl atoms, which we shall discuss in some deatil in the next paragraphs.

The Amherst EDM Experiment on Cs. Owing to its large mass and its unpaired valence electron, Cesium is a good candidate to search for an atomic electric dipole moment induced by an electron EDM with an enhancement factor of R=120 (see 2.2.A). Although the enhancement factor of cesium is almost 6 times smaller than the corresponding value for thallium, cesium EDM experiments offer several technical advantages: the relatively high vapour pressure avoids the use of high temperature ovens and the wavelength of the resonance transition is accessible to low-cost laser diodes. Fig.14 shows the terms of the Cs atom relevant to the Amherst EDM experiment.

The experiment was done in a Cs vapour cell using an all optical technique involving no oscillating fields. The spin flip in this case is induced by a precession in a transverse static magnetic field, which may be interpreted as the $\omega_{rf} \to 0$ limit of conventional magnetic resonance. This technique is a variant of level-crossing spectroscopy called ground-state Hanle effect. The principal layout of the Amherst experiment is shown in Fig. 13 and involves two laser beams, a pump beam and a probe beam. The pump beam is circularly polarized and is used to optically pump the vapour, giving it a magnetization

$$M_z = g_F \, \mu_B \, P_z = g_F \, \mu_B < F_z >,$$

---

[8] If the upper state is a $P_{3/2}$ state, the system has no dark state and optical pumping is less efficient

proportional to the polarization P_z, and oriented along the k-vector of the pump beam. Under the action of the torque $\vec{M} \times \vec{B}$ exerted by a static magnetic field along Oy the magnetization precesses in the x-z plane at the frequency ω_{prec}. If one assumes that the magnetization $\vec{M}$ is produced by the optical pumping process at a constant (pump) rate γ_p and that all the components M_i relax exponentially during the precession with equal time constants τ, a simple calculation shows that the time averaged values of the magnetization components in the x-z plane are given by the "Hanle"-curves

$$<M_z>_t = M_z(\infty) \frac{1}{1+x^2} \tag{22}$$

and

$$<M_x>_t = M_z(\infty) \frac{x}{1+x^2} \tag{23},$$

where $M_z(\infty)$ is the steady state magnetization obtained under the combined action of pumping and relaxation, and x is the dimensionless "Hanle" parameter $x = \omega_{prec}\tau$.

The transverse magnetization thus has a dispersively shaped dependence on x. i.e. on the magnetic field. It is precisely the shift of the zero crossing of this resonance under the action of a static electric field which is used to probe the existence of an atomic EDM. The transverse magnetization is monitored by measuring a circular dichroism of the probe beam: the helicity is periodically reversed and the difference of the transmitted intensities $I_\pm$ measured. It is easy to show that for an optically thin medium the normalized circular dichroism $CD \equiv \dfrac{I_+ - I_-}{I_+ + I_-}$ is equal to the transverse polarization

$$CD = <P_x>_t = \frac{<M_x>_t}{g_F \mu_B} = P_z(\infty) \frac{x}{1+x^2} \tag{24},$$

so that near the center of the resonance, for x<<1 one has $CD = P_z(\infty) x = P_z(\infty) \omega_{prec}\tau$.

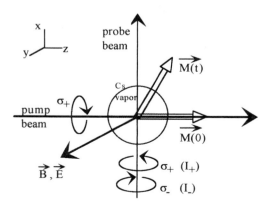

Figure 13. Principle of the Amherst EDM experiment in Cs

Let us recall that the precession frequency is given by

$$\omega_{prec} = -\frac{\vec{\mu}\cdot\vec{B} + \vec{d}\cdot\vec{E}}{\hbar}.$$

If the circular dichroism CD($\pm$|E|) is recorded by reversing the electric field, while keeping B=0, one thus obtains the experimental asymmetry

$$A = CD(+) - CD(-) = 2\, g_F\, d_A\, |E|\frac{\tau}{\hbar}, \tag{25}$$

from which the atomic EDM, d_A, can be extracted. The Landé g-factor in alkali ground states is given by $g_{F=I\pm1/2} = \pm\dfrac{2}{2I+1}$; for Cs I = 7/2.

A remark on the relaxation time τ: The linewidth (HWHM) of the resonance given by Eq. 23, expressed in units of magnetic field is given by $\Delta B = \dfrac{\hbar}{g_F\,\mu_B}\dfrac{1}{\tau}$, or after inserting numerical factors $\Delta B(G) = 0.45\,\dfrac{1}{\tau(\mu\,\text{sec})}$. In low pressure vapors the ground-state spin polarization is completely destroyed upon collisions with the cell walls, and the effective relaxation time is determined by the transit time of the atoms through the light fields, which for beam diameters in the mm to cm range, is 5 - 50 µsec, which translates to linewidths in the 10-100 mG range. As was first shown by Dehmelt (Happer, 1972), the relaxation time can be significantly enhanced by the addition of an inert diamagnetic buffer gas. By making multiple velocity changing collisions with the buffer gas atoms, the Cs atoms diffuse only slowly out of the light beams. If the buffer gas pressure is not too high the spin depolarization during the collisions may be neglected (Corney, 1979).

The Amherst experiment uses 250 Torr of N_2 as buffer gas giving an effective spin relaxation time of 17 msec and a resonance linewidth of 27 µG. The collisions strongly broaden the optical transition. The hyperfine structure of the excited state can no longer be resolved, so that the absorption spectrum consists of the resolved lines $6S_{1/2}$ (F=3) $\rightarrow$ $6P_{1/2}$ (F'=3,4) and $6S_{1/2}$ (F=4) $\rightarrow$ $6P_{1/2}$ (F'=3,4). The pump laser is tuned to the former transition, and empties the F=3 ground state by optical pumping. All atoms are thus transferred to the upper hyperfine level in the ground state, and a polarization $P_z = <F_z>_{F=4}$ of 70% is build up. The circularly polarized probe laser is tuned to the $6S_{1/2}$ (F=4) $\rightarrow$ $6P_{1/2}$ transition and allows a high contrast detection of the ground state polarization.

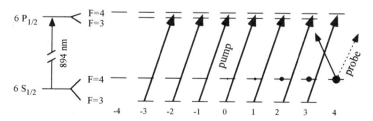

Figure 14. Cs terms relevant to the Amherst EDM experiment

A serious disadvantage of using vapor cells in EDM experiments is their relatively low electrical break-down voltage. In the Amherst experiment fields of 4 kV/cm were used. The choice of the buffer gas and its pressure is a compromise between maximizing the breakdown voltage and minimizing the spin depolarization rate. A considerable advantage of using an atomic vapor rather than an atomic beam is the absence of the motional B-field effect (see 3.3), due to the isotropic distribution of the atomic velocities in the vapor. A major source of systematic uncertainty are magnetic fields reversing synchronously with the electric field. Such fields may for example be generated by leakage currents between the electric field electrodes. Furthermore great care has to be taken to minimize time varying stray magnetic fields. Multiple layers of μ-metal shields, as well as an active field stabilization reduce the field in the vapor cell to below 100 nG. Any residual magnetic field induced effects are eliminated by using two stacked cells with opposite electric fields of equal magnitude. When subtracting the asymmetries recorded in the two cells, effects from an EDM add; magnetic effects on the other hand cancel, assuming that no field gradients are present.

With this apparatus the Amherst team has measured the EDM of the Cs atom to be (Murthy et al., 1989)

$$d_{Cs} = (-1.8 \pm 6.7 \pm 1.8) \ 10^{-24} \ e \ cm,$$

and deduced an upper limit on the electron EDM of

$$d_e = (-1.5 \pm 5.5 \pm 1.5) \ 10^{-26} \ e \ cm.$$

The total integration time was 220 hours. At the time of publication this result presented a 25-fold improvement over previous measurements. The first error represents the statistical uncertainty, while the second error gives the sytematic uncertainties. An improved version of the experiment is currently in preparation at Amherst college.

The Berkeley EDM Experiment on Tl. The Berkeley EDM experiment on Tl has set in the past few years the lowest limit to date on the electric dipole moment of the electron. The apparatus used in this experiment is closely related to the ones used in the early neutron and early atomic EDM experiments. It is basically an atomic beam machine using Ramsey's technique of separated oscillatory fields. Several major changes/additions characterize the apparatus:

- the preparation and analysis of the atomic spin polarization are done by optical pumping and laser induced fluorescence respectively,
- the intrinsic symmetric structure of the Ramsey technique, together with a symmetric arrangement of the polarization and analysis sections allows to run alternatively two counter-propagating atomic beams in the apparatus,
- the atomic beams propagate along the vertical direction.

An initially unpolarized Tl beam (ground state configuration $6^2P_{1/2},F=0,1$) is optically pumped with a resonant light beam polarized parallel to the quantization axis defined by the static magnetic field. The laser is tuned to the $6^2P_{1/2},F=1 \rightarrow 7^2S_{1/2},F'=1$ transition. As the Clebsch-Gordan coefficient for the $|F=1, M=0\rangle \rightarrow |F'=1, M=0\rangle$ under π-excitation vanishes, the M=±1 Zeeman levels are selectively depopulated and a longitudinal alignment[9] $A_{zz}=1-3p_0$ is build up in the F=1 ground state. The absorption coefficient for

---

[9] The populations p_M in the $|F=1 \ M\rangle$ ground state sublevels are normalized, i.e. $p_{-1} + p_0 + p_{+1} = 1$.

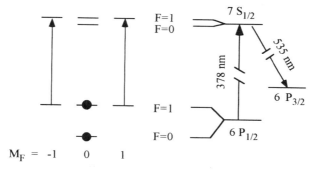

$M_F = -1 \quad 0 \quad 1$

Figure 15. *Relevant terms of Tl EDM experiment*

linearly polarized light - monitored by the fluorescence in the $7^2S_{1/2} \to 6^2P_{3/2}$ branch (50% branching ratio) - is proportional to $p_{+1}+p_{-1} = 2-A_{zz}$, and is maximal for $p_0 =1/3$ (no alignment) and vanishes for $p_0 = 1$ (maximal alignment). The state of maximal alignment is thus characterized by a minimum of the fluorescence in the probe region.

A magnetic resonance is driven in the aligned atoms using a conventional two-zone Ramsey resonance arrangement (Fig. 16). With a static field of 260 mG the resonance frequency is 120 kHz. On resonance the r.f. field drives simultaneously the IF=1, M=0> $\to$ IF=1, M=±1> transitions, thus destroying the alignment and restoring the fluorescence. By chosing the relative phase of the two zones to be ±3π/4 a dispersive shaped resonance signal is generated and the signal change on resonance induced by the static electric field E is proportional to d_{Tl} E and is thus a measure for the atomic EDM d_{Tl}. The actual EDM precession region has a length of 1 m and a field of 110 kV/cm was used.

The major systematic effect here is, as in all beam experiments, the effect of motional magnetic fields induced by a non-parallelism of E and B. This effect is eliminated by running the same experiment with two counterpropagating Tl beams, the motional field induced fringe shift being linear in the atomic velocity will reverse, whereas the EDM signal proper, being independent of v, will not. This cancellation holds however only if the velocity

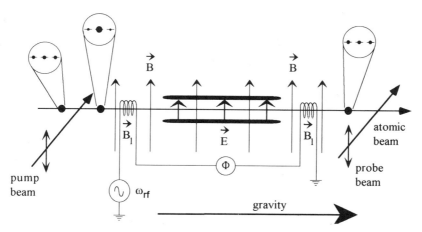

Figure 16. Principle of Berkeley EDM experiment on [205]Tl. Fluorescence detection at intersection points of atomic and laser beams as well as second, counterpropagating atomic beam are not shown. Enlargements on top show populations of Zeeman levels in F=1 state

distributions in the upward and downward running beams are identical. The use of a vertical beam machine avoids curvatures of the atomic trajectories under the influence of gravity, by which the two beams are bound to traverse different regions of space. An additional systematic effect arises from the Stark shift of the optical transition frequency. As this shift is quadratic in the applied electric field it cancels to first order and contributes only if the magnitude of the electric field changes upon reversing its orientation. Finally let us mention yet another systematic effect which may be interpreted in terms of a geometric phase (Commins et al., 1994), and whose origin is the adiabatic evolution of the atomic angular momentum F in the total (= applied + motional) magnetic field in presence of a gradient of transverse components. The importance of this effect is comparable to the motional B-field effect.

First results on the atomic and electronic EDM's were published in 1990 (Abdullah et al., 1990) and yielded

$$d_{Tl} = (1.6 \pm 5.0) \ 10^{-24} \text{ e cm,}$$

and

$$d_e = (-2.7 \pm 8.3) \ 10^{-27} \text{ e cm.}$$

respectively. After several improvements of the apparatus this limit was recently (Commins et al., 1994) pushed to

$$d_e = (1.8 \pm 1.2 \pm 1.0) \ 10^{-27} \text{ e cm,}$$

from which the authors, using the enhancement factor of -585 calculated by Liu and Kelly (1992), give the upper limit on the electron EDM of

$$|d_e| < \ 4 \cdot 10^{-27} \text{ e cm,}$$

which is the lowest limit set so far on the electron EDM.

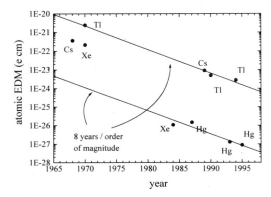

Figure 17. History of experimental upper limits on atomic EDM's. Upper and lower lines represent average lowering of limit in paramagnetic (Tl, Cs) and diamagnetic (Hg, Xe) atoms respectively.

What have we learned from Atomic EDM Experiments?

The history of the experimental search for atomic EDM's is now close to 30 years old. Figure 17 shows the temporal evolution of the experimental sensitivity to atomic EDM's. The lowest upper limits on atomic EDM's come from experiments on the diamagnetic atoms Xe and Hg. The lowest experimental limit for the permanent electric dipole moment of *any* system is the current upper limit on the EDM of the ^{199}Hg atom (Jacobs et al., 1993),

$$|d_{Hg}| < 1.3 \cdot 10^{-27} \text{ e cm}$$

which has recently been lowered (Lamoreaux, 1995) to

$$|d_{Hg}| < 9 \cdot 10^{-28} \text{ e cm}.$$

It is the extremely long nuclear spin relaxation time which is, besides the experimenters skills, at the origin of this extraordinarily low bound. The lines in Figure 17 show the history of the experimental sensitivity to atomic EDM's, which is well represented by an exponential improvement rate. For both the diamagnetic (Hg, Xe) and paramagnetic (Cs, Tl) atoms the time required for an order of magnitude improvement is approximately 8 years; the ratio of sensitivities obtained with the two species is approximately 2000, which is basically the ratio of electronic and nuclear magnetic moments.

Despite the low upper bounds on atomic EDM's in diamagnetic atoms, our current best knowledge of the maximal possible value for the EDM of the electron comes from experiments on atoms with unpaired electron spins, Cs and Tl, as discussed in the previous section. This is visualized in Fig. 18, which shows how the upper bound on the electron EDM has been lowered in the past 30 years. The time period needed for a one order of magnitude improvement in sensitivity, as measured by the results from the paramagnetic atoms Cs and Tl, is, again, approximately eight years.

In the theory section we have discussed several other P- and T- violating effects contributing to atomic EDM's. From the above experiments in paramagnetic atoms, as well as from experiments in diamagnetic atoms and polar molecules, upper limits for these contributions may be obtained. The EDM value of the diamagnetic ^{199}Hg atom sets the best present upper limit on the tensor coupling constant of the nucleon-electron interaction C_T

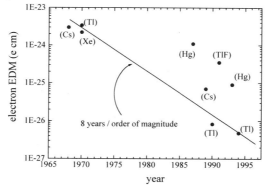

Figure 18. History of experimental upper limits on electron EDM's. Symbols in brackets denote atoms from EDM of which limits were deduced.

and on the Schiff moment Q_S. The Tl experiment, besides giving the best upper limit for the electron EDM, has also set the best upper limit on C_S, the scalar part of the P-,T-violating electron-nucleon interaction, as well as on the β–parameters of the P-even, T-odd model (Commins et al., 1994)discussed by Conti and Khriplovich (1992). The lowest upper limit on the EDM of the proton comes from an experiment performed in a molecular beam of TlF and is

$$|d_p| < 3.7(6.3) \ 10^{-23} \text{ e cm.}$$

Several authors have recently prepared tables comparing the present upper limits for various T-violating processes (Barr, 1993; Jacobs et al., 1993; Mårtensson-Pendrill, 1992; Ramsey, 1994).

New Generation of Atomic EDM Experiments

Although the various teams working in the experimental search for atomic EDM's have produced spectacular upper bounds on the latter and are actively pursuing the development of their techniques toward increased sensitivity and better control of systematic effects, it is nevertheless worthwhile to consider novel experimental approaches to the problem of the electron dipole moment. Before giving a more detailed description of alternative EDM experiments currently under development, we start by giving some general considerations of the experimental sensitivities.

Criteria for Sensitive Experiments: Systematics.   Sensitive experiments require a large phase angle $\Phi = \omega_{prec} T = \dfrac{\vec{d} \cdot \vec{E}}{\hbar} T$ to be accumulated by the precession of the EDM in the static field, implying a large value of the field-time product ET. The precession time T has different interpretations, depending on the experiment. In Ramsey-type experiments it is the time between successive exposures of the particle to the oscillating fields, and in confined samples, such as atomic vapors it is the effective spin relaxation time as defined by collisions or the finite interaction time with the electromagnetic fields. The measured asymmetry is proportional to Φ

$$\text{Asymmetry} \propto \Phi \propto R\, d_e\, ET \quad ,$$

where R is the enhancement factor. It is of course advantageous to design an experiment which yields a large asymmetry, as the level at which EDM simulating systematic effects have to be eliminated depends directly on the latter.

Statistics.   The statistical limit of the sensitivity is given by the following argument. For a resonance linewidth $\Delta \omega_0 = T^{-1}$, the uncertainty $\delta \omega$ in the determination of the center frequency, and hence in the detection of any shift of this frequency is of the order

$$\delta \omega \approx \frac{\Delta \omega_0}{(S/N)} \quad ,$$

which, for a shot-noise limited count rate R_0 and an integration time T_{int}, yields

$$\delta \omega \approx \frac{\Delta \omega_0}{T \sqrt{R_0 T_{int}}} \quad .$$

Interpreting $\delta\omega$ as an EDM induced shift, the sensitivity to the electron EDM is given by

$$\delta d_e \approx \frac{\hbar}{R\,E\,T\,\sqrt{R_0\,T_{int}}}\quad.$$

The last equation has to be applied with care as in certain experiments the inverse linewidth 1/T and the count rate R_0 may not be independent. In case of an optically detected magnetic resonance the count rate is proportional to the product of the number of atoms N_{At} and the otical pumping rate γ_p. For large pump rates the linewidth will be determined by power broadening

$$\Delta\omega = \Delta\omega_o\,\sqrt{1+S}\;,$$

where $S = \gamma_p/\Delta\,\omega_o$ is the saturation parameter of the optical pumping pocess, and $\Delta\omega_0 = 1$ / T the intrinsic linewidth[10]. The sensitivity in the strongly power broadened limit (S>>1) then reads

$$\Delta\,d_e \approx \frac{\hbar}{R\,E\,\sqrt{N_{At}\,T\,T_{int}}}\quad,$$

Sensitive experiments thus call for samples which sustain large electric field strengths, while having long intrinsic spin relaxation T. As we have seen above these requirements are often mutually exclusive; atomic beam EDM experiments can be carried out in rather large fields of the order of 100kV/cm, but suffer from a relatively short time T, on the order of ms, limited by the thermal atomic velocity. On the other hand spin relaxation times of up to 20 ms can be achieved in atomic vapours in presence of an inert buffer gas. Metallic vapors however have a rather low break-down voltage, allowing the use of fields of only a few kV/cm. The Berkeley Tl and the Amherst Cs experiments are good examples for the above arguments. Another criterion, not considered so far, is detection efficiency. In this respect it seems that experiments measuring the medium susceptibility directly by forward scattering are superior to experiments using fluorescence detection; in the latter case limitations are imposed by the finite solid angle, whereas in the former case every single information carrying photon can in principle be detected [11]. Last but not least, a straightforward way to enhance the sensitivity to elctron EDM's is the choice of atoms/molecules with a larger enhancement factor R.

We conclude this paper by giving a brief discussion of three novel experimental approaches to search for an electron EDM:

- Paramagnetic, polar molecules: these molecules may have enhancement facors exceeding the ones of atoms by several orders of magnitude.
- Faraday-Ramsey spectrocopy: a variant of the Berkeley atomic beam experiment using forward scattering and no oscillating fields.
- Atoms in cryogenic matrices: this sample combines the properties of sustaining large electric fields, while allowing long intrinsic spin relaxation times.

---

[10] For these qualitative arguments we do not distinguish between longitudinal and transverse relaxation rates.
[11] In optically very thin samples however inefficient fluorescence collection may still be superior to forward scattering.

Polar Molecules. The polar molecule TlF has a long history (Cho et al., 1991; Sandars, 1967) in the experimental search for T-violating processes in the atomic/molecular sector, but is less sensitive to contributions from an electron EDM than the heavy paramagnetic atoms Cs and Tl. The paramagnetic, polar molecule YbF, on the other hand, seems to be an extremely promising candidate to search for electron EDM induced effects (Kozlov and Ezhov, 1994). In this molecule, close lying rotational levels of opposite parity are very strongly mixed by even modest electric fields of the order of 10kV/cm, so that the mixing parameter ε (see Eqs. 5,6) , which is $\ll 1$ in atoms, becomes of order unity. The correponding enhancement factor is of the order 10^6.

E. Hinds and his collaborators at Yale University are currently setting up a molecular beam experiment on YbF aimed at a one to two orders of magnitude improvement on the electron EDM (Hinds, 1994).

Faraday-Ramsey spectroscopy. Faraday-Ramsey (FR) spectroscopy is an extension of the nonlinear Faraday effect which has been extensively studied in vapor cells (see e.g. (Kanorsky et al., 1993; Weis et al., 1993) and references therein), to an atomic beam geometry with spatially separated pump and probe regions (Ramsey geometry).

Atoms are optically pumped with linearly polarized resonance radiation and an alignment (second rank tensor in spin space) builds up in the ground state. Upon traversing a homogeneous B-field the alignment precesses and its orientation is probed downstream by forward scattering of a weak linearly polarized probe beam.

The technique is related to both the Berkeley Tl and the Amherst Cs expriments; it may be viewed, on one hand as the $B_0 \to 0$ limit of the conventional Ramsey configuration with orthogonal static (B_0) and oscillating ($B_1 \cos\omega_{rf}t$) magnetic fields, in which case the resonance condition $\omega_{rf} \to 0$ implies B_1 to be a static field (compare Figs.16 and 19).

On the other hand the relation to the Amherst experiment (cf Fig. 13) is obvious: pump and probe beams are separated in space and atoms are bound to traverse both regions by using an atomic beam. A major difference, however, is the use of linearly polarized light in FR spectroscopy. The relevant alignment component created by pumping with *linearly* polarized light is a superposition of ground state Zeeman levels with magnetic quantum numbers m and m±2 ($\Delta m = 2$ coherence). The Larmor precession frequency of the alignment is hence twice the conventional Larmor frequency describing the evolution of vector spin polarization (rank 1 tensor) in a magnetic field and the sensitivity to EDM's is doubled. We have performed preliminary tests of the method in a thermal Rb beam using pump probe separations of 5 and 30 cm (Schuh et al., 1993; Weis et al., 1993) and find that the statistical

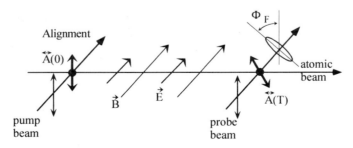

Figure 19. Principle of Faraday-Ramsey spectroscopy

sensitivity of the FR-method, when scaled to the experimental conditions of the current best result from paramagnetic atoms (Commins et al., 1994), is quite comparable.

Atoms in Cryogenic ^{4}He Matrices. The use of atoms embedded in a cryogenic noble gas matrix, matrix isolated atoms, has been proposed in the past to search for EDM's (Pryor and Wilczek, 1987). This proposal was based on conventional ESR (electron spin resonance) techniques and asked for unconventionally large matrices of 10 cm^3. In heavy noble gas matrices the much more sensitive method of optically detected magnetic resonance cannot be applied. For this reason S.I. Kanorsky suggested in 1991 the use of atoms immersed in superfluid ^{4}He to perform a highly sensitive EDM search. In collaboration with S.I. Kanorsky we started such a cryogenic experiment at the Max-Planck-Institute for Quantum Optics in 1992. First experiments based on techniques developed earlier by the pioneering work of the Heidelberg group (Reyher et al., 1986)[12] were mainly concerned with the development of implantation techniques and the measurement of atomic residence times (Arndt et al., 1993). Since 1993 we have extended our experiments to solid ^{4}He (sHe) samples and were able, for the first time, to implant atoms into such a matrix (Kanorsky et al., 1994) and to observe their laser induced resonance fluorescence. We have furthermore investigated the perturbation of the optical excitation and emission spectra by the He matrix (Kanorsky et al., 1994) and were able to optically pump Cs atoms, obtaining a spin polarization in excess of 95% and a spin relaxation time, T_1, in the range of one to two seconds in fields as low as 10 mG (Arndt et al., 1995). As a next step towards an EDM experiment we have performed zero-field level crossing experiments (longitudinal and transverse ground state Hanle effect) as well as optical-r.f. (Arndt et al., 1995) and optical-microwave double resonance experiments. High contrasts were observed in all of these experiments. The linewidths so far do not reflect the extremely long spin relaxation time, but were rather limited by magnetic field inhomogeneities.

The advantage of using paramagnetic atoms imbedded in sHe is the fact that the He matrix is an isotropic[13] trap, in which the atoms may be trapped for minutes up to hours. The long observation times, together with the non-magnetic character of the matrix, He has neither nuclear nor electronic magnetic moments, and an electrical break-down voltage of several 10^5 V/cm are expected to make such a sample ideally suited for an EDM search in paramagnetic atoms. Extrapolating our present experimental parameters towards a realistic EDM experiment we think that the method is likely to provide a one to two orders of magnitude improvement in the statistical sensitivity over state-of-the-art experiments in paramagnetic atoms (Weis et al., 1995). The optimum experimental strategy for an EDM experiment in sHe is still an open question: the transverse Hanle resonance, as well as the optical-r.f. resonances, or even the hyperfine transition could serve as the narrow magnetic resonance structure whose shift under the influence of a static electric field gives the sought signature of an atomic EDM. From the point of view of systematic effects the sample presents two major advantages: there are no motional magnetic field effects and the cryogenic surrounding is ideally suited for the implementation of superconducting magnetic shields.

---

[12] The interested reader is referred to the forthcoming proceedings of the first workshop on" Ions and Atoms in Superfluid Helium" held in Heidelberg in 1994 (to be published in Z.Phys.B).
[13] Provided the atoms are implanted into the bcc phase of the matrix.

APPENDIX: Relation between electron EDM and atomic EDM signal

Both the Cs and Tl EDM experiments are made in atomic ground states within a level of well defined total angular momentum F. As the electron EDM is defined for the free electron and the enhancement factor is usually given for atomic states ignoring hyperfine structure the question arises of how the experimental signal - usually proportional to an atomic precession frequency in a well defined hyperfine state - is related to the electron EDM. In this Appendix we discuss this relation.

The enhancement factor R is defined as the ratio of expectation values of the dipole operator d_z in the states with maximal value of the z-component of the relevant angular momentum, i.e.

$$R \equiv \frac{d_A}{d_e} = \frac{< n \, L_J \; m_j = j | d_z | n \, L_J \; m_j = j >}{< s_z = 1/2 | d_z | s_z = 1/2 >}$$

Defining d_j via

$$\vec{d}_A \equiv d_j \vec{J} \; ,$$

the atomic EDM d_A is given, for j=1/2 states (Cs, Tl), by

$$d_A = < 1/2, 1/2 | J_z | 1/2, 1/2 > d_j = \frac{d_j}{2} .$$

Both the Tl and the Cs experiments measure an atomic precession frequency. In the regime of a linear Zeeman effect the associated energy is defined as the difference of the atomic energies in adjacent Zeeman levels

$$\hbar \omega_{prec} = < \vec{d}_A \cdot \vec{E} >_{F,M} - < \vec{d}_A \cdot \vec{E} >_{F,M-1}) = E_z d_j \{ \langle F, M | J_z | F, M \rangle - \langle F, M - 1 | J_z | F, M - 1 \rangle \}$$

Using the relation (Sobelman, 1979)

$$\langle J, F, M | J_z | J, F, M \rangle \equiv \frac{g_F}{g_J} M$$

the precession frequency is thus given by

$$\omega_{prec} = 2 \frac{g_F}{g_J} \frac{d_A E_z}{\hbar} .$$

For the Tl $6P_{1/2}(F=1)$ state (I=1/2) the ratio of Lande factors is $g_F/g_J = 1/2$ and the precession frequency becomes

$$\omega_{prec} = \frac{d_A E_z}{\hbar} = \frac{R_{Tl} d_e E_z}{\hbar} ,$$

whereas for the Cs $6S_{1/2}(F=I\pm1/2)$ states (I=7/2) the precession frequency becomes

$$\omega_{prec} = \pm \frac{2}{2I+1} \frac{d_A E_z}{\hbar} = \pm \frac{R_{Cs} d_e E_z}{4\hbar} .$$

ACKNOWLEDGEMENTS

I dedicate this paper to Stephen B. Ross, who died in a tragic accident on May 25, 1996. In his Ph.D. thesis work Stephen set the lowest current limit on the electric dipole moment of the electron. As a post-doc at Max-Planck-Institut für Quantenoptik Stephen collaborated for one year in our project «Atoms in solid ^{4}He». Our team mourns the loss of a vital physicist and a good friend.

I thank my colleagues Stephen B. Ross, Sergej Kanorsky, Krzysztof Pachucki, and Markus Arndt for their critical reading of the manuscript and many constructive comments.

REFERENCES

Abdullah, K., Carlberg, C., Commins, E.D., Gould, H., and Ross, S.B., 1990, New Experimental Limit on the Electron Electric Dipole Moment, *Phys.Rev.Lett.* 65:2347.

Altarev, I.S. *et al.*, 1986, Search For an Electric Dipole Moment of the Neutron, *JETP Lett.* 44:460.

Altarev, I.S. *et al.*, 1992, New Measurement of the Electric Dipole Moment of the Neutron, *Phys.Lett.* B276:242.

Arndt, M., Kanorsky, S.I., Weis, A., and Haensch, T.W., 1993, Can Paramagnetic Atoms in Superfluid Helium be Used to Search for Permanent Electric Dipole Moments?, *Phys.Lett.* A174:298.

Arndt, M., Kanorsky, S.I., Weis, A., and Haensch, T.W., 1995, Long Electronic Spin Relaxation Times of Cs Atoms in solid ^{4}He, *Phys.Rev.Lett.* 74:1359.

Arnison, G. and al., e., 1983, Experimental Observation of Isolated Large Transverse Energy Electrons with Associated Missing Energy at $s^{1/2}$=540 GeV, *Phys.Lett.* 122B:103.

Arnison, G. and al., e., 1983, Experimental Observation of Lepton Pairs of Invariant Mass Around 95 GeV/c^2 at the CERN SPS Collider, *Phys.Lett.* 126B:398.

Aubert, B. and al., e., 1974, Further Observation of Muonless Neutrino-Induced Inelastic Interactions, *Phys.Rev.Lett.* 32:1454.

Banner, M. and al., e., 1983, Observation of Single Isolated Electrons of High Transverse Momentum in Events with Missing Transverse Energy at the CERN p(bar)p Collider, *Phys.Lett.* 122B.

Barr, S.M., 1993, A Review of CP Violation in Atoms, *Int.J.Mod.Phys.* A8:209.

Barr, S.M., 1994, Atomic Electric Dipole Moments and CP Violation, in *XXIXth Rencontres de Moriond*, Editions Frontieres, Villars/Ollon, Switzerland.

Benvenuti, A. and al., e., 1974, Observation of Muonless Neutrino-Induced Inelastic Interactions, *Phys.Rev.Lett.* 32:800.

Bernreuther, W. and Suzuki, M., 1991, The Electric Dipole Moment of the Electron, *Rev.Mod.Phys.* 63:313.

Bouchiat, M.A. and Bouchiat, C.C., 1974, I. Parity Violation Induced by Weak Neutral Currents in Atomic Physics, *J.Physique* 35:899.

Bouchiat, M.A. and Bouchiat, C.C., 1974, Weak Neutral Currents in Atomic Physics, *Phys.Lett.* 48B:111.

Bouchiat, M.A., Guena, J., Hunter, L., and Pottier, L., 1982, Observation of a Parity Violation in Cesium, *Phys.Lett.* B 117:358.

Cho, D., Sangster, K., and Hinds, E.A., 1991, Search for Time-Reversal-Symmetry Violation in Thallium Fluoride using a Jet Source, *Phys.Rev.A* 44:2783.

Christenson, J.H., Cronin, J.W., and Fitch, V.L., 1964, Evidence for the 2π Decay of the K_2 Meson, *Phys.Rev.Lett.* 13:138.

Commins, E.D., 1993, Atomic Parity Nonconservation and Electric Dipole Moment Experiments - a 1992 Review, *Phys.Script.* t46:92.

Commins, E.D., 1993, Resource Letter ETDTSTS-1: Experimental Tests of the Discrete Space-Time Symmetries, *Am.J.Phys.* 61:778.

Commins, E.D., Ross, S.B., DeMille, D., and Regan, B.C., 1994, Improved Experimental Limit on the Electric Dipole Moment of the Electron, *Phys.Rev.A* 50:2960.

Conti, R.S. and Khriplovich, I.B., 1992, New Limits on T-Odd, P-Even Interactions, *Phys.Rev.Lett.* 68:3262.

Corney, A., 1979, *Atomic and Laser Spectroscopy.* Oxford University Press, Oxford

Flambaum, V.V., 1976, *Yad.Fiz* 24:383.

Fortson, E.N. and Lewis, L.L., 1984, Atomic Parity Nonconservation Experiments, *Phys.Rep.* 113:289.

Friedman, J.I. and Telegdi, V.L., 1957, Nuclear Emulsion Evidence for Parity Nonconservation in the Decay Chain π^+-μ^+-e^+, *Phys.Rev.* 105:1681.

Garwin, R.L., Lederman, L.M., and Weinrich, M.W., 1957, Observations of the Failure of Conservation of Parity and Charge Conjugation in Meson Decays: the Magnetic Moment of the Free Muon, *Phys.Rev.* 105:1415.

Golub, R., 1993, New Application of the Superthermal Ultra-Cold Neutron Source. I - The Search for the Neutron Electric Dipole Moment, *J.Physique* 44:L321.

Golub, R. and Lamoreaux, S.K., 1994, Neutron Electric Dipole Moment, Ultracold Neutrons and Polarized ^{3}He, *Phys.Rep.* 237:2.

Gould, H., 1970, Search for an Electric Dipole Moment in Thallium, *Phys.Rev.Lett.* 24:1091.

Happer, W., 1972, Optical Pumping, *Rev.Mod.Phys.* 44:169.

Hartley, A.C., Lindroth, E., and Mårtensson-Pendrill, A.M., 1990, Parity Non-Conservation and Electric Dipole Moments in Caesium and Thallium, *J.Phys.B.:At.Mol.Phys.* 23:3417.

Hasert, F.J. and al., e., 1973, Observation of Neutrino-Like Interactions Without Muon or Electron in the Gargamelle Neutrino Experiment, *Phys.Lett.* 46B:138.

Hinds, E., private communication (1994)

Hinds, E.A. and Sandars, P.G.H., 1980, Electric Dipole Hyperfine Structure of TlF, *Phys.Rev.A* 21:471.

Hinds, E.A., 1988, Experiments on T-Symmetry and Parity Violation in Atomic Physics, in *Atomic Physics 11*, World Scientific, Paris.

Hunter, L.R., 1991, Tests of Time-Reversal Invariance in Atoms, Molecules, and the Neutron, *Science* 252:73.

Jackson, J.D., Treiman, S.B., and Wyld, H.W.J., 1957, Possible Tests of Time Reversal Invariance in Beta Decay, *Phys.Rev.* 106:517.

Jacobs, J.P., Klipstein, W.M., Lamoreaux, S.K., Heckel, B.R., and Fortson, E.N., 1993, Testing Time-Reversal Symmetry using ^{199}Hg, *Phys.Rev.Lett.* 71:3782.

Johnson, W.R., Guo, D.S., Idrees, M., and Sapirstein, J., 1985, Weak Interaction Effects in Heavy Atomic Systems, *Phys.Rev.A* 32:2093.

Johnson, W.R., Guo, D.S., Idrees, M., and Sapirstein, J., 1986, Weak Interaction Effects in Heavy Atomic Systems II, *Phys.Rev.A* 34:1043.

Jost, R., 1960, *Theoretical Physics in the Twentieth Century*. Interscience, New York

Kanorsky, S.I., Arndt, M., Dziewior, R., Weis, A., and Hänsch, T.W., 1994, Optical Spectroscopy of Atoms Trapped in Solid Helium, *Phys.Rev.B* 49:3645.

Kanorsky, S.I., Arndt, M., Dziewior, R., Weis, A., and Hänsch, T.W., 1994, Pressure Shift and Broadening of the Resonance Line of Barium Atoms Trapped in Liquid Helium, *Phys.Rev.B* 50:6296.

Kanorsky, S.I., Weis, A., Wurster, J., and Hänsch, T.W., 1993, Quantitative Investigation of the Resonant Nonlinear Faraday Effect under Conditions of Optical Hyperfine Pumping, *Phys.Rev.A* 47:1220.

Kastler, A., 1950, Quelques Suggestions Concernant la Production Optique et la Detection Optique d'une Inegalite de Population des Niveaux de Quantification Spatiale des Atomes. Application a l'Experience de Stern-Gerlach et a la Resonance Magnetique, *J.Phys.Rad.* 11:255.

Khriplovich, I.B., 1988, T-Invariance Violation, can it be observed in Atoms?, in *Atomic Physics 11*, World Scientific, Paris.

Khriplovich, I.B., 1991, *Parity Nonconservation in Atomic Phenomena*. Gordon and Breach Science Publishers, Amsterdam

Kobayashi, M. and Maskawa, T., 1973, CP-Violation in the Renormalizable Theory of Weak Interaction, *Prog.Theor.Phys.* 49:652.

Kozlov, M.G. and Ezhov, V.F., 1994, Enhancement of the Electric Dipole Moment of the Electron in the YbF Molecule, *Phys.Rev.A* 49:4502.

Lamoreaux, S.K., private communication (1995)

Landau, L., 1957, On the Conservation laws for weak Interactions, *Nucl.Phys.* 3:127.

Lee, T.D. and Yang, C.N., 1956, Question of Parity Nonconservation in Weak Interactions, *Phys.Rev.* 104:254.

Liu, Z.W. and Kelly, H.P., 1992, Ananlysis of Atomic Electric Dipole Moment in Thallium by All-Order Calculation in Many-Body Perturbation Theory, *Phys.Rev.A* 45:4210.

Mårtensson-Pendrill, A.-M., 1992, *Calculation of P- and T-Violating Properties in Atoms and Molecules*. Plenum Press, New York

Michel, F.C., 1965, Neutral Weak Interaction Currents, *Phys.Rev.B* 138:408.

Moriond, 1995, *XXXth Rencontres de Moriond*, Editions Frontieres, Villars/Ollon, Switzerland.

Murthy, S.A., Krause, D., Li, Z.L., and Hunter, L.R., 1989, New Limits on the Electron Electric Dipole Moment from Cesium, *Phys.Rev.Lett.* 63:965.

Noecker, M.C., Masterson, B.P., and Wieman, C.E., 1988, Precision Measurement of Parity Nonconservation in Atomic Cesium, a Low Energy Test of the Electroweak Theory, *Phys.Rev.Lett.* 61:310.

Pendlebury, J.M., 1992, Steps to Improve the Measurement of the Neutron Electric Dipole Moment, *Nucl.Phys.* A546:359.

Pendlebury, M., 1995, The Neutron EDM Experiment in Preparation at the ILL, in *XXXth Rencontres de Moriond*, Villars/Ollon (CH).

Prescott, C.Y. and al., e., 1978, Parity Non-Conservation in Inelastic Electron Scattering, *Phys.Lett.* B77:347.

Pryor, C. and Wilczek, F., 1987, "Artificial Vacuum" for T-Violation Experiment, *Phys.Lett.* B194:137.

Purcell, E.M. and Ramsey, N.F., 1950, On the Possibility of Electric Dipole Moments for Elementary Particles and Nuclei, *Phys.Rev.* 78:807.

Ramsey, N.F., 1990, Electric Dipole Moment of the Neutron, *Am.Rev.Nucl.Part.Sci.* 40:1.

Ramsey, N.F., 1994, Electric Dipole Tests of Time Reversal Symmetry, in *Atomic Physics 14*, Boulder.

Ramsey, N.F., 1956, *Molecular Beams*. Oxford Univ. Press, Clarendon

Ramsey, N.F., 1958, Time Reversal, Charge Conjugation, Magnetic Pole Conjugation, and Parity, *Phys.Rev.* 109:225.

Reyher, M.J., Bauer, H., Huber, C., Mayer, R., Schäfer, A., and Winnacker, A., 1986, Spectroscopy of Barium Ions in HeII, *Phys.Lett.* A115:238.

Sandars, P.G.H., , in *Atomic Physics*, (World Scientific, Singapore, 1984).

Sandars, P.G.H., 1965, The Electric Dipole Moment of an Atom, *Phys.Lett.* 14:194.

Sandars, P.G.H., 1966, Enhancement Factor for the Electric Dipole Moment of the Valence Electron in an Alkali Atom, *Phys.Lett.* 22:290.

Sandars, P.G.H., 1967, Measurability of the Proton Electric Dipole Moment, *Phys.Rev.Lett.* 19.

Sandars, P.G.H. and Sternheimer, R.M., 1975, Electric Dipole Moment Enhancement Factor for the Thallium Atom, and a New Upper Limit on the Electric Dipole Moment of the Electron, *Phys.Rev.A* 11:473.

Sandars, P.G.H., 1984, *Atomic Physics 9*, World Scientific, Singapore,

Schiff, L.I., 1963, Measurability of Nuclear Electric Dipole Moments, *Phys.Rev.* 132:2194.

Schubert, K.R. and al, e., 1970, The Phase of v_{00} and the Invariances CPT and T, *Phys.Lett.* 31B:662.

Schuh, B., Kanorsky, S.I., Weis, A., and Haensch, T.W., 1993, Observation of Ramsey Fringes in Nonlinear Faraday Rotation, *Opt.Commun.* 100:451.

Shukla, A., Das, B.P., and Andriessen, J., 1994, Relativistic Many-Body Calculation of the Electric Dipole Moment of Atomic Rubidium due to Parity and Time-Reversal Violation, *Phys.Rev.A* 50:1155.

Slichter, C.P., 1963, *Principles of Magnetic Resonance*. Harper&Row, New York

Smith, J.H., Purcell, E.M., and Ramsey, N.F., 1957, Experimental Limit to the Electric Dipole Moment of the Neutron, *Phys.Rev.* 108:120.

Smith, K.F. *et al.*, 1990, A Search for the Electric Dipole Moment of the Neutron, *Phys.Lett.* B234:191.

Sobelman, I.I., 1979, *Atomic Spectra and Radiative Transitions*. Springer, Berlin

Sternheimer, R.M., 1969, Electronic Polarizybilities of the Alkali Atoms. II, *Phys.Rev.* 183:112.

Streater, R.F. and Wightman, A.S., 1964, *PCT, Spin, Statistics, and All That*. Benjamin, New York

Weinberg, S., 1974, Recent Progress in Gauge Theories of the Weak, Electromagnetic, and Strong Interactions, *Rev.Mod.Phys.* 46:255.

Weis, A., Schuh, B., Kanorsky, S.I., and Hänsch, T.W., 1993, Use of Faraday-Ramsey Spectroscopy to Search For Permanent Atomic Electric Dipole Moments, in *EQEC '93*, EPS, Firenze.

Weis, A., Kanorsky, S.I., Arndt, M., and Hänsch, T.W., 1995, Spin Physics in Solid Helium: Experimental Results and Applications, *Z. Physik* B98:359 .

Weis, A., Wurster, J., and Kanorsky, S.I., 1993, Qualitative Interpretation of the Nonlinear Farady Effect as a Hanle Effect of a Light-Induced Birefringence, *JOSA B* 10:716.

Weisskopf, M.C., Carrico, J.P., Gould, H., Lipworth, E., and Stein, T.S., 1968, Electric Dipole Moment of the Cesium Atom. A New Upper Limit to the Electric Dipole Moment of the Electron, *Phys.Rev.Lett.* 21:1645.

Wu, C.S., Ambler, E., Hayward, R.W., Hoppes, D.D., and Hudsom, R.P., 1957, Experimental Test of Parity Conservation in Beta Decay, *Phys.Rev.* 105:1413.

Zeldovich, Y.B., 1959, Parity Nonconservation in the First Order in the Weak-Interaction Constant in Electron Scattering and Other Effects, *ZETF* 36:964.

DYNAMICAL ASPECTS OF CLASSICAL ELECTRON THEORY

D. Bambusi, A. Carati, L. Galgani, D. Noja, J. Sassarini

Dipartimento di Matematica dell'Università
Via Saldini 50
20133 Milano, Italy

Dedication. *It occurred to the five of us to come in contact with Asim Barut in the very short period from june to september 1994, between a conference in Erice and "his" conference in Edirne, where he was so kind to invite all of us and for which the present notes are written down. It was a joyful experience to share with him the sensation that classical electrodynamics had not been fully exploited, and the hope that a full appreciation of it might prove relevant even for the foundations of quantum mechanics. Certainly there will be other people among his pupils that will take up his heritage, but we very willingly aknowledge that, although working along already estabilished lines, a great support comes to us from the sensation of continuing also his work.*

INTRODUCTION

By classical electron theory we mean what is in principle a very simple thing, namely the Maxwell–Lorentz system, which consists of Maxwell equations with sources due to a point particle, and the relativistic Newton equation for the particle, with Lorentz force due to the electromagnetic field. The unknowns are then the fields $\mathbf{E}(\mathbf{x}, t)$, $\mathbf{B}(\mathbf{x}, t)$ and the particle motion $\mathbf{q}(t)$, governed by the equations

$$\operatorname{div}\mathbf{E} = \rho \qquad \operatorname{div}\mathbf{B} = 0$$

$$\operatorname{rot}\mathbf{E} = -\frac{1}{c}\frac{\partial \mathbf{B}}{\partial t} \qquad \operatorname{rot}\mathbf{B} = \frac{1}{c}\left(\mathbf{j} + \frac{\partial \mathbf{E}}{\partial t}\right)$$

$$\frac{\mathrm{d}}{\mathrm{d}t}\left(\frac{m_0}{\sqrt{1 - \dot{q}^2/c^2}}\dot{\mathbf{q}}\right) = e\left(\mathbf{E}(\mathbf{q}, t) + \frac{1}{c}\dot{\mathbf{q}} \times \mathbf{B}(\mathbf{q}, t)\right)$$

$$\rho(\mathbf{x}, t) = e\delta(\mathbf{x} - \mathbf{q}(t)) \qquad \mathbf{j}(\mathbf{x}, t) = \frac{e}{c}\dot{\mathbf{q}}\delta(\mathbf{x} - \mathbf{q}(t))$$

from which the continuity equation $\frac{\partial \rho}{\partial t} + \operatorname{div}\mathbf{j} = 0$ follows; here c is the speed of light while e is the particle's charge and m_0 its (bare!) mass. The intent would be to

Electron Theory and Quantum Electrodynamics: 100 Years Later
Edited by Dowling, Plenum Press, New York, 1997

study such a system as a dynamical system in the standard sense, namely by looking at the Cauchy problem, also taking into proper account (but we are not able to do it, at present) the nonlinearities, which appear for example in the definitions of the charge density ρ and of the current density $\mathbf{j}$, and in the Lorentz force. The main mathematical problem comes from the presence of the delta function in the definition of the current, which leads to a singularity in the Lorentz force at the right hand side of the equation governing the particle's motion (think of the Coulumb force in the static case). This problem can be dealt with by suitable regularizations; a standard procedure, well familiar from quantum field theory, consists in introducing cutoffs in the Fourier transforms of the fields, and studying the limit when the cutoffs are removed. In this connection we like to quote one of our favorite authors, namely E. Nelson, [1] who, in his book "Quantum Fluctuations" (page 65) says: "*With suitable ultraviolet and infrared cutoffs, this is a dynamical system of finitely many degrees of freedom, and we have global existence and uniqueness... . Is it an exaggeration to say that nothing whatever is known about the behavior of the system as the cutoffs are removed, and there is not one single theorem that has been proved ?*". Though incredible, this is just the actual situation: nothing whatever seemed to be known rigorously for the motion of a point particle in interaction with the electromagnetic field, when the latter is not assumed to be assigned in advance.

We believe that we were finally able to provide at least some preliminary results concerning the limit in which the cutoffs are removed, on which we will report below. The results might at first sight appear to be almost trivial, because we essentially confirmed the validity of the famous Abraham–Lorentz (or AL) equation for the particle in the so–called dipole approximation. But on the other hand this required a strong conceptual effort, because we had to become convinced of a deep fact, which paradoxically was well known, but on the other hand was essentially removed by the scientific community (see however [2]). Namely, the fact that classical electrodynamics, when extended to microscopic bodies, is radically different from the macroscopic one, due to the fact that it requires negative bare masses, and so leads for generic initial data to absurd runaway solutions; and these can be removed by some prescription *à la* Dirac, which leads to a conceptually different theory, exhibiting nonlocal aspects. By the way, as foreseen by Nelson too, this turns out to have strong implications on the relations between classical physics and quantum mechanics.

These facts will be illustrated, though in a rather sketchy way, in the present notes. In this introduction we would like, however, to mention some authorities, in support of the significance of our studies. Indeed, if it turns out that microscopic classical electrodymanics has so many and great complications, why at all to insist on it and not just abandon it ? Here the quotation we like most is taken from the beautiful chapter of the Feynman lectures devoted to the electromagnetic mass (page 28.10), namely: "*....it might be a waste of time to straighten out the classical theory, because it could turn out that in quantum electrodynamics the difficulties will disappear or may be resolved in some other fashion. But the difficulties do not disappear in quantum electrodynamics.... The Maxwell theory still has the difficulties after the quantum mechanics modifications are made.*". Essentially the same opinion is expressed in some works of Dirac and Haag; in particular, in the introduction to his famous work on the selfinteraction of the electron [3] Haag says: " *This often discussed subject will be here reconsidered in light of the difficulties of quantum field theory*". For other relevant contributions see [4], [5] (see also [6]).

RIGOROUS RESULTS FOR THE LINEARIZED MAXWELL–LORENTZ SYSTEM

We give here a short review of some results recently found by two of us (see [7]), which somehow constitute the culmination of a long line of research (see [8], [9] and also [10] from a numerical point of view, and [11], [12] from an analytical point of view; however, in all these works the bare mass was kept positive, because it took much time to understand the necessity of negative bare masses for microscopic particles). Other aspects of the problem were discussed in the very recent work [13]; finally, in a work still in preparation [14] for the first time the closed equation was found which is obeyed by the field in the limit in which the cutoffs are removed.

The aim is thus to have information on the dynamics when the regularization is removed. In fact, the preliminary results illustrated here are restricted to the projection of the solution on the particle variables, i.e. concern only the motion of the particle, which is induced by the solution of the full system. However, the main drawback is that such results are obtained not for the original Maxwell–Lorentz system, but only for its linearization about the equilibrium point defined by the particle at rest at the origin and vanishing field (a mechanical linear restoring force acting on the particle is also included); in particular, within such an approximation (often called the "dipole approximation") the system is nonrelativistic; finally (but this is an unessential technical point), the regularization is performed not by imposing cutoffs on the fields, but by considering, as we use to say, a "fat" particle, i.e. by substituting the δ function appearing in the charge and current densities by a smooth charge distribution (or form factor) ρ. Because of this, the limit in which the cutoffs are removed will be called here the "point limit".

We chose to work in the Coulomb gauge. Let us recall that in such a gauge the dynamically relevant parts of the fields depend just on the vector potential $\mathbf{A} = \mathbf{A}(\mathbf{x}, t)$ (subject to the constraint $\operatorname{div} \mathbf{A} = 0$), and that the complete Maxwell–Lorentz system takes the form

$$
\frac{1}{c^2} \ddot{\mathbf{A}} - \triangle \mathbf{A} = \frac{4\pi e}{c} \Pi(\dot{\mathbf{q}} \delta_q) ,
$$

$$
\frac{\mathrm{d}}{\mathrm{d}t} \left(\frac{m_0 \dot{\mathbf{q}}}{\sqrt{1 - (\dot{q}/c)^2}} \right) = -\frac{e}{c} \dot{\mathbf{A}}(\mathbf{q}, t) + \frac{e}{c} \dot{\mathbf{q}} \times \operatorname{rot} \mathbf{A}(\mathbf{q}, t) - \alpha \mathbf{q} ,
\tag{1}
$$

where $\alpha > 0$ is a constant characterizing the external linear force, while $\delta_{\mathbf{q}}$ is the delta function translated by $\mathbf{q}$ with respect to the δ function centered at the origin, i.e. $\delta_q(\mathbf{x}) := \delta(\mathbf{x} - \mathbf{q})$. Finally Π is the projector on the subspace of vector fields with vanishing divergence, i.e. $\Pi(\mathbf{j})$ is the so called transversal part of the current $\mathbf{j}$, often denoted by $\mathbf{j}_{\mathrm{tr}}$. Now we take the so called dipole approximation, i.e. linearize the system about the equilibrium point $\mathbf{q} = \dot{\mathbf{q}} = \mathbf{A} = \dot{\mathbf{A}} = 0$, and regularize it by substituting the delta function by a smooth normalized (in L^1) charge distribution ρ. So we obtain the system

$$
\frac{1}{c^2} \ddot{\mathbf{A}} - \triangle \mathbf{A} = \frac{4\pi e}{c} \Pi(\dot{\mathbf{q}} \rho) ,
$$

$$
m_0 \ddot{\mathbf{q}} = -\frac{e}{c} \int_{\mathbb{R}^3} \rho(\mathbf{x}) \dot{\mathbf{A}}(\mathbf{x}, t) \mathrm{d}^3 \mathbf{x} - \alpha \mathbf{q} ,
\tag{2}
$$

which is the one actually studied; the appropriate configuration space $\mathcal{Q}_0$ is

$$
\mathcal{Q}_0 := \mathcal{S}_*(\mathbb{R}^3, \mathbb{R}^3) \oplus \mathbb{R}^3 \ni (\mathbf{A}, \mathbf{q}) ,
$$

where $\mathcal{S}_*$ denotes the subset of the vector fields belonging to the Schwartz space $\mathcal{S}$ (C^∞ functions decaying at infinity faster than any power) having vanishing divergence.

Concerning the form factor ρ one assumes that it is C^∞, decays at least exponentially fast at infinity, is spherically symmetric, and its Fourier transform $\hat\rho$, defined by

$$\rho(\mathbf{x}) = \frac{1}{(2\pi)^{3/2}} \int_{\mathbb{R}^3} \hat\rho(\mathbf{k}) e^{i\mathbf{k}\mathbf{x}} d^3\mathbf{k} \;,$$

is everywhere non–vanishing (*i.e.* $\hat\rho(\mathbf{k}) \neq 0$, $\forall \mathbf{k} \in \mathbb{R}^3$); finally, in order to simplify the discussion of the point limit one assumes that ρ has the form

$$\rho_R(x) := \frac{1}{R^3} \mathcal{D}\left(\frac{x}{R}\right) \;, \tag{3}$$

where $\mathcal{D}$ is a positive, normalized (in L^1) function, and $R > 0$ a parameter characterizing the "radius" of the particle. For any $R > 0$ the Cauchy problem for system 2 is well posed in the phase space $\mathcal{Q}_0 \times \mathcal{Q}_0$, and the problem is to discuss the limit of the particle's motion as the "radius" R of the charge distribution tends to zero.

The first result is obtained for the special class of initial data such that the particle is at rest in some position $q_0 \neq 0$, and the field vanishes. One has the

Proposition. *Having fixed $m_0 > 0$ and $R > 0$ denote by $\mathbf{q}^{(R)}(t)$ the solution of 2 with form factor 3, corresponding to initial data $\mathbf{A}_0 = \dot{\mathbf{A}}_0 = \dot{\mathbf{q}}_0 = 0$, and $\mathbf{q}_0 \neq 0$. Assume that $\mathbf{q}^{(R)}(\cdot)$ converges weakly to a distribution $\mathbf{q}(\cdot)$ as $R \to 0$. Then there exists a constant vector $\bar{\mathbf{q}}$ such that $\mathbf{q}(t) \equiv \bar{\mathbf{q}}$.*

This means that, for fixed positive bare mass, the limit dynamics of the point particle is trivial, if it exists. This result would be obvious for physicists of the old generation, and should be obvious even for undergraduates that are familiar with the Feynman lectures on physics, where he explains that there is an electromagnetic contribution to mass, which diverges in the point limit;[1] in fact, the above proposition just proves that a point particle is unaffected by the presence of a force $(-\alpha\mathbf{q}_0)$ no matter how large it is, so that it behaves as if its actual mass were infinite. From the mathematical point of view this is seen as follows. The proof of all results discussed here is based on the use of a representation formula for the solution of the Cauchy problem of 2 through normal modes, and it turns out that in such a formula the bare mass m_0 appears always summed to the quantity

$$m_{\mathrm{em}} := \frac{32}{3}\pi^2 \frac{e^2}{c^2} \int_0^\infty |\hat\rho(k)|^2 dk = \frac{1}{R}\left[\frac{32}{3}\pi^2 \frac{e^2}{c^2} \int_0^\infty \left|\hat{\mathcal{D}}(k)\right|^2 dk\right] \;, \tag{4}$$

which is just the electromagnetic mass corresponding to the given charge distribution; and this is in agreement with the expectation that the particle should behave as if

---

[1] This is seen in the simplest way (see the quoted chapter of Feynman's book) as follows. Consider a "fat" particle, in the form of a sphere of radius R, in uniform rectilinear motion with velocity v. Compute by the familiar formulae of retarded potentials the fields $\mathbf{E}$ and $\mathbf{B}$ "created" by the particle, and the corresponding Poynting vector, proportional to $\mathbf{E} \times \mathbf{B}$, and integrate in the domain outside the particle, thus obtaining the total momentum $\mathbf{p}_{em}$ of the electromagnetic field dragged along by the particle. In the nonrelativistic approximation one immediately finds

$$\mathbf{p}_{em} = \tfrac{2}{3} m_{em} v \;,$$

in terms of the electromagnetic mass m_{em} defined by $m_{em} = \frac{e^2}{Rc^2}$.

its experimental mass m were the sum of m_0 and m_{em}: $m = m_0 + m_{em}$. Notice that $m_{em} = m_{em}(R) \to +\infty$ as $R \to 0$, so that the effective mass also $\to \infty$ as $R \to 0$ if the bare mass m_0 is kept fixed.

So, in order to obtain meaningful results, one has to renormalize mass, i.e. to consider the bare mass m_0 as a function of R, by putting

$$m_0(R) := m - m_{em}(R)\,, \tag{5}$$

where m is a fixed phenomenological parameter to be identified with the physical mass of the particle, and which does not appear at all in the original Maxwell–Lorentz system. This point shoul be particularly emphasized. If one considers the original regularized Maxwell–Lorentz system, and takes the limit $R \to 0$ with the prescription 5, one is in fact defining a new system, which can be said to be just suggested by the original one. This is particularly evident, if one remarks that, according to 5 there exists a threshold radius $\bar{R}$ such that

$$m_0(\bar{R}) = 0\,, \quad m_0(R) < 0 \quad \text{if } R < \bar{R}\,,$$

and $m_0(R) \to -\infty$ as $R \to 0$. The critical radius $\bar{R}$ is called the "classical electron radius" Concerning the behaviour of the system in the limit $R \to 0$ one has the

Proposition. *Consider the Cauchy problem for system 2 with form factor ρ_R given by 3, m_0 given by 5, and initial data $\mathbf{A}_0 = \dot{\mathbf{A}}_0 = \dot{\mathbf{q}}_0 = 0$, and $\mathbf{q}_0 \neq 0$; let $\mathbf{q}^{(R)}(t)$ be the corresponding particle's motion. Then, for any $T > 0$, as $R \to 0$ the function $\mathbf{q}^{(R)}(.)$ converges in $C^1([-T,T], \mathbb{R}^3)$ to a non constant function.*
So, the particular solution corresponding to the above initial data has a point limit which is nontrivial, provided mass is renormalized.

Consider now the case where the initial particle's velocity too is different from zero; this is a nontrivial generalization, because it requires that one be familiar with the notion of the field "adapted" to the given initial velocity. This is illustrated by the following result which shows that, with $\dot{\mathbf{q}}_0 \neq 0$, if one takes a vanishing initial field the trajectory of the particle turns out to have no sensible point limit. Precisely one has the

Proposition. *Consider the Cauchy problem for system 2 with form factor ρ_R given by 3, m_0 given by 5, and initial data $\mathbf{A}_0 = \dot{\mathbf{A}}_0 = \mathbf{q}_0 = 0$, and $\dot{\mathbf{q}}_0 \neq 0$; let $\mathbf{q}^{(R)}(t)$ be the corresponding particle's motion. Then, for any $T > 0$, as $R \to 0$ one has*

$$|\mathbf{q}^{(R)}(t)| \to \infty\,, \quad \forall t \in [-T,T] \backslash \mathcal{N}\,,$$

where $\mathcal{N}$ is a finite (possibly empty) set.
It is not difficult to prove that the same happens also if one takes as initial data for the field any regular function (i.e. without singularities).

This result should be not completely astonishing, because it is very well known (and was also recalled above) that, in the case of a uniform motion, a particle drags with it a field which in the case of a point particle has a singularity at the particle's position. So, it seems natural to study the particular class of initial data such that a particle with a non vanishing velocity is accompanied by such a "proper or adapted field".[2] In order to give a precise statement we recall that[15,11] in the non–linear

---

2   Within our group, the awareness of this elementary fact was obtained through the work [8], where it was shown that such a field produces a Lorentz force vanishing exactly at the instantaneous particle position; by the way, it is just this fact that allows dynamically for the existence of uniform

Maxwell–Lorentz system a free particle can move uniformly with velocity $\mathbf{v}$, only if accompanied by a field $\mathbf{X}$, which is defined as the solution of equation

$$\triangle \mathbf{X} - \frac{1}{c^2} \sum_{i,l=1}^{3} v_i v_l \frac{\partial^2}{\partial x_i \partial x_l} \mathbf{X} = -4\pi \frac{e}{c} \Pi(\rho \mathbf{v})$$

vanishing at infinity. In the dipole approximation, i.e. after a linearization in the velocity and in the field, such an equation reduces to

$$\triangle \mathbf{X} = -4\pi \frac{e}{c} \Pi(\rho \mathbf{v}) . \tag{6}$$

We denote by $\mathbf{X_v}$ the unique solution of equation 6 vanishing at infinity, and study the point limit of the solutions of the Cauchy problem corresponding to initial data of the form $\dot{\mathbf{q}}_0 \neq 0$, $\mathbf{A}_0 = \mathbf{X}_{\dot{\mathbf{q}}_0}$. Initial data of the form $(\mathbf{q}_0, \dot{\mathbf{q}}_0, \mathbf{X}_{\dot{\mathbf{q}}_0}, 0)$ will be called of "congruent type". One has then the

Proposition. *Consider the Cauchy problem for system 2 with form factor ρ_R given by 3, m_0 given by 5, and initial data $(\mathbf{q}_0, \dot{\mathbf{q}}_0, \mathbf{A}_0, \dot{\mathbf{A}}_0) = (\mathbf{q}_0, \dot{\mathbf{q}}_0, \mathbf{X}_{\dot{\mathbf{q}}_0}, 0)$; let $\mathbf{q}^{(R)}(t)$ be the corresponding particle's motion. Then, for any $T > 0$, as $R \to 0$ the function $\mathbf{q}^{(R)}(\cdot)$ converges in $C^1([-T, T], \mathbb{R}^3)$ to a non constant function.*

Analogously, for the case of general initial data for the field, one has the following

Theorem. *Consider the Cauchy problem for system 2 with form factor ρ_R given by 3, m_0 given by 5, and initial data $(\mathbf{q}_0, \dot{\mathbf{q}}_0, \mathbf{A}_0, \dot{\mathbf{A}}_0) = (\mathbf{q}_0, \dot{\mathbf{q}}_0, \mathbf{X}_{\dot{\mathbf{q}}_0} + \mathbf{A}_0', \dot{\mathbf{A}}_0)$ with $(\mathbf{A}_0', \dot{\mathbf{A}}_0) \in \mathcal{S}_*(\mathbb{R}^3, \mathbb{R}^3) \times \mathcal{S}_*(\mathbb{R}^3, \mathbb{R}^3)$; let $\mathbf{q}^{(R)}(t)$ be the corresponding particle's motion. Then, for any $T > 0$, as $R \to 0$ the function $\mathbf{q}^{(R)}(\cdot)$ converges in $C^0([-T, T], \mathbb{R}^3)$. Moreover, the limiting particle's motion depends continuously on*

$$(\mathbf{q}_0, \dot{\mathbf{q}}_0, \mathbf{A}_0', \dot{\mathbf{A}}_0) \in \mathbb{R}^3 \times \mathbb{R}^3 \times \mathcal{S}_*(\mathbb{R}^3, \mathbb{R}^3) \times \mathcal{S}_*(\mathbb{R}^3, \mathbb{R}^3) .$$

So the above theorem shows that the dynamics of a point particle is well defined in the point limit, at least for initial fields which are regular modifications of congruent fields. Moreover, the Cauchy problem is well posed in the sense of Hadamard. By the way this existence result could be extended to the case of initial fields $(\mathbf{A}_0', \dot{\mathbf{A}}_0)$ which are only C^0, and decay at infinity faster than $r^{-3/2}$.

In the case of congruent initial data it is possible to calculate explicitly the point limit of the solution of the Maxwell-Lorentz system. Let us introduce some notations: define

$$\omega_0^2 := \frac{\alpha}{m} , \quad \tau := \frac{2}{3} \frac{e^2}{mc^3} ,$$

consider the equation

$$\tau \nu^3 - \nu^2 - \omega_0^2 = 0 , \tag{7}$$

and denote by ν_r, $\nu_+ = \nu_3 + i\nu_2$, $\nu_- = \nu_3 - i\nu_2$ its three solutions $(\nu_2, \nu_3 > 0)$. One has then the

---

motions (a circumstance that Abraham used to qualify as "consistency of electrodynamics with the inertia principle"). Moreover, numerical integrations with a positive bare mass and a cutoff on the field showed that, if the initial data are nonvanishing particle velocity and vanishing field, then the solution of the complete system is such that the proper field of the particle tends to be created. As this fact was discovered by Lia Forti, in our jargon we use to call the proper field "il campo della Lia", i.e. Lia's field.

Theorem. *The point limit of the particle's motion corresponding to the solution of the Cauchy problem for the linearized Maxwell–Lorentz system 2 with initial data of congruent type* $(\mathbf{q}_0, \dot{\mathbf{q}}_0, \mathbf{A}_0, \dot{\mathbf{A}}_0) = (\mathbf{q}_0, \dot{\mathbf{q}}_0, \mathbf{X}_{\dot{q}_0}, 0)$ *is given by*

$$
\mathbf{q}(t) = \begin{cases} e^{-\nu_3 t}[\mathcal{A}_1^+ \cos(\nu_2 t) + \mathcal{A}_2^+ \sin(\nu_2 t)] + \mathcal{A}_3^+ e^{\nu_r t} , & \text{if } t > 0 \\ e^{\nu_3 t}[\mathcal{A}_1^- \cos(\nu_2 t) + \mathcal{A}_2^- \sin(\nu_2 t)] + \mathcal{A}_3^- e^{-\nu_r t} , & \text{if } t < 0 \end{cases} , \tag{8}
$$

where $\mathcal{A}_1^\pm$, $\mathcal{A}_2^\pm$, $\mathcal{A}_3^\pm$ *are real vector constants depending on the initial data and on e, m,* ω_0*. Moreover one has the asymptotics*

$$
\begin{cases} \nu_r = \dfrac{\omega_0}{\epsilon} + O(\epsilon) \\ \nu_2 = \omega_0 + O(\epsilon^2) \\ \nu_3 = \omega_0 \epsilon/2 + O(\epsilon^2) \end{cases} , \qquad \begin{cases} \mathcal{A}_1^\pm = \mathbf{q}_0 + O(\epsilon^2) \\ \mathcal{A}_2^\pm = \dfrac{\dot{\mathbf{q}}_0}{\omega_0} + O(\epsilon^2) \\ \mathcal{A}_3^\pm = O(\epsilon) \end{cases} \tag{9}
$$

for $\epsilon \to 0$*, in terms of the dimensionless parameter* $\epsilon := \omega_0 \tau$*.*

We can now compare 8 with the solutions of the celebrated Abraham–Lorentz (AL) equation

$$
m\tau \dddot{\mathbf{q}} = m\ddot{\mathbf{q}} + \alpha \mathbf{q} , \tag{10}
$$

which was known since a century, but the deduction of which from the Maxwell–Lorentz system should be considered as a heuristic one (see [16] or [17]). In this connection one has the

Theorem. *The point limit 8 of the particle's motion in the Maxwell–Lorentz system is also a solution of the problem*

$$
\begin{aligned} -m\tau \dddot{\mathbf{q}} &= m\ddot{\mathbf{q}} + \alpha \mathbf{q} , \quad t < 0 , \\ m\tau \dddot{\mathbf{q}} &= m\ddot{\mathbf{q}} + \alpha \mathbf{q} , \quad t > 0 . \end{aligned}
$$

So, for initial data of congruent type, the particle's motion satisfies exactly the AL equation for positive times, and the corresponding one with the substitution $\tau \to -\tau$ for negative times. Notice that the fact that different equations occur for negative and positive times is not a particular feature of the present model; a classical case where this happens is that of the Boltzmann equation, and a simple and enlightening mechanical model where this phenomenon occurs was given by Lamb at the beginning of the century.

A HEURISTIC APPROACH TO THE ABRAHAM–LORENTZ EQUATION

It was shown above that the motion of the particle in the point limit for the linearized Maxwell–Lorentz system with the prescription of mass renormalization, for a special class of initial data satisfies exactly the Abraham–Lorentz equation 10. Now, the rigorous deduction reported above is based on an explicit representation formula using of the normal modes of the complete system, which does not allow one to really understand what is going on. So we intend to devote the present section to the illustration of a heuristic deduction, which allows to see by eyes what is occurring (see [18]). A rigorous treatment along similar lines can be found in [13], where it was also shown how the initial acceleration for the AL equation is defined by the initial field for the complete system.

In order to study the linearized Maxwell–Lorentz system 2, it is convenient to introduce the space Fourier transform of the vector potential by

$$\mathbf{A}(\mathbf{x},t) = \frac{1}{(2\pi)^{3/2}} \int_{\mathbb{R}^3} \hat{\mathbf{A}}(\mathbf{k},t) e^{i\mathbf{k}\cdot\mathbf{x}} d^3\mathbf{k} \ ,$$

so that the system becomes

$$m_0 \ddot{\mathbf{q}}(t) = -\frac{e}{c} \int \hat{\rho}^*(\mathbf{k}) \frac{\partial \hat{\mathbf{A}}}{\partial t}(\mathbf{k},t) \, d^3k - \alpha\mathbf{q} + \mathbf{F}_{\text{ext}}(\mathbf{q})$$

$$\frac{d^2}{dt^2}\hat{\mathbf{A}}(\mathbf{k},t) + k^2 c^2 \hat{\mathbf{A}}(\mathbf{k},t) = 4\pi \, e \, c \, \hat{\rho}(\mathbf{k}) \left[\dot{\mathbf{q}}(t) - \frac{(\dot{\mathbf{q}}(t)\cdot\mathbf{k})\mathbf{k}}{k^2}\right] \ ,$$

(11)

where $*$ denotes complex conjugate, and a generic external force $\mathbf{F}_{\text{ext}}$ additional to the linear one $-\alpha\mathbf{q}$ has also been introduced. By the variation of constants formula, the second equation can also be written in the integral form

$$\hat{\mathbf{A}}(\mathbf{k},t) = \hat{\mathbf{A}}_{\text{hom}}(\mathbf{k},t) + \frac{e\hat{\rho}(\mathbf{k})}{k} \int_0^t \left[\dot{\mathbf{r}}(t') - \frac{(\dot{\mathbf{r}}(t')\cdot\mathbf{k})\mathbf{k}}{k^2}\right] \sin[kc(t-t')]dt' \ ,$$

where $\hat{\mathbf{A}}_{\text{hom}}$ is the solution of the corresponding homogeneous problem. By substitution in the equation for the particle, exchanging the order of the integrations and performing the integrations on the angular variables (by exploiting the spherical simmetry of the form factor) one then obtains an integro–differental equation for the particle, namely

$$m_0\ddot{\mathbf{q}}(t) = -\frac{32\pi^2}{3} e^2 \int_0^t dt' \dot{\mathbf{q}}(t') \int_0^\infty dk \, k^2 \, |\tilde{\rho}(k)|^2 \cos[kc(t-t')]$$
$$+ \mathbf{F}_{\text{hom}} - \alpha\mathbf{q}(t) + \mathbf{F}_{\text{ext}}(\mathbf{q}) \ ,$$

where $\mathbf{F}_{\text{hom}}$ represents the Lorentz force on the particle due to the free evolution of the initial field; in particular, $\mathbf{F}_{\text{hom}}$ vanishes if the initial field does. From now on we wiil simply denote $\mathbf{F}_{\text{ext}} + \mathbf{F}_{\text{hom}}$ by $\mathbf{F}_{\text{ext}}$.

An equivalent interesting form for the particle equation is obtained by performing two integrations by parts on the variable t'. Indeed this leads to

$$(m_0 + m_{\text{em}})\ddot{\mathbf{q}}(t) = -\alpha\mathbf{q} + \mathbf{F}_{\text{ext}} +$$
$$+ \frac{32\pi^2 e^2}{3c^3} \int_0^t dt'\dddot{\mathbf{q}}(t') \int_0^\infty dk \, \left|\tilde{f}(k)\right|^2 \cos[kc(t-t')] +$$
$$- \frac{32\pi^2 e^2}{3c^3} \int_0^\infty \left|\tilde{f}(k)\right|^2 \left\{\dot{\mathbf{q}}(0) \, k \, \sin(kct) - \frac{\ddot{\mathbf{q}}(0)}{c}\cos(kct)\right\} dk$$

where it appears that the bare mass m_0 occurs only summed to the electromagnetic mass m_{em} defined by 4.

We now concentrate our attention on a particularly convenient form factor, namely the sharp one corresponding to a cutoff $\mathcal{K}$ in momentum space, defined by

$$\tilde{\rho}(\mathbf{k}) = \frac{1}{(2\pi)^{3/2}} \, \theta(\mathcal{K} - k)$$

where θ is the usual step function; this is somehow equivalent to considering an extended particle of spatial dimension (or "radius") R with $R \simeq \mathcal{K}^{-1}$. Correspondingly, we affix an index $(\mathcal{K})$ to the quantities of interest, and we will be looking for the limit as $\mathcal{K} \to \infty$. In particular, for the electromagnetic mass one has then

$$m_{\text{em}}(\mathcal{K}) = \frac{4\,e^2}{3\,\pi\,c^2}\,\mathcal{K}\,.$$

Introduce now time Fourier transforms, so that one has for example, for the particle acceleration $\mathbf{a}^{(\mathcal{K})}(\mathbf{t}) \equiv \ddot{\mathbf{q}}^{(\mathcal{K})}(\mathbf{t})$,

$$\mathbf{a}^{(\mathcal{K})}(t) = \frac{1}{(2\pi)^{1/2}} \int_{-\infty}^{\infty} \tilde{\mathbf{a}}^{(\mathcal{K})}(\omega)\,e^{i\omega t}\,d\omega\,,$$

while the time Fourier transform of the external force is given by

$$\hat{\mathbf{F}}_{\text{ext}}^{(\mathcal{K})}(\omega) = \frac{1}{(2\pi)^{1/2}} \int_{-\infty}^{\infty} \mathbf{F}_{\text{ext}}(\mathbf{q}^{(\mathcal{K})}(t), t)\,e^{i\omega t}\,dt\,,$$

and turns out to be a functional of the particle motion $\mathbf{q}^{(\mathcal{K})}(t)$.

By computing some trivial integrals, the equation for the particle then takes the form

$$\tilde{\mathbf{a}}^{(\mathcal{K})}(\omega) = \frac{\tilde{\mathbf{F}}_{\text{ext}}^{(\mathcal{K})}(\omega)}{m_0 + m_{\text{em}}(\mathcal{K}) + \frac{2e^2}{3\pi c^3}\,\omega \log \frac{\omega - c\,\mathcal{K}}{\omega + c\,\mathcal{K}} + \frac{\alpha}{\omega^2}}\,. \tag{12}$$

By the way, it is of interest to remark that for $\omega = 0$ equation 12 reduces to

$$\tilde{\mathbf{a}}^{(\mathcal{K})}(0) = \frac{\tilde{\mathbf{F}}_{\text{ext}}^{(\mathcal{K})}(0) - \alpha\tilde{\mathbf{q}}(0)}{m}$$

with $m = m_0 + m_{\text{em}}$; this relation is consistent with a kind of "correspondence principle" because it is analogous to the mechanical equation $m\ddot{\mathbf{q}} = \mathbf{F}_{\text{ext}}$ for the integrated quantities $\tilde{\mathbf{a}}^{(\mathcal{K})}(0) = \int_{-\infty}^{+\infty} \mathbf{a}^{(\mathcal{K})}(t)dt$ and $\hat{\mathbf{F}}_{\text{ext}}^{(\mathcal{K})}(0) = \int_{-\infty}^{+\infty} \mathbf{F}_{\text{ext}}(\mathbf{q}^{(\mathcal{K})}(t))dt$.

Equation 12 for the Fourier transform of the particle acceleration allows in a rather simple way, through a control of its poles, to obtain an information on the function $\ddot{\mathbf{q}}(t)$ itself. Clearly there are poles due to the particular form chosen for the external force, but the most relevant ones are those that are independent of it, namely those corresponding to the zeroes of the denominator, i.e. the zeroes of the function

$$\zeta^{(\mathcal{K})}(\omega) = m_0 + m_{\text{em}}(\mathcal{K}) + \frac{2e^2}{3\pi c^3}\,\omega \log \frac{\omega - c\,\mathcal{K}}{\omega + c\,\mathcal{K}} + \frac{\alpha}{\omega^2}\,.$$

One easily proves that:
* if $m_0 > 0$, then the function $\zeta(\omega)$ has two real zeroes $\pm\bar{\omega}$, with $\bar{\omega} > c\mathcal{K}$;
* if $m_0 < 0$, then the function $\zeta(\omega)$ has two conjugate imaginary zeroes.

This result allows one to deduce several interesting consequences. The first one is that if one goes to the point limit $\mathcal{K} \to \infty$ while keeping the bare mass positive, then the system presents oscillations of increasingly high frequencies, diverging with the cutoff. The second remark concerns the case in which the point limit is taken while performing a coherent renormalization procedure, i.e. one considers, for any cutoff $\mathcal{K}$, a corresponding bare mass $m_0(\mathcal{K})$ such that $m_0(\mathcal{K}) + m_{\text{em}}(\mathcal{K}) = m$ where m is a fixed positive mass; in particular this requires to take $m_0 < 0$ if $\mathcal{K} > \bar{k}$ where $\bar{k} \simeq \bar{R}^{-1}$,

$\bar{R} = e^2/mc^2$ being the so called "classical radius". In such a case one has that the real poles escape to infinity, reaching it for $\mathcal{K} = \bar{k}$; then they reappear on the imaginary axis, tending to $\pm i\tau^{-1}$, where τ is the familiar parameter defined by $\tau = e^2/mc^3$. In particular this has the fundamental consequence that for large enough cutoff the solutions for generic initial data have runaway character, i.e. diverge exponentially fast with time t as $t \to +\infty$, due to the presence of the pole in the lower half plane, and also for $t \to -\infty$, due to the other pole in the upper half plane.

In the literature it appears that the existence of oscillatory motions for "fat" particles with positive bare mass was put into evidence particularly by Bohm and Weinstein [19], after the classical works of Schott. [20] The idea of such authors was to find a relation between the bare mass m_0 and the form factor in order to have exactly periodic (and thus non radiating) solutions for the complete system; moreover they were stressing that such oscillations *"do not constitute a form of instability, as does the self-acceleration of the Dirac classical electron"*. What we have shown here for the case of the sharp cutoff is that the presence of nonradiating oscillatory motions for "fat" particles (i.e. for particles with $R > \bar{R}$) is just a premonition of the appearence of runaway motions for "thin" particles (i.e. for particles with $R < \bar{R}$), and in particular also in the point limit. The classical radius $\bar{R}$ appears thus as a *critical radius* or a threshold: for "thin" particles of radius $R < \bar{R}$ the Maxwell–Lorentz system presents in general absurd runaway solutions which are not present for larger radii. In fact for $R \gg \bar{R}$ the motions are just the smooth ones to which we are accustomed in macroscopic electrodynamics, and the transition to the absurd microscopic electrodymanics occurs with the premonition of the high frequency nonradiating oscillations of Schott and Bohm–Weinstein.

So, microscopic classical electrodynamics is qualitatively completely different from the macroscopic one, and one could be tempted to decide to throw it out, because it could be *"a waste of time to try to straighten it"*. Another possible attitude is to take the Maxwell–Lorentz equations seriously even for "thin" particles, and to try to give sense to them by restricting the attention to a special class of initial data, as was suggested by Dirac and will be recalled in the next section. This is the attitude we are trying to pursue.

As a final comment we indicate how equation 12 allows one to understand qualitatively that the particle motion $\mathbf{q}(t)$ in the point limit has to satisfy approximately the Abraham–Lorentz equation for times larger than the characteristic time τ. Indeed, defining

$$\tilde{\mathbf{a}}(\omega) = \lim_{\mathcal{K} \to +\infty} \tilde{\mathbf{a}}^{(\mathcal{K})}(\omega) ,$$

one finds

$$\tilde{\mathbf{a}}(\omega) = \frac{\tilde{\mathbf{F}}_{\text{ext}}(\omega)}{m + im\tau\omega\,\text{Sign}[\text{Im}(\omega)] + \frac{\alpha}{\omega^2}} . \tag{13}$$

So the function $\tilde{\mathbf{a}}(\omega)$ has a cut on the real axis and two poles of the first order at the points $\pm i\tau^{-1}$. Thus, for positive times the upper pole gives an exponentially small contribution which can be neglected after a convenient time. In such a limit equation 13 reduces to

$$(m - i\omega m\tau + \frac{\alpha}{\omega^2})\tilde{\mathbf{a}}(\omega) = \tilde{\mathbf{F}}_{\text{ext}}(\omega) ,$$

which, by performing the inverse Fourier transformation, gives the Abraham–Lorentz equation

$$m\ddot{\mathbf{q}} - m\tau\dddot{\mathbf{q}} = \mathbf{F}_{\text{ext}}(\mathbf{q}) - \alpha\mathbf{q} .$$

In an analogous way, for negative sufficiently large times one gets approximately a similar equation with $-\tau$ instead of $+\tau$, namely

$$m\ddot{\mathbf{q}} + m\tau\dddot{\mathbf{q}} = \mathbf{F}_{\text{ext}}(\mathbf{q}) - \alpha\mathbf{q} \ ,$$

in agreement with the time reversal symmetry of the full problem.

We do not have time to report here on some numerical computations (see [18]) which illustrate very well the role eof the electromagnetic mass as a function of the cutoff.

QUANTUM–LIKE ASPECTS OF CLASSICAL ELECTRON THEORY

In his paper of the year 1938, where the relativistic version of the Abraham–Lorentz equation was introduced, Dirac[21] pointed out that generic solutions of such an equation are absurd, presenting the so called runaway character. Apparently, the scientific community has not yet really digested this fact, which in our opinion is of great importance; so we concentrate now on it and on its qualitative implications.

The runaway solutions were already mentioned in the previous section. But the simplest example in which they occur is the nonrelativistic Abraham–Lorentz equation for the free particle; indeed, in such a case the equation is a closed one for the acceleration $\mathbf{a}(t)$, namely (for $t > 0$) $\tau\dot{\mathbf{a}} = \mathbf{a}$. The general solution is then $\mathbf{a}(t) = \mathbf{a}_0 e^{t/\tau}$, where $\mathbf{a}_0$ is the initial acceleration. So the solution for the free particle explodes (i.e. diverges exponentially) for positive times, unless one takes the initial datum $\mathbf{a}_0 = 0$, which gives the "physical solution" $\mathbf{a}(t) = 0$. It is rather easy to understand by qualitative arguments that runaway solutions occur generalically for the AL equation with an external force (see [22] and also [23]). Moreover, as should be clear from the discussion of the previous section, the generic runaway character is a property of the complete system and not just of the particle equation. Thus, for generic initial data microscopic classical electrodynamics is absurd.

One has then the following general mathematical problem. Given an initial "mechanical state" $(\mathbf{q}_0, \mathbf{v}_0)$, does there exist an initial acceleration $\mathbf{a}_0$ (in the case of the AL equation) or an initial field (in the case of the complete system) such that the corresponding solution does not have runaway character? For example, in the case of a scattering problem, the nonrunaway character could be defined by the "final condition" $\mathbf{a}(t) \to 0$ as $t \to \infty$. In geometrical terms (considering for example of the AL equation) one asks whether there exists in the complete phase space $(\mathbf{q}, \mathbf{v}, \mathbf{a})$ a surface constituted of trajectories not having runaway character. Moreover, one has the problem whether such a surface can be expressed in the explicit form $\mathbf{a} = \mathbf{a}(\mathbf{q}, \mathbf{v})$ (one speaks in such a case of "uniqueness" of the runaway solutions, in the sense that the nonrunaway solution is then uniquely determined by the mechanical state). The idea of Dirac was to considee electrodynamics of point particles as defined only for initial data restricted to such a "physical surface".

From the mathematical point of view, the problem of the existence of the physical surface in the case of the AL equation was solved positively by Hale and Stokes[24] in the year 1962, for external forces of quite general a type. But in general there is no uniqueness. In fact this was already known long ago to Bopp[25] and Haag,[3] who gave an example with two solutions; but this was apparently forgotten, being not even mentioned in the most popular handbook on electron theory.[26] However, this nonuniqueness property was encountered again quite recently,[22] and even understood in terms of familiar concepts of the theory of dynamical systems. The case dealt with is the one–dimensional scattering of a particle by a barrier, and the nonuniqueness is described as follows: there exists an interval of energies for which there are any number of distinct nonrunaway solutions, that turn out to be divided into two classes. The first class corrsponds to solutions reflected by the barrier, and the second one to solutions transmitted beyond the barrier (for the same initial mechanical data!); so one is here in presence of a classical effect qualitatively analogous to the tunnel effect.

Now, is this fact, namely the occuring of a phenomenon qualitatively analogous to a quantum one already within the framework of classical electron theory, just an accidental fact? In our opinion it is not so. The deep reason is that classical electron theory in the sense of the "physical solutions" *à la* Dirac described above is not at all the standard classical electron theory. As we tried to show, classical electrodynamics loses sense for point particles (or more precisely alreday below the so called "classical radius"), and so one could just abandon it. Another possibility is to try to keep it in some extended sense, for example just by taking the Dirac's point of view of restricting it to the "physical surface". But notice that such a prescription is a kind of nonlocal one, because it is a prescription on the "final time", which mathematically leads to a kind of Sturm–Liouville problem. So it is not astonishing to find that such a new classical electrodynamics presents peculiar properties with respect to the standard (or macroscopic) one, which deals instead with "fat" particles, having dimensions larger the so–called classical radius.

We quote in passing two examples on which some of us are presently working. First,[27] on can show that the microscopic classical electrodynamics *à la* Dirac discussed here leads to violations of the Bell inequalities (think of the initial acceleration as the hidden parameter, and take into account the nonlocality – or passive locality in the sense of Nelson – implied by the Dirac prescription). Second,[28] in the problem of the scattering of electrons by an uncharged conducting sphere one finds diffraction patterns corresponding to electron wavelengths of the order of one hundredth of the de Broglie wavelength.

This second fact might appear astonishing at first sight, but is in fact quite obvious. Indeed, for what concerns the wavelike properties of the classical electron, they are just due to the wave properties of the field which is intrinsically related to the particle (think of the complete Maxwell–Lorentz system). For what concerns the appearence of an action of the order of one hundredth of Planck's constant, this is just due to the fact that in classical electron theory one has available the action e^2/c, which is just equal to $(1/137)\hbar$.

Now, the fact that the action e^2/c naturally enters in problems of classical electrodynamics is very well illustrated by the following example, where reference is made to elementary formulas that can be found in the Landau handbook but, as far as we know, were never understood previously in this way.[29] The problem is the scattering of an electron by a nucleus, with atomic number Z. In the Landau handbook one finds the formula for the emitted energy ΔE (computed in first order approximation by Larmor's formula along the trajectory of the purely mechanical approximation); the formula is a complicated one depending on Z. One also finds a formula for the emitted spectrum, which was known since always (particularly by Kramers) to decay exponentially at the high frequencies, so that there is a corresponding cut–off frequency $\bar{\omega}$; the formula for $\bar{\omega}$ is also a complicated one depending on Z. But for the ratio $\Delta E/\bar{\omega}$, which is an action, one finds the simple formula $\Delta E/\bar{\omega} = \frac{e^2}{c}(v/c)^2$, which is independent of Z and exhibits indeed e^2/c, namely essentially a hundredth of $\hbar$.

In conclusion, we hope we were able to show that classical electron theory can be extended to point (or microscopic) particles in an interesting way, leading to a theory which presents some aspects reminiscent of quantum aspects.

REFERENCES

[1] E. Nelson, *Quantum fluctuations*, Princeton U.P. (Princeton, 1985).
[2] S. Coleman, R.E. Norton, *Runaway modes in model field theories*, Phys. Rev. **125**, 1422–1428 (1962).
[3] R. Haag, *Z. Naturforsch.* **10 A**, 752 (1955).
[4] A. Kramers, Contribution to the 1948 Solvay Conference, in *Collected scientific papers*, North–Holland (Amsterdam, 1956), page 845.
[5] N.G. van Kampen, *Mat. Fys. Medd. K. Dansk. Vidensk. Selsk* **26**, 1 (1951).
[6] J. Kijowski, *Electrodynamics of moving particles*, GRG **26**, 167–201 (1994).
[7] D. Bambusi and D. Noja, *Classical electrodynamics of point particles and mass renormalization. Some preliminary results. Lett. Math. Phys.* in print (1966).
[8] L. Galgani, C. Angaroni, L. Forti, A. Giorgilli and F. Guerra, *Phys. Lett.* **A139**, 221 (1989).
[9] G. Arioli and L. Galgani, *Numerical studies on classical electrodynamics*, Phys. Lett. **A162**, 313–322 (1992).
[10] G. Benettin and L. Galgani, *J. Stat. Phys.* **27**, 153 (1982).
[11] D. Bambusi and L. Galgani, *Some rigorous results on the Pauli–Fierz model of classical electrodynamics*, Ann. Inst. H. Poincaré, Physique théorique, **58**, 155–171 (1993).
[12] D. Bambusi, *A Nekhoroshev–type theorem for the Pauli–Fierz model of classical elctrodynamics*, Ann. Inst. H. Poincaré, Physique théorique, **60**, 339–371 (1994).
[13] D. Bambusi, *A proof of the Lorentz–Dirac equation for charged point particles*, preprint (1988).
[14] A. Posilicano, D. Noja, *The wave equation with one point interaction and the linearized classical electrodynamics of a point particle*, preprint (1996)
[15] M. Abraham, *Ann. d. Phys.*, **10**, 105 (1903).
[16] H.A. Lorentz, *The theory of electrons*, Dover (New York, 1952); first edition 1909.
[17] G. Morpurgo, *Introduzione alla fisica delle particelle*, Zanichelli (Bologna, 1987).
[18] J. Sassarini, Doctoral thesis, University of Milano (1995).
[19] D. Bohm, M. Weinstein, *The self oscillations of a charged particle*, Phys. Rev, **74**, 1789–1798 (1948).
[20] G.A. Schott, *Phil. Mag.* **29**, 49–62 (1915).
[21] P.A.M. Dirac, *Proc. Royal Soc. (London)* A**167**, 148–168 (1938).
[22] A. Carati, P. Delzanno, L. Galgani, J. Sassarini, *Nonuniqueness properties of the physical solutions of the Lorentz–Dirac equation*, Nonlinearity, **8** 65–79 (1995).
[23] A. Carati and L. Galgani, *Asymptotic character of the series of classical electrodynamics, and an application to bremsstrahlung*, Nonlinearity **6**, 905–914 (1993).
[24] J.K. Hale and A.P. Stokes, *J. Math. Phys.* **3**, 70 (1962).
[25] F. Bopp, *Ann. der Phys.* **42**, 573–608 (1943).
[26] F. Rohrlich, *Classical charged particles*, Addison–Wesley (Reading, 1965).
[27] A. Carati, *Bell inequalities in classical electrodynamics*, in preparation.
[28] A. Carati, L. Galgani, *Wave–like properties of classical charged particles*, in preparation.
[29] A. Carati, unpublished.

CLIFFORD ALGEBRAS, SUPERCALCULUS, AND SPINNING PARTICLE MODELS

Jayme Vaz, Jr.[1] and Waldyr A. Rodrigues, Jr.[2]

[1]DFESCM - Instituto de Física "Gleb Wataghin"
Universidade Estadual de Campinas
CP 6165, 13081-970, Campinas, S.P., Brazil
[2]Departamento de Matemática Aplicada - IMECC
Universidade Estadual de Campinas
CP 6065, 13081-970, Campinas, S.P., Brazil

INTRODUCTION

Spinning particles have been studied by different authors and in different occasions (see [1] for some references), and actually the interest on spinning particles models revived specially due to the interest on spinning strings. Notwithstanding, producing a classical spinning particle model that after quantization gives Dirac equation has always been a very appealing idea which has been the subject of several interesting papers containing new physical insights and beautiful mathematics. In this paper we plan to study in details some aspects of spinning particle models, and in particular the one proposed by Barut and Zanghi [2], and some of its possible generalizations.

There are different approaches to spinning particles, and among those an important one [3] uses Grassmann variables – the so called supercalculus [4]. Although this approach has the advantage of being near the approach of some modern theories like supersymmetry, it has the obvious disadvantage of being unnatural, in the sense that it does not show the meaning of that inner degree of freedom which is the spin. On the other hand, Grassmann and Clifford algebras are closely related (see, v.g., [5]), but Clifford algebras have the advantage of being "more geometrical" than Grassmann algebra, in the sense that geometrical operations have a natural counterpart in terms of algebraic operations in Clifford algebra. In our opinion it is desirable to have a description of spinning particles only in terms of spacetime than in terms of an additional superspace structure. The models we plan to study are such that they have a natural description in terms of Clifford Algebras and Spacetime.

The algebraic structure suitable for our purposes is the Clifford algebra of spacetime, or spacetime algebra (STA), which we briefly review in sec.2. In order to use STA to describe a spinning particle, we must first say what we mean by a spinning particle. It is well-known that relativity imposes some dificults in defining a rigid body, which is a concept that restrict the internal degrees of freedom to be rotations, and consequently identifying spin with the angular momentum of such rotations. We shall assume a spinning particle to be a body of infinitesimal dimensions, but with internal rotation degrees of freedom. In this way, to a spinning particle we attach a tetrad to describe its internal motions. Such a tetrad will be called a moving frame, and STA enable us to relate it to a particular inertial reference frame frame, and consequently describe its internal motions. In other words, we model a spinning particle by a pair (ς, e_μ) where ς is a timelike curve pointing to the future and $\{e_\mu\}$ is a moving frame along ς such that $\varsigma_* = e_0$.

The spinning particle model we want to discuss is the so-called Barut and Zanghi (BZ) model [2]. Although already studied by us in other occasions [6, 7], we want now to pursue our analysis further by considering some cases we have not taken into account in those occasions. In fact, our analysis in other occasions have been limited to particular solutions of BZ model; our idea now is to study it by focusing our attention on its whole structure, and study the particle's worldline according to this model by means of their geometrical characteristics given by their curvatures. This will be done in sec.3, where we show that all worldlines in that model have vanishing third curvature. In sec.4 we discuss a class of generalizations of BZ model, and one of these models will be of particular interest. In fact, BZ model can be seen as a classical analogue of Dirac theory [8], and one of those models we will study appears as a classical analogue of Heisenberg's fundamental field equation, which is the basis of Heisenberg's program of a unified field theory of elementary particles [9] (sec.5). Then in sec.6 we generalize BZ model in order to include radiation reaction.

THE SPACETIME ALGEBRA AND SPINNING PARTICLES

Let $\mathbb{R}_{1,3}$ denote the STA, which is the Clifford algebra of $V = \mathbb{R}^{1,3}$ (Minkowski vector space) with quadractic form $Q(v) = g(v,v)$ and let $\{\gamma_\mu\}$ ($\mu = 0,1,2,3$) be a basis of $\mathbb{R}^{1,3}$ such that $g(\gamma_\mu, \gamma_\nu) = g_{\mu\nu} = \mathrm{diag}(1,-1,-1,-1)$. As a vector space $\mathbb{R}_{1,3}$ is 16-dimensional and a basis is $\{1; \gamma_0, \cdots, \gamma_3; \gamma_{01}, \cdots, \gamma_{23}; \gamma_{012}, \cdots, \gamma_{123}; \gamma_5\}$ (that is, as a vector space $\mathbb{R}_{1,3} \simeq \bigwedge(\mathbb{R}^{1,3})$) and where $\gamma_{\mu\nu} = \gamma_\mu \gamma_\nu = \gamma_\mu \wedge \gamma_\nu$ ($\mu \neq \nu$) and $\gamma_5 = \gamma_{0123} = \epsilon$ is the volume element. The Clifford (or geometrical) product is denoted by justaposition, and we have for $a \in \mathbb{R}^{1,3} \subset \mathbb{R}_{1,3}$ and $X \in \mathbb{R}_{1,3}$ that

$$aX = a \lrcorner X + a \wedge X, \tag{1}$$

where by $\lrcorner$ we denoted the contraction and by $\wedge$ the external (or wedge or Grassmann) product; left contraction $\lrcorner$ is defined by $G(A \lrcorner B, C) = G(B, \tilde{A} \wedge C)$, $\forall C \in \mathbb{R}_{1,3} \simeq \bigwedge(\mathbb{R}^{1,3})$, and the right contraction $\llcorner$ is defined by $G(A \llcorner B, C) = G(A, C \wedge \tilde{B})$, $\forall C \in \mathbb{R}_{1,3} \simeq \bigwedge(\mathbb{R}^{1,3})$, where G is the extension of g to $\bigwedge(\mathbb{R}^{1,3})$ [10]; we shall in what follows simplify the notation by denoting $a \cdot X = a \lrcorner X$, but the dot must not be confused with internal product (although, when $A, B \in \mathbb{R}^{1,3} \subset \mathbb{R}_{1,3}$ we have $a \lrcorner b = a \llcorner b = g(a,b) = a \cdot b$. We denote by $\tilde{}$ the main antiautomorphism (called reversion) $(AB)\tilde{} = \tilde{B}\tilde{A}$ with $\tilde{A} = A$ for A being a scalar or a vector, and by $\langle \ \rangle_k$ we denote the k-vector part of a multivector (or Clifford number), that is, for $A = A_0 + A_1 + \cdots + A_n$ ($A_k \in \bigwedge^k(V)$) we have $\langle A \rangle_k = A_k$. We also note that $A_r \lrcorner B_s = (-1)^{r(s-r)} B_s \llcorner A_r$ ($s > r$). The reciprocal basis $\{\gamma^\mu\}$ is such that $\gamma^\mu \cdot \gamma_\nu = \delta^\mu_\nu$. For more details see [10, 11].

Let $\mathbb{R}^+_{1,3}$ denote the even sub-algebra of STA $\mathbb{R}_{1,3}$; an element of $\mathbb{R}^+_{1,3}$ will be called an operator spinor. An operator spinor ψ has therefore the general form

$$\psi = A + A^{\mu\nu}\gamma_{\mu\nu} + A^{0123}\gamma_5. \tag{2}$$

The spin group $\mathrm{Spin}_+(1,3)$ is given by $\mathrm{Spin}_+(1,3) = \{R \in \mathbb{R}^+_{1,3} \mid R\tilde{R} = \tilde{R}R = 1\}$, and $\mathrm{Spin}_+(1,3)/\{\pm 1\} \simeq SO_+(1,3)$. Any $R \in \mathrm{Spin}_+(1,3)$ can be written in the form

$R = \pm e^B$ with $B \in \wedge^2(\mathbb{R}^{1,3})$ being a 2-vector, and B can be chosen in such a way to the sign be positive (except in the particular case where $B^2 = 0$ and $R = -e^B$). If B is a spatial 2-vector ($B^2 < 0$) then R describes a spatial rotation, while if B is a temporal 2-vector ($B^2 > 0$) then R describes a boost. It is well known that R can be decomposed in a product of a spatial rotation U and a boost L.

Now, consider an operator spinor $\psi \in \mathbb{R}^+_{1,3}$. Since $\psi \in \mathbb{R}^+_{1,3}$ we have that $\psi\tilde{\psi}$ has scalar and pseudo-scalar (4-vector) parts, i.e.,

$$\psi\tilde{\psi} = \sigma + \gamma_5\omega, \tag{3}$$

where

$$\sigma = \langle\psi\tilde{\psi}\rangle_0, \tag{4}$$

$$\omega = -\langle\psi\gamma_5\tilde{\psi}\rangle_0. \tag{5}$$

If $\psi\tilde{\psi} = 0$, i.e., $\sigma = \omega = 0$, then ψ is a singular spinor; however, if $\psi\tilde{\psi} \neq 0$, i.e., $\sigma \neq 0$ and/or $\omega \neq 0$, then ψ is a non-singular spinor, with $\psi^{-1} = (\psi\tilde{\psi})^{-1}\tilde{\psi}$, and if we define

$$\rho = \sqrt{\sigma^2 + \omega^2}, \tag{6}$$

$$\tan\beta = \frac{\omega}{\sigma}, \tag{7}$$

in such a way that

$$\psi\tilde{\psi} = \rho e^{\gamma_5\beta} = \rho\cos\beta + \gamma_5\sin\beta, \tag{8}$$

then one can write the following expression for ψ:

$$\psi = \sqrt{\rho}e^{\gamma_5\beta/2}R, \tag{9}$$

where $R \in \text{Spin}_+(1,3)$. Such ψ will be called a Dirac-Hestenes spinor, and the angle β appearing in this canonical decomposition of ψ will be called Yvon-Takabayasi angle. The usual covariant Dirac spinor $|\,\Psi\rangle \in \mathbb{C}^4$ is related to the algebraic spinor $\Psi = \psi\frac{1}{2}(1+\gamma_0)\frac{1}{2}(1+i\gamma_{12}) \in (\mathbb{C}\otimes\mathbb{R}_{1,3})\frac{1}{2}(1+\gamma_0)\frac{1}{2}(1+i\gamma_{12})$ (see [10, 12], but there is no need for the complexification $\mathbb{C}\otimes\mathbb{R}_{1,3}$ of STA $\mathbb{R}_{1,3}$ in order to formulate Dirac theory [13]).

Since a Clifford algebra is isomorphic to a particular matrix algebra, its elements can be represented by matrices; in the particular case of Dirac-Hestenes spinors one representation is ($*$ denotes complex conjugation):

$$\psi = \begin{pmatrix} \psi_1 & -\psi_2^* & \psi_3 & \psi_4^* \\ \psi_2 & \psi_1^* & \psi_4 & -\psi_3^* \\ \psi_3 & \psi_4^* & \psi_1 & -\psi_2^* \\ \psi_4 & -\psi_3^* & \psi_2 & \psi_1^* \end{pmatrix}, \quad \tilde{\psi} = \begin{pmatrix} \psi_1^* & \psi_2^* & -\psi_3^* & -\psi_4^* \\ -\psi_2 & \psi_1 & -\psi_4 & \psi_3 \\ -\psi_3^* & -\psi_4^* & \psi_1^* & \psi_2^* \\ -\psi_4 & \psi_3 & -\psi_2 & \psi_1 \end{pmatrix}. \tag{10}$$

Note that in the same representation the idempotent $e = \frac{1}{2}(1+\gamma_0)\frac{1}{2}(1+i\gamma_1\gamma_2)$ is

$$e = \begin{pmatrix} 1 & 0 & 0 & 0 \\ 0 & 0 & 0 & 0 \\ 0 & 0 & 0 & 0 \\ 0 & 0 & 0 & 0 \end{pmatrix}, \tag{11}$$

and fot the algebraic spinor $\Psi = \psi e$

$$\psi = \begin{pmatrix} \psi_1 & 0 & 0 & 0 \\ \psi_2 & 0 & 0 & 0 \\ \psi_3 & 0 & 0 & 0 \\ \psi_4 & 0 & 0 & 0 \end{pmatrix}, \tag{12}$$

which exhibit the equivalence between those notions of spinors and the usual covariant one.

We have now to define the Clifford bundle since our calculations within STA are to be performed over spacetime. Let (M, D, g) be a relativistic spacetime [14, 15], D being the Levi-Civita connection of g, and the tangent space at $x \in M$ is $T_x M \simeq \mathbb{R}^{1,3}$. $Cl(M, g)$ is the Clifford bundle over the spacetime (M, D, g), that is: M is the base manifold, the STA $\mathbb{R}_{1,3}$ is the typical fiber and $\text{Spin}_+(1,3)$ is the structural group. Multivector fields are sections of $Cl(M, g)$. These objects are isomorphic to the superfields introduced by Salam and Strathdee [16]. Note that since the Dirac-Hestenes spinor $\psi \in \mathbb{R}_{1,3}^+$ is an (even) multivector field[0], it is a superfield (see below). In what follows all fields used are sections of $Cl(M, g)$ or are Clifford fields over the map $\varsigma : \mathbb{R} \supset I \rightarrow M$, where ς is a timelike vector curve pointing to the future. For what follows we need some definitions, which are adapted from definitions appearing originally in [15]. An *observer* in (M, D, g) is a timelike curve $\varsigma : \mathbb{R} \supset I \rightarrow M$ pointing to the future, such that the assignment $\tau \mapsto \varsigma_* \tau$ is by definition the tangent vector field of ς, denoted simply by ς_* and $g(\varsigma_*, \varsigma_*) = 1$. An *instantaneous observer* is a pair (z, Z), $z \in M$ and $Z \in T_z M$ is a timelike vector field pointing to the future. A *reference frame* in (M, D, g) is a timelike vector field defined in $U \subset M$ such that each of its integral lines is an observer. A *moving system* for $x \in M$ is an orthonormal basis for $T_x M$. A moving system for all points $x \in \varsigma(\tau)$ is called a *comoving frame* for ς.

Now, consider a particle moving along a worldline $x : \tau \mapsto x(\tau)$, where τ is an invariant time parameter. If this particle is spinning, then it must be characterized, besides by the pair of conjugate variables (x, p), by the motion of a comoving frame $e_\mu = e_\mu(\tau)$ $(\mu = 0, 1, 2, 3)$. Since the comoving frame $\{e_\mu\}$ is orthonormal, it can be related to the orthonormal frame $\{\gamma_\mu\}$ (which we associate with the observer frame) by

$$e_\mu = R \gamma_\mu \tilde{R}, \tag{13}$$

where $R = R(\tau) \in \text{Spin}_+(1,3)$. In view of eq.(9), and since γ_5 anticommutes with vectors, we can write

$$\rho e_\mu = \psi \gamma_\mu \tilde{\psi}. \tag{14}$$

A spinning particle can therefore be characterized by (x, p) and $(\psi, \bar{\psi})$ (where $\bar{\psi}$ is the spinor canonically conjugated to ψ ($\bar{\psi} = \gamma_{21} \tilde{\psi}$) [7]).

From eq.(13) we have that

$$\dot{e}_\mu = \Omega \cdot e_\mu = \frac{1}{2}[\Omega, e_\mu], \tag{15}$$

where $\dot{e}_\mu = \frac{De_\mu}{d\tau}$, and

$$\dot{R} = \frac{1}{2}\Omega R, \tag{16}$$

with $\Omega \in \bigwedge^2(\mathbb{R}^{1,3})$, which is the angular velocity 2-vector or the Darboux 2-vector. Note that Ω can be written as

$$\Omega = \dot{e}_\mu \wedge e^\mu = \dot{e}_\mu e^\mu = \omega_{\mu\nu} e^\nu e^\mu, \tag{17}$$

where $\omega_{\mu\nu} = -\omega_{\nu\mu}$. We shall see later that we can call eq.(15) the superparticle equation of motion.

---

[0]More precisely a Dirac-Hestenes spinor field is an equivalence class of even sections of the Clifford bundle – see [12] for details.

Another frame of particular interest is the Frenet frame $\{f_\mu\}$. The Frenet frame is defined in such a way that the Darboux 2-vector assumes the form

$$\Omega = \kappa_1 f^1 \wedge f^0 + \kappa_2 f^2 \wedge f^1 + \kappa_3 f^3 \wedge f^2, \tag{18}$$

where $\kappa_1, \kappa_2, \kappa_3$ are called the first, second and third curvatures of the worldline. The equations for the motion of $\{f_\mu\}$ are $\dot{f}_\mu = \Omega \cdot f_\mu$, which in this case are explicitly

$$\dot{f}_0 = \kappa_1 f^1, \tag{19}$$

$$\dot{f}_1 = -\kappa_1 f^0 + \kappa_2 f^2, \tag{20}$$

$$\dot{f}_2 = -\kappa_2 f^1 + \kappa_3 f^3, \tag{21}$$

$$\dot{f}_3 = -\kappa_3 f^2. \tag{22}$$

Note the careful distinction that must be made between the comoving frame $\{e_\mu\}$ and the Frenet frame $\{f_\mu\}$. The comoving frame has a "dynamical meaning" in the sense that its motion along the worldline has to be given by the equations of motion of a given model, that is, from the equation for the spinor ψ that gives e_μ by eq.(14). On the other hand, the Frenet frame has a "kinematical meaning" related to any given curve. By an element of $\mathrm{Spin}_+(1,3)$ it can be related at each point $x \in M$ to any other frame $\{e'_\mu\}$, that is, $f_\mu = U' e'_\mu \tilde{U}'$, and in particular to the comoving frame $\{e_\mu\}$ by $f_\mu = U e_\mu \tilde{U}$. In this case we can study a spinning particle by means of $\{f_\mu\}$.

Finally, suppose we are moving along the worldline $x = x(\tau)$. We must now define a criterium which says when we have rotation in the local rest frame of the particle. Let $v = \frac{d}{d\tau}$ be a tangent vector to $x = x(\tau)$, and D a connexion on spacetime. If the particle is not acelerated we say that two vectors X and X' have the same spatial direction if X' is the parallel transport of X along $x(\tau)$. However, if the particle is acelerated, i.e., $D_v v \neq 0$, then this criterium is no longer valid; we have now to introduce the so-called Fermi-Walker connexion $\mathcal{F}$ given by [15]

$$\mathcal{F}_v X = D_v X - [(D_v v) \wedge v] \cdot X, \tag{23}$$

or in the case of Minkowski spacetime:

$$\mathcal{F}_v X = \dot{X} - (\dot{v} \wedge v) \cdot X. \tag{24}$$

The gyroscopic axes $\{X_\mu\}$ are defined by $\mathcal{F}_v X_\mu = 0$, and we say that a vector X is rotating along the worldline $x = x(\tau)$ if it is rotating in relation to those gyroscopic axes.

Now, let us compare our approach with the usual one that uses Grassmann variables. The calculus based on Grassmann variables is known as "supercalculus" [4]; it is used in the so called "pseudo-classical" mechanics (a Grassmann variant of Classical Mechanics) which describes the dynamics of the so called "superparticle". Let ξ_i $(i = 1, \cdots, n)$ be the generators of the Grassmann algebra $\mathcal{G}_n$, i.e.,

$$\xi_i \xi_j = -\xi_j \xi_i. \tag{25}$$

A general element of $\mathcal{G}_n$ is therefore of the form

$$f^0 + f^i \xi_i + \frac{1}{2!} f^{ij} \xi_i \xi_j + \cdots + \frac{1}{n!} f^{i_1 \cdots i_n} \xi_{i_1} \cdots \xi_{i_n} = f(\xi_1, \cdots, \xi_n) = f(\xi), \tag{26}$$

which defines a Grassmann function $f(\xi)$. Then Berezin[3] introduced the operations of differentiation and integration. Left and right derivations are given by the rules

$$\frac{\vec{\partial}}{\partial \xi_l}(\xi_{k_1} \cdots \xi_{k_\nu}) = \delta_{k_1 l}\xi_{k_2} \cdots \xi_{k_\nu} - \delta_{k_2 l}\xi_{k_1}\xi_{k_3} \cdots \xi_{k_\nu} + \cdots + (-1)^\nu \delta_{k_\nu l}\xi_{k_1} \cdots \xi_{k_{\nu-1}}, \quad (27)$$

$$(\xi_{k_1} \cdots \xi_{k_\nu})\frac{\overleftarrow{\partial}}{\partial \xi_l} = \delta_{k_\nu l}\xi_{k_1} \cdots \xi_{k_{\nu-1}} - \cdots + (-1)^\nu \delta_{k_1 l}\xi_{k_2} \cdots \xi_{k_\nu}, \quad (28)$$

and integration by the rules

$$\int 1 d\xi_l = 0, \quad \int \xi_l d\xi_l = 1, \quad (29)$$

with multiple integration defined by means of iteration. It follows that

$$\int \xi_{k_1} \cdots \xi_{k_\nu} d\xi_\nu \cdots d\xi_1 = \epsilon_{k_1 \cdots k_\nu}, \quad (30)$$

$$\int f(\xi) d\xi_n \cdots d\xi_1 = \epsilon_{k_1 \cdots k_n} f^{k_1 \cdots k_n}. \quad (31)$$

From these results we can easily see that integration is equivalent to right derivation, i.e.,

$$\int f(\xi) d\xi_n \cdots d\xi_1 = f(\xi)\frac{\overleftarrow{\partial}}{\partial \xi_n} \cdots \frac{\overleftarrow{\partial}}{\partial \xi_1}. \quad (32)$$

The superfields are fields depending also on ξ, and according to eq.(26) a superfield $\Phi(x,\xi)$ is of the form

$$\Phi(x,\xi) = \Phi_0(x) + (\Phi_1(x))^i \xi_i + \frac{1}{2!}(\Phi_2(x))^{ij} \xi_i\xi_j + \cdots + \frac{1}{n!}(\Phi_n(x))^{i_1 \cdots i_n} \xi_{i_1} \cdots \xi_{i_n}. \quad (33)$$

The relationship between our approach and the supercalculus can be easily established. Consider the Clifford product of two vectors a and b, i.e.,

$$ab = a \cdot b + a \wedge b. \quad (34)$$

Comparing eq.(25) and eq.(34) it follows that the Grassmann algebra $\mathcal{G}_n$ can be seen as the Clifford algebra of the isotropic vector space V_n with basis $\{\xi_1, \cdots, \xi_n\}$ ($\xi_i \cdot \xi_j = 0, \forall i, j$). However, this is not a good relationship for our purposes since we have in this case no possibility of translating the operations of supercalculus in terms of *algebraic* operations in Clifford algebra, which is our objective. Consider a vector space V_n endowed with an inner product g; choose a basis $\{e_1, \cdots, e_n\}$ such that $g(e_i, e_j) = e_i \cdot e_j = g_{ij} = \mathrm{diag}(+1, \cdots, +1; -1, \cdots, -1)$, and let $\{e^i\}$ be the reciprocal basis: $e^i \cdot e_j = \delta^i_j$ (obviously there is no need for introducing the reciprocal basis if (V_n, g) is euclidian). The product given by eq.(25) in $\mathcal{G}_n$ can be related to the external product $\wedge$ in the Clifford algebra of (V_n, g) according to

$$\xi_i\xi_j \leftrightarrow e_i \wedge e_j. \quad (35)$$

Note that we can use now the contraction in the Clifford algebra in order to define the operations of supercalculus. In fact, left and right derivations are related to left and right contractions according to the rules:

$$\frac{\vec{\partial}}{\partial \xi_k} \leftrightarrow e^k \lrcorner, \quad (36)$$

$$\frac{\overleftarrow{\partial}}{\partial \xi_k} \leftrightarrow \mathsf{L} e^k. \tag{37}$$

It is easy to obtain from the properties of the contractions the results of eqs.(27,28). Integration, according to eq.(32), is related to

$$\int f(\xi)d\xi_n \cdots d\xi_1 \leftrightarrow (\cdots ((f(e)\,\mathsf{L}\,e^n)\,\mathsf{L}\,e^{n-1})\cdots \mathsf{L}\,e^1) = f(e)\,\mathsf{L}\,(e^n e^{n-1}\cdots e^1) = f(e)\,\mathsf{L}\,\eta^{-1}, \tag{38}$$

where $e = (e_1, \cdots, e_n)$ and η^{-1} is the inverse of the volume element $\eta = e_1 \cdots e_n$ of V_n. We see therefore that *the operations of derivation and integration of supercalculus are related to the notions of orthogonality and orientation of the vector space V_n*. The above identifications show us that superfields can be regarded as sections of an appropriate Clifford bundle. Dirac-Hestenes spinors are therefore examples of (even) superfields.

One can claim that supercalculus over Grassmann variables is more general than Clifford algebra since the former does not involves the definition of an inner product in V_n. This is, however, not the case! The inner product is needed in order to define orthogonality and to give in a canonical way a correlation between the vector space V_n and its dual V_n^*, which enable us to write the reciprocal basis $\{e^i\}$ of $\{e_i\}$. If we do not have an inner product on V_n we can take a basis $\{e_i\}$ of V_n and a basis $\{\theta^i\}$ of V_n^* such that $\theta^j(e_i) = e_i \lrcorner \theta^j = \delta_j^i$ and define an algebra over $\bigwedge(V) \oplus \bigwedge(V^*)$ according to the rules

$$e_i e_j = -e_j e_i, \tag{39}$$

$$\theta^i \theta^j = -\theta^j \theta^i, \tag{40}$$

$$\theta^i e_j + e_j \theta^i = e_j \lrcorner \theta^i = \theta^i(e_j) = \delta_j^i, \tag{41}$$

and the identifications given by eqs.(35,36,37) now read

$$\xi_i \xi_j \leftrightarrow e_i e_j, \qquad \frac{\overrightarrow{\partial}}{\partial \xi_k} \leftrightarrow \theta^k \lrcorner, \qquad \frac{\overleftarrow{\partial}}{\partial \xi_k} \leftrightarrow \mathsf{L} \theta^k. \tag{42}$$

If an inner product g is then defined in V_n we have a canonical way to define θ^i such that $\theta^i(e_j) = \delta_j^i$, that is, $\theta^i = g^{ij} e_j$, and after defining $E_i^{\pm}$ by

$$E_i^{\pm} = (e_i \pm g_{ij} \theta^j) \tag{43}$$

the above algebra is just the sum of two Clifford algebras, one for the vector space (V_n, g) and other for $(V_n, -g)$ (which carries essentialy the same information!). The algebra defined by the rules (39), (40) and (41) was first considered by Schönberg [17], and more recently by Witten[18].

Finally, we observe that in [19] we showed that eq.(15), which we called the superparticle equation of motion, can be put in correspondence with the famous Berezin-Marinov model using the identifications above. The formalism use the concept of multivector lagrangians fully developed in [20, 21].

BARUT AND ZANGHI MODEL

The BZ model [2], translated in terms of STA [1, 22], is based on the following lagrangian:

$$\mathcal{L} = \langle \dot{\tilde{\psi}} \psi \gamma_{21} + p(\dot{x} - \psi \gamma_0 \tilde{\psi}) + eA\psi \gamma_0 \tilde{\psi} \rangle_0, \tag{44}$$

where A denotes the electromagnetic potential and p is the canonical momentum, which follows from the fact that BZ model is a hamiltonian system [23, 7], with hamiltonian

$$\mathcal{H} = \langle (p - eA)\psi\gamma_0\tilde{\psi}\rangle_0, \tag{45}$$

and conjugate variables (x, p) and $(\psi, \gamma_{21}\tilde{\psi})$. The equations of motion are

$$\dot{\psi}\gamma_{21} = \pi\psi\gamma_0, \tag{46}$$

$$\dot{x} = \psi\gamma_0\tilde{\psi}, \tag{47}$$

$$\dot{\pi} = eF \cdot \dot{x}, \tag{48}$$

where $\pi = p - eA$ is the kinetic momentum and $F = \partial \wedge A$ is the electromagnetic field. In the following we will denote $\dot{x} = v$.

Let us analyse BZ model for the free case $(A = 0)$, where the equations are

$$\dot{\psi}\gamma_{21} = p\psi\gamma_0, \tag{49}$$

$$v = \dot{x} = \psi\gamma_0\tilde{\psi}, \tag{50}$$

$$\dot{p} = 0. \tag{51}$$

In their original paper BZ looked directly for a solution of the above system, which in terms of STA reads

$$\psi(\tau) = \cos{(m\tau)}\psi(0) + \sin{(m\tau)}\frac{p}{m}\psi(0)\gamma_0\gamma_{12}, \tag{52}$$

where $p^2 = m^2$; it follows for this solution that

$$v(\tau) = \dot{x}(\tau) = \frac{p}{m} + \left[v(0) - \frac{p}{m}\right]\cos 2m\tau + \frac{\dot{v}(0)}{2m}\sin 2m\tau, \tag{53}$$

whose integration gives a worldline $x = x(\tau)$ which is a cilindrical helix, the classical analogue of the quantum phenomenum called zitterbewegung.

Although extensively studied in other occasions [6, 7], these studies were limited to that particular solution given by eq.(52). Our objective now is to give a general analysis of BZ model by centering our attention on the whole structute of BZ model.

We start by supposing that the spinor ψ is non-singular. In this case it admits the canonical decomposition given by eq.(9). From that expression we have

$$\dot{\psi} = \frac{1}{2}\left(\frac{\dot{\rho}}{\rho} + \gamma_5\beta + \Omega\right)\psi, \tag{54}$$

where Ω was defined by eq.(16). Using eq.(54) into eq.(49) we obtain, after some manipulations,

$$\frac{1}{2}\left(\frac{\dot{\rho}}{\rho} + \gamma_5\beta + \Omega\right) = \cos\beta\, p \wedge e_0 \wedge e_1 \wedge_2 + \cos\beta\, p \cdot (e_0 \wedge e_1 \wedge e_2) +$$
$$+ \sin\beta\, p \wedge e_3 + \sin\beta\, p \cdot e_3, \tag{55}$$

where $\{e_\mu\}$ are given by eq.(13). If we write

$$p = \alpha^\mu e_\mu, \tag{56}$$

or $\alpha_\mu = p \cdot e_\mu$, then we obtain from eq.(55):

$$\frac{1}{2}\left(\frac{\dot{p}}{\rho} + \gamma_5\beta + \Omega\right) = -\cos\beta\,\alpha_3 e_0 \wedge e_1 \wedge e_2 \wedge e_3 + \cos\beta\,\alpha_0 e_1 \wedge e_2 - \cos\beta\,\alpha_1 e_0 \wedge e_2 +$$

$$+ \cos\beta\,\alpha_2 e_0 \wedge e_1 + \sin\beta\,(\alpha_0 e_0 - \alpha_1 e_1 - \alpha_2 e_2) \wedge e_3 + \sin\beta\,\alpha_3,(57)$$

and after splitting eq.(57) into its scalar, 2-vector and pseudo-scalar parts, we have:

$$\dot{\rho} = 2\rho\sin\beta\,\alpha_3, \tag{58}$$

$$\dot{\beta} = 2\cos\beta\,\alpha_3, \tag{59}$$

$$\Omega = 2\cos\beta\,\alpha_2 e_0 \wedge e_1 - 2\cos\beta\,\alpha_1 e_0 \wedge e_1 + 2\sin\beta\,\alpha_0 e_0 \wedge e_3 +$$
$$+ 2\cos\beta\,\alpha_0 e_1 \wedge e_2 - 2\sin\beta\,\alpha_1 e_1 \wedge e_3 - 2\sin\beta\,\alpha_2 e_2 \wedge e_3. \tag{60}$$

From eq.(60) for Ω and from eq.(15) we can obtain expressions for $\dot{e}_\mu$; the result is

$$\dot{e}_0 = -2\cos\beta\,\alpha_2 e_1 + 2\cos\beta\,\alpha_1 e_2 - 2\sin\beta\,\alpha_0 e_3, \tag{61}$$

$$\dot{e}_1 = -2\cos\beta\,\alpha_2 e_0 + 2\cos\beta\,\alpha_0 e_2 - 2\sin\beta\,\alpha_1 e_3, \tag{62}$$

$$\dot{e}_2 = 2\cos\beta\,\alpha_1 e_0 - 2\cos\beta\,\alpha_0 e_1 - 2\sin\beta\,\alpha_2 e_3, \tag{63}$$

$$\dot{e}_3 = -2\sin\beta\,\alpha_0 e_0 + 2\sin\beta\,\alpha_1 e_1 + 2\sin\beta\,\alpha_2 e_2. \tag{64}$$

Another important result of BZ model can be obtained after multiplying eq.(49) on the right by $\tilde{\psi}$ and subtracting from it the reverse of eq.(49) multiplied on the left by ψ; the result is

$$\dot{S} = p \wedge v, \tag{65}$$

where we defined

$$S = \frac{1}{2}\psi\gamma_{21}\tilde{\psi} = \frac{1}{2}(\sigma e_{21} + \omega e_{30}). \tag{66}$$

Since $\dot{p} = 0$, it follows from eq.(65) that

$$\frac{d}{d\tau}(S + x \wedge p) = 0. \tag{67}$$

We note that $\dot{p} = 0$ imposes some conditions on α^μ in eq.(56). Since $\dot{p} = \dot{\alpha}^\mu e_\mu + \alpha^\mu \dot{e}_\mu = 0$, after using eq.(61-64) one obtains

$$\dot{\alpha}_0 = -2\sin\beta\,\alpha_0\alpha_3, \tag{68}$$

$$\dot{\alpha}_1 = -2\sin\beta\,\alpha_1\alpha_3, \tag{69}$$

$$\dot{\alpha}_2 = -2\sin\beta\,\alpha_2\alpha_3, \tag{70}$$

$$\dot{\alpha}_3 = -2\sin\beta\,(\alpha_0^2 - \alpha_1^2 - \alpha_2^2). \tag{71}$$

Now we are able to find the Frenet tetrad $\{f_\mu\}$ and the curvatures $\kappa_1, \kappa_2, \kappa_3$. Taking $f_0 = e_0$ one obtains, after some calculations, that

$$f_0 = e_0, \tag{72}$$

$$f_1 = \sqrt{\frac{\alpha_1^2 + \alpha_2^2}{\alpha_1^2 + \alpha_2^2 + \tan^2\beta\,\alpha_0^2}}\left(b_2 + \frac{\tan\beta\,\alpha_0}{\sqrt{\alpha_1^2 + \alpha_2^2}}e_3\right), \tag{73}$$

$$f_2 = \sqrt{\frac{\alpha_1^2 + \alpha_2^2 + \tan^2\beta\,\alpha_0^2}{\alpha_1^2 + \alpha_2^2 + \alpha_3^2 + \tan^2\beta\,\alpha_0^2}} \left[-b_1 - \frac{\alpha_3\sqrt{\alpha_1^2 + \alpha_2^2}}{\alpha_1^2 + \alpha_2^2 + \tan^2\beta\,\alpha_0^2}\left(e_3 - \frac{\tan\beta\,\alpha_0}{\sqrt{\alpha_1^2 + \alpha_2^2}}b_2 \right) \right],$$

(74)

$$f_3 = \sqrt{\frac{\alpha_1^2 + \alpha_2^2}{\alpha_1^2 + \alpha_2^2 + \tan^2\beta\,\alpha_0^2}} \left[-\left(e_3 - \frac{\tan\beta\,\alpha_0}{\sqrt{\alpha_1^2 + \alpha_2^2}}b_2 \right) + \frac{\alpha_3}{\sqrt{\alpha_1^2 + \alpha_2^2}}b_1 \right],$$

(75)

where we defined

$$b_1 = \frac{\alpha_1 e_1 + \alpha_2 e_2}{\sqrt{\alpha_1^2 + \alpha_2^2}}, \quad b_2 = \frac{\alpha_2 e_1 - \alpha_1 e_2}{\sqrt{\alpha_1^2 + \alpha_2^2}},$$

(76)

and for the curvatures

$$\kappa_1 = 2\cos\beta\sqrt{\alpha_1^2 + \alpha_2^2 + \tan^2\beta\,\alpha_0^2},$$

(77)

$$\kappa_2 = \frac{2\alpha_0}{\cos\beta}\frac{\sqrt{\alpha_1^2 + \alpha_2^2}\sqrt{\alpha_1^2 + \alpha_2^2 + \alpha_3^2 + \tan^2\beta\,\alpha_0^2}}{(\alpha_1^2 + \alpha_2^2 + \tan^2\beta\,\alpha_0^2)},$$

(78)

$$\kappa_3 = 0.$$

(79)

It is an interesting and unexpected fact that *for all worldlines in BZ model the third curvature vanishes*. It must be observed, since BZ model after quantization gives Dirac equation, that one can ask from this result if the same result (i.e., $\kappa_3 = 0$) holds in Dirac theory. In his zitterbewegung interpretation of quantum mechanics Hestenes [13] showed that particles' worldlines in Dirac theory have $\kappa_3 = 0$, which is a very interesting fact when compared with our results.

From the knowledge of the curvatures one can obtain informations about the particles' worldline. Since $\kappa_3 = 0$ we have, of course, that the ratio, say, κ_3/κ_2, is constant along the worldline. Now, for a helix we have the ratios κ_3/κ_2 and κ_1/κ_2 constants. Let us calculate $\frac{d}{d\tau}(\kappa_1/\kappa_2)$. Using eqs.(77- 78) we obtain

$$\frac{d}{d\tau}\left(\frac{\kappa_1}{\kappa_2}\right) = \frac{6\cos^2\beta\,\sin\beta\,\alpha_3(\alpha_0^2 - \alpha_1^2 - \alpha_2^2)\sqrt{\alpha_1^2 + \alpha_2^2 + \tan^2\beta\,\alpha_0^2}}{\alpha_0\sqrt{\alpha_1^2 + \alpha_2^2}\sqrt{\alpha_1^2 + \alpha_2^2 + \alpha_3^2 + \tan^2\beta\,\alpha_0^2}}.$$

(80)

Let us analyse the situations where κ_1/κ_2 is constant. The first possibility is $\cos\beta = 0$ or $\sin\beta = 0$. If $\cos\beta = 0$ we have problems with $\tan\beta$, so we take $\sin\beta = 0$ as our first possibility. The other one is $\alpha_3 = 0$, or $p \cdot e_3 = 0$. The other one, $\alpha_0^2 - \alpha_1^2 - \alpha_2^2 = 0$, implies $p^2 = -\alpha_3^2 \leq 0$, which corresponds to a tachyon when $\alpha_3 > 0$ or to a luminal particle (luxon) when $\alpha_3 = 0$ [26]; we shall not consider the case $\alpha_0^2 - \alpha_1^2 - \alpha_2^2 = 0$ since we are interested in sub-luminal case. So, we have that

$$\frac{d}{d\tau}\left(\frac{\kappa_1}{\kappa_2}\right) = 0 \implies \begin{cases} \sin\beta = 0 \\ \text{or} \\ \alpha_3 = 0 \end{cases},$$

(81)

or, in other words, *whenever $\sin\beta = 0$ or/and $\alpha_3 = 0$ the worldlines in BZ model are helices.*

One interesting fact that one can see when calculating (80) is that

$$\frac{d}{d\tau}(\alpha_1^2 + \alpha_2^2 + \alpha_3^2 + \tan^2\beta\,\alpha_0^2) = 0.$$

(82)

But $\alpha_1^2 + \alpha_2^2 + \alpha_3^2 = \alpha_0^2 - p^2$, so from eq.(82) we have

$$\frac{d}{d\tau}\left[\left(\frac{\alpha_0}{\cos\beta}\right)^2 - p^2\right] = 0.$$

(83)

In order to see the meaning of this result let us denote

$$\alpha_0 = m_0, \quad p^2 = m^2. \tag{84}$$

Then we have that

$$m^2 - \left(\frac{m_0}{\cos\beta}\right)^2 = \text{cte.,} \tag{85}$$

and if the constant is chosen as zero we have

$$m = \frac{m_0}{\cos\beta} = m_0\sqrt{1 + \frac{\omega^2}{\sigma^2}}, \tag{86}$$

where σ and ω were given by eq.(4-5), and related to ρ and β by eq.(6-7). It is a remarkable fact that if we suppose m_0 constant, then eq.(86) is just *de Broglie - Vigier formula* for the variable mass m in de Broglie et al. interpretation of quantum mechanics (see, for example, [27]). If this is a mere coincidence, or not, is certainly a subject of further investigations which may be of physical relevance. However, the choice of the constant in eq.(85) to be zero is very restrictive; in fact, from $p^2 = \alpha_0^2 - \alpha_1^2 - \alpha_2^2 - \alpha_3^2$ one have, in eq.(85),

$$-\tan^2\beta\, m_0^2 = \text{cte.} + \alpha_1^2 + \alpha_2^2 + \alpha_3^2, \tag{87}$$

that is, the constant must be less or equal to $-(\alpha_1^2 + \alpha_2^2 + \alpha_3^2)$; denoting

$$\mu^2 = \alpha_1^2 + \alpha_2^2 + \alpha_3^2 \geq 0, \tag{88}$$

we have that

$$m^2 + \mu^2 = \left(\frac{m_0}{\cos\beta}\right)^2. \tag{89}$$

Let us come back to our analysis of the worldlines. In the case of helices we must have either $\alpha_3 = 0$ or $\sin\beta = 0$. Let us consider $\alpha_3 = 0$. In this case, for all α_μ we have, from eqs.(68-71), that $\dot{\alpha}_\mu = 0$; but, in particular, eq.(71) implies that

$$\sin\beta = 0 \quad \text{or} \quad \alpha_0^2 - \alpha_1^2 - \alpha_2^2 = 0. \tag{90}$$

But, since we are supposing $\alpha_3 = 0$, the second case, i.e., $\alpha_0^2 - \alpha_1^2 - \alpha_2^2 = 0$ implies that $p^2 = 0$, that is, to a light-like case. The first possibility is just the one we left to consider later. On the other hand, if we consider $\sin\beta = 0$, that is, $\beta = 0$ or $\beta = \pi$, then we have $\dot{\beta} = 0$; but eq.(59) implies that $\cos\beta\,\alpha_3 = 0$, and since we are supposing $\sin\beta = 0$, we must have $\alpha_3 = 0$. In summary, for a helical worldline in BZ model we have two possibilities:

$$(\text{A}) \quad \alpha_3 = 0 \text{ and } \sin\beta = 0, \tag{91}$$

$$(\text{B}) \quad \alpha_3 = 0 \text{ and } p^2 = 0. \tag{92}$$

Case (A) is easy to handle. First, since for the curvatures we have in this case

$$\kappa_1 = \pm 2\sqrt{\alpha_1^2 + \alpha_2^2}, \tag{93}$$

$$\kappa_2 = \pm 2\alpha_0, \tag{94}$$

$$\kappa_3 = 0, \tag{95}$$

where the plus/minus sign corresponds to $\beta = 0/\beta = \pi$, then p can be written as

$$p = \alpha_0\left(f_0 + \frac{\kappa_1}{\kappa_2}f_2\right), \tag{96}$$

since $f_0 = e_0$ and $f_2 = -b_1$. Using eq.(20) we have

$$\frac{p}{\alpha_0} = \left[1 - \left(\frac{\kappa_1}{\kappa_2}\right)^2\right] f_0 - \frac{\kappa_1}{\kappa_2^2} \dot{f}_1, \tag{97}$$

or, using eqs.(93,94) and the notation (84):

$$\left(\frac{m^2}{m_0^2}\right) f_0 = \frac{p}{m_0} \pm \frac{\sqrt{1 - m^2/m_0^2}}{2m_0} \dot{f}_1, \tag{98}$$

which, since $f_0 = e_0 = \dot{x}/\rho$, can be immediatly integrated to obtain $x = x(\tau)$:

$$x = \left(\frac{m_0}{m}\right) \frac{p}{m} \tau \pm \frac{\sqrt{m_0^2/m^2 - 1}}{2m} f_1 + x_0, \tag{99}$$

where we put $\rho = $ cte. $= 1$ by identifying τ with the center of mass proper time. Note that the above worldline is a helix with radius r_0 given by

$$r_0 = \frac{1}{2m}\sqrt{\left(\frac{m_0}{m}\right)^2 - 1}. \tag{100}$$

When $m = m_0$ we have $r_0 = 0$, and $x = (p/m)\tau$ as expected. However, as we shall see later, even in this limit the particle is spinning.

However, when $m^2 \to 0$ (lightlike case), which is case (B), our approach fails. Case (B) cannot be treated in a similar way to case (A). In fact, in case (B) we have

$$\kappa_1 = \kappa_2 = 2\alpha_0, \quad \kappa_3 = 0, \tag{101}$$

and eq.(97) gives

$$\dot{f}_1 = -2p. \tag{102}$$

From this equation we have $f_1 = -2p\tau + c_1$, and from eq.(19) $\dot{f}_0 = -4\alpha_0 p\tau + c_1$, which gives $f_0 = -2\alpha_0 p\tau^2 + c_1\tau + c_2$, and $x = -(2/3)\alpha_0 p\tau^3 + (c_1/2)\tau^2 + c_2\tau + c_3$, which is not a desirable solution in the sense of being a limit case of the one previously considered. It is interesting to note, however, that this solution depends on three constants, which suggests us to ask if there is some relationship here with self-interactions as in Lorentz-Dirac equation (part of this question will be discussed later).

The only way we have to scape from the above problem is to question one of our hypothesis. This hypothesis can only be that one that the spinor ψ in BZ model is non-singular. So, we shall suppose that in the limit $p^2 \to 0$ the spinor ψ is singular. This case has been considered in [6], where ψ was supposed to be a Majorana spinor, and we saw in that occasion that the worldline is indeed a helix with radius equal to half Compton radius.

Finally, let us calculate the Fermi-Walker derivative of $\{e_\mu\}$ and $\{f_\mu\}$. We have

$$\mathcal{F}_v f_0 = \mathcal{F}_v e_0 = 0, \tag{103}$$

$$\mathcal{F}_v f_3 = \mathcal{F}_v e_3 = 0, \tag{104}$$

as expected, but

$$\mathcal{F}_v f_1 = -\kappa_2 f_2, \tag{105}$$

$$\mathcal{F}_v f_2 = \kappa_2 f_1, \tag{106}$$

that is, f_1 and f_2 rotate in the plane spanned by f_1 and f_2 (or e_1 and e_2) with frequency $\omega = \kappa_2$; and for e_1 and e_2 we have

$$\mathcal{F}_v e_1 = \kappa_2 e_2, \tag{107}$$

$$\mathcal{F}_v e_2 = -\kappa_2 e_1, \tag{108}$$

that is, e_1 and e_2 rotates in the same plane with the same frequency. In other words, we have an intrinsic rotation, which we identify with spin. *This analysis shows that BZ model exhibit an intrinsic spin, which is not due to any helical motion.*

As a final comment, we remember that solution (99) is given in terms of the co-moving frame $\{e_\mu\}$ (or the Frenet frame $\{f_\mu\}$). In order to write it in terms of the observer frame $\{\gamma_\mu\}$ one has to solve explicitly the equations of BZ model in order to obtain $R \in \mathrm{Spin}_+(1,3)$. Once we find R we can decompose it as $R = LU$, where L is a boost (i.e., $e_0 = L\gamma_0\tilde{L}$) and U is a spatial rotation (i.e., $\gamma_0 = L\gamma_0\tilde{L}$). Then, a convenient analysis of BZ model can be performed in the inertial system $E_\mu = U\gamma_\mu\tilde{U}$. In this system we have $\dot{U} = \frac{1}{2}wU$, where w can be decomposed [28] as $w = w_T + w_L$, where w_T is the angular velocity for Thomas precession and w_L is the angular velocity for Larmor precession. Since $\Omega = \kappa_1 f^1 f^0 + \kappa_2 f^2 f^1$, it is immediate that $w_L = 2\cos\beta\,\alpha_0\tilde{L}e_{21}L = 2\cos\beta\,\alpha_0 E_1 \wedge E_2$, but the knowledge of w_T depends explicitly on the knowledge of the boost L. Anyway, our analysis in the comoving frame enable us to understand some of the main general aspects of BZ model.

GENERALIZED BARUT-ZANGHI MODEL

The quantities σ and ω, defined by eqs.(4,5), are just the invariants of Dirac theory (with the obvious difference that in Dirac theory ψ is a spinor field, while here ψ is defined along the particle's worldline). Being invariant quantities, one can think of generalizing BZ model by adding to the lagrangian $\mathcal{L}_{\mathrm{BZ}}$ a term that depends on σ and ω. In this way, we shall study a model based on the lagrangian

$$\mathcal{L} = \langle \dot{\tilde{\psi}}\psi\gamma_{21} + p(\dot{x} - \psi\gamma_0\tilde{\psi}) + eA\psi\gamma_0\tilde{\psi}\rangle_0 + \lambda F(\sigma,\omega), \tag{109}$$

where $F(\sigma,\omega)$ is a scalar function of σ and ω, and λ is a constant, while the other symbols have the same meaning as in BZ model. Since

$$\partial_{\tilde{\psi}}F = \left(\frac{\partial F}{\partial\sigma} - \gamma_5\frac{\partial F}{\partial\omega}\right)\psi \tag{110}$$

the equations for this generalized BZ (GBZ) model are

$$\dot{\psi}\gamma_{21} = \pi\psi\gamma_0 - \lambda\left(\frac{\partial F}{\partial\sigma} - \gamma_5\frac{\partial F}{\partial\omega}\right)\psi, \tag{111}$$

$$\dot{x} = \psi\gamma_0\tilde{\psi}, \tag{112}$$

$$\dot{\pi} = e(F \cdot \dot{x}), \tag{113}$$

where $\pi = p - eA$ is the kinetic momentum. In the free case ($A = 0$) we have

$$\dot{\psi}\gamma_{21} = p\psi\gamma_0 - \lambda\left(\frac{\partial F}{\partial\sigma} - \gamma_5\frac{\partial F}{\partial\omega}\right)\psi, \tag{114}$$

$$\dot{x} = v = \psi\gamma_0\tilde{\psi}, \tag{115}$$

213

$$\dot{p} = 0. \tag{116}$$

First of all, it is important to note that, if we define S like in BZ model, i.e., by eq.(66), then from eq.(114) it follows that

$$\dot{S} = p \wedge v, \tag{117}$$

just like in BZ model.

Now, let us suppose that ψ is non-singular. Then eq.(9) holds, and as a consequence we have eq.(54), which when introduced in eq.(111) gives

$$\frac{1}{2}\left(\frac{\dot{\rho}}{\rho} + \gamma_5\dot{\beta} + \Omega\right) = \cos\beta\, p \wedge e_0 \wedge e_1 \wedge e_2 + \cos\beta\, p \cdot (e_0 \wedge e_1 \wedge e_2) +$$

$$+ \sin\beta\, p \wedge e_3 + \sin\beta\, p \cdot e_3 - \lambda\left(\frac{\partial F}{\partial\sigma} - \gamma_5\frac{\partial F}{\partial\omega}\right)e_1 \wedge e_2. \tag{118}$$

If we write $p = \alpha^\mu e_\mu$ as in eq.(56) and then split eq.(118) into its scalar, 2-vector and pseudo-scalar parts, we obtain

$$\dot{\rho} = 2\rho\sin\beta\,\alpha_3, \tag{119}$$

$$\dot{\beta} = 2\cos\beta\,\alpha_3, \tag{120}$$

$$\Omega = 2\cos\beta\,\alpha_2 e_0 \wedge e_1 - 2\cos\beta\,\alpha_1 e_0 \wedge e_2 + 2\left(\sin\beta\,\alpha_0 - \lambda\frac{\partial F}{\partial\omega}\right)e_0 \wedge e_3 +$$

$$+ 2\left(\cos\beta\,\alpha_0 - \lambda\frac{\partial F}{\partial\sigma}\right)e_1 \wedge e_2 - 2\sin\beta\,\alpha_1 e_1 \wedge e_3 - 2\sin\beta\,\alpha_2 e_2 \wedge e_3. \tag{121}$$

Using this expression for Ω we can write the equations for $\dot{e}_\mu = \Omega \cdot e_\mu$; the result is:

$$\dot{e}_0 = -2\cos\beta\,\alpha_2 e_1 + 2\cos\beta\,\alpha_1 e_2 - 2\left(\sin\beta\,\alpha_0 - \lambda\frac{\partial F}{\partial\omega}\right)e_3, \tag{122}$$

$$\dot{e}_1 = -2\cos\beta\,\alpha_2 e_0 + 2\left(\cos\beta\,\alpha_0 - \lambda\frac{\partial F}{\partial\sigma}\right)e_2 - 2\sin\beta\,\alpha_1 e_3, \tag{123}$$

$$\dot{e}_2 = 2\cos\beta\,\alpha_1 e_0 - 2\left(\cos\beta\,\alpha_0 - \lambda\frac{\partial F}{\partial\sigma}\right)e_1 - 2\sin\beta\,\alpha_2 e_3, \tag{124}$$

$$\dot{e}_3 = -2\left(\sin\beta\,\alpha_0 - \lambda\frac{\partial F}{\partial\omega}\right)e_0 + 2\sin\beta\,\alpha_1 e_1 + 2\sin\beta\,\alpha_2 e_2. \tag{125}$$

From $\dot{p} = 0 = \dot{\alpha}^\mu e_\mu + \alpha^\mu\dot{e}_\mu$ we obtain equations for $\dot{\alpha}_\mu$, which are:

$$\dot{\alpha}_0 = -2\left(\sin\beta\,\alpha_0\alpha_3 - \lambda\frac{\partial F}{\partial\omega}\alpha_3\right), \tag{126}$$

$$\dot{\alpha}_1 = -2\left(\sin\beta\,\alpha_1\alpha_3 + \lambda\frac{\partial F}{\partial\sigma}\alpha_2\right), \tag{127}$$

$$\dot{\alpha}_2 = -2\left(\sin\beta\,\alpha_2\alpha_3 - \lambda\frac{\partial F}{\partial\sigma}\alpha_1\right), \tag{128}$$

$$\dot{\alpha}_3 = -2\left[\sin\beta\left(\alpha_0^2 - \alpha_1^2 - \alpha_2^2\right) - \lambda\frac{\partial F}{\partial\omega}\alpha_0\right]. \tag{129}$$

Now we can proceed as in BZ model and calculate from eqs.(122-125) the Frenet frame $\{f_\mu\}$ and the curvatures $\kappa_1, \kappa_2, \kappa_3$. However, the calculations in the general case, although straightforward, are too long, and we may ask ourself what we want from such calculus. Our objective is to compare the GBZ model with BZ one, and we saw in the preceeding section that the situations of interest (that is, helical worldlines) require $\alpha_3 = 0$. So, if we suppose now that $\alpha_3 = 0$ then the calculations will not be so long, and we will keep pur goals. Moreover, for the sake of comparison, there is no need to consider a whole class of GBZ models, each one specified by a different function $F(\sigma, \omega)$. We can restrict our interest in a particular function, and the most interesting one seems to be

$$F(\sigma, \omega) = \frac{1}{2}(\sigma^2 + \omega^2). \tag{130}$$

In this way, if we define the quantity Δ by

$$\Delta = \frac{\sin \beta \, \alpha_0 - \lambda \omega}{\cos \beta \sqrt{\alpha_1^2 + \alpha_2^2}} = \frac{\sin \beta (\alpha_0 - \lambda \rho)}{\cos \beta \sqrt{\alpha_1^2 + \alpha_2^2}}, \tag{131}$$

then the Frenet tetrad and the curvatures for the case $\alpha_3 = 0$ and $F(\sigma, \omega) = \frac{1}{2}(\sigma^2 + \omega^2)$ can be written as

$$f_0 = e_0, \tag{132}$$

$$f_1 = \frac{b_2 + \Delta e_3}{\sqrt{1 + \Delta^2}}, \tag{133}$$

$$f_2 = -b_1, \tag{134}$$

$$f_3 = \frac{-e_3 + \Delta b_2}{\sqrt{1 + \Delta^2}}, \tag{135}$$

where b_1 and b_2 are given by eq.(76), and

$$\kappa_1 = 2 \cos \beta \sqrt{\alpha_1^2 + \alpha_2^2} \sqrt{1 + \Delta^2}, \tag{136}$$

$$\kappa_2 = \frac{2\alpha_0}{\cos \beta} \frac{(1 - \sin \beta \, \lambda \omega / \alpha_0)}{\sqrt{1 + \Delta^2}}, \tag{137}$$

$$\kappa_3 = \frac{2(\sin \beta \sqrt{\alpha_1^2 + \alpha_2^2} - \cos \beta \, \alpha_0 \Delta)}{\sqrt{1 + \Delta^2}}. \tag{138}$$

It seems that now we have a model which permits worldlines with $\kappa_3 \neq 0$ when $\alpha_3 = 0$. However, as we shall see, *we have indeed* $\kappa_3 = 0$. Anyway, a calculation shows that

$$\frac{d}{d\tau}\left(\frac{\kappa_1}{\kappa_2}\right) = 0, \quad \frac{d}{d\tau}\left(\frac{\kappa_3}{\kappa_2}\right) = 0, \tag{139}$$

that is, *the worldlines in GBZ model with $\alpha_3 = 0$ and $F(\sigma, \omega) = \frac{1}{2}(\sigma^2 + \omega^2)$ are helices.*

We must observe now that the condition $\alpha_3 = 0$ requires a "compatibility condition" due to eq.(129). Putting $\alpha_3 = 0$ in eq.(129) we obtain the following condition that has to be satisfied:

$$2 \sin \beta \, (p^2 - \lambda \rho \alpha_0) = 0, \tag{140}$$

that is, either

$$\sin \beta = 0 \tag{141}$$

or

$$p^2 = \lambda \rho \alpha_0. \tag{142}$$

215

Within condition (141) we obtain case (A) in BZ model (eq.91), and of course we would obtain BZ results since for $\sin \beta = 0$ we have $\Delta = 0$. The interesting condition to be satisfied is therefore condition (142). When $\lambda = 0$ we obtain case (B) of BZ model (eq.92), so we will suppose $\lambda \neq 0$. Supposing also that $\alpha_0 = m_0 \neq 0$ and with $p^2 = m^2$ (notations of eq.84) we have

$$\lambda \rho = \frac{m^2}{m_0}. \tag{143}$$

Quantity Δ can be rewritten now as

$$\Delta = \tan \beta \sqrt{1 - \frac{m^2}{m_0^2}}, \tag{144}$$

where we used $\alpha_1^2 + \alpha_2^2 = m_0^2 - m^2$. Note that, since $m_0^2 \geq m^2$, quantity $\xi = 1 - m^2/m_0^2$ has to assume values $0 \leq \xi \leq 1$.

Now, consider the numerator of κ_3 (eq.138). We can rewrite it as

$$\frac{\sin \beta}{\sqrt{\alpha_1^2 + \alpha_2^2}} \left[\alpha_1^2 + \alpha_2^2 - \alpha_0(\alpha_0 - \lambda \rho)\right] = \frac{\sin \beta}{\sqrt{\alpha_1^2 + \alpha_2^2}} \left(-p^2 + \lambda \rho \alpha_0\right) = 0 \tag{145}$$

due to condition (142). So, even in GBZ model we have

$$\kappa_3 = 0. \tag{146}$$

It is indeed remarkable that *both in BZ model and in the present GBZ model all helical worldlines have vanishing third curvature.*

Eq.(143) enable us to rewrite the quantity $(1 - \sin \beta \, \lambda \omega / \alpha_0)$ as $\cos^2 \beta \, (1 + \Delta^2)$. Using it we can rewrite the curvatures κ_1 and κ_2 as

$$\kappa_1 = 2 \cos \beta \sqrt{\alpha_1^2 + \alpha_2^2} \sqrt{1 + \Delta^2}, \tag{147}$$

$$\kappa_2 = 2 \cos \beta \, \alpha_0 \sqrt{1 + \Delta^2}. \tag{148}$$

Another interesting property of GBZ model is that eq.(83) of BZ model still holds true, i.e.,

$$\frac{d}{d\tau} \left[\left(\frac{m_0}{\cos \beta}\right)^2 - m^2 \right] = 0. \tag{149}$$

Let us look now for the worldline $x = x(\tau)$. Since we will take τ as the center of mass proper time, in what follows the constant ρ will be taken as $\rho = 1$. Note now that

$$p = \alpha_0 \left(f_0 + \frac{\kappa_1}{\kappa_2} f_2 \right), \tag{150}$$

and using eq.(20) we obtain eq.(97) again. Introducing the values of κ_1 and κ_2 we obtain

$$\left(\frac{m^2}{m_0^2}\right) f_0 = \frac{p}{m} + \frac{\sqrt{1 - m^2/m_0^2}}{2m_0} \frac{1}{\cos \beta \sqrt{1 + \Delta^2}} \dot{f}_1, \tag{151}$$

which can be integrated to obtain

$$x = \left(\frac{m_0}{m}\right) \frac{p}{m} \tau + \frac{\sqrt{m_0^2/m^2 - 1}}{2m \cos \beta \sqrt{1 + \Delta^2}} f_1 + x_0. \tag{152}$$

This is a helix with radius r_0 given by

$$r_0 = \frac{\sqrt{m_0^2/m^2 - 1}}{2m \, |\cos\beta| \, \sqrt{1 + \Delta^2}}. \tag{153}$$

Now we calculate the Fermi-Walker derivative of $\{f_\mu\}$ and $\{e_\mu\}$. For $\{f_\mu\}$ we would obtain, of course, the same results of BZ model; but for $\{e_\mu\}$ we have

$$\mathcal{F}_v e_0 = 0, \tag{154}$$

$$\mathcal{F}_v e_1 = \frac{\kappa_2(1 - m^2/m_0^2)}{\sqrt{1 + \Delta^2}} e_2 - 2\sin\beta\,\alpha_1 e_3, \tag{155}$$

$$\mathcal{F}_v e_2 = \frac{-\kappa_2(1 - m^2/m_0^2)}{\sqrt{1 + \Delta^2}} e_1 - 2\sin\beta\,\alpha_2 e_3, \tag{156}$$

$$\mathcal{F}_v e_3 = \frac{\kappa_1 \tan\beta}{\sqrt{1 + \Delta^2}} b_1. \tag{157}$$

Finally, we remark that, as in BZ model, explicit additional informations can be obtained only after solving the equation for ψ. In next section we will make some remarks about those equations.

SOME REMARKS ABOUT THE EQUATIONS

Let us make some comments about the equations for ψ both in BZ and GBZ model. First, consider BZ model. The parameter τ appearing in BZ lagrangian (44) is an invariant time parameter which is not to be identified a priori with proper time. In fact, if τ is identified a priori with proper time then the term $(\dot{x} - \psi\gamma_0\tilde{\psi})$ has to be replaced by $\dot{x} - (1/\rho)\psi\gamma_0\tilde{\psi})$. In this case the term $p\psi\gamma_0$ in eq.(49) would be replaced by $(p/\rho)\psi\gamma_0 - (p/\rho^2)\psi\gamma_0\tilde{\psi}\partial_{\tilde{\psi}}\rho = 0$ since $\partial_{\tilde{\psi}}\rho = \exp(-\gamma_5\beta)\psi$, and eq.(49) becomes $\dot{\psi} = 0$. The parameter τ can be identified as a proper time of a center of mass, but not for the whole helical motion.

In relation to GBZ model the same comment above applies. However, an interesting fact occurs in GBZ model if we identify τ with proper time from the beginning. Indeed, if the term $(\dot{x} - \psi\gamma_0\tilde{\psi})$ is replaced by $(\dot{x} - (1/\rho)\psi\gamma_0\tilde{\psi})$ in GBZ lagrangian (109) with $F(\sigma, \omega) = (1/2)(\sigma^2 + \omega^2)$, we would obtain the folowing equation for ψ:

$$\dot{\psi}\gamma_{21} = -\lambda(\sigma - \gamma_5\omega)\psi, \tag{158}$$

which can be rewritten as

$$\dot{\psi}\gamma_{21} + \lambda\gamma_0\tilde{\psi}\psi\gamma_0\psi = 0 \tag{159}$$

or also as

$$\dot{\psi}\gamma_{21} + \lambda\gamma_0\gamma_5\tilde{\psi}\psi\gamma_0\gamma_5\psi = 0, \tag{160}$$

which we recognize as the classical analogue of Heisenberg fundamental field equation [9]. We have therefore a classical model for Heisenberg program of description of elementary particles by a unified field.

When τ is not identified a priori with proper time, the equation we obtained for GBZ model was

$$\dot{\psi}\gamma_{21} = p\psi\gamma_0 - \lambda\gamma_0\tilde{\psi}\psi\gamma_0\psi, \tag{161}$$

which can be rewritten, after using $\lambda\rho = m^2/m_0$, as

$$\dot{\psi}\gamma_{21} = P\psi\gamma_0, \tag{162}$$

where

$$P = p - \frac{m^2}{m_0} e_0. \tag{163}$$

This equation looks like BZ one, but there is one problem in solving GBZ equation due to the fact that now we have $\dot{P} \neq 0$ in general, while in BZ model it appears only p which satisfies $\dot{p} = 0$. Indeed, until now we do not succeed in finding an interesting solution for eq.(162).

A MODEL INCLUDING RADIATION REACTION

Eq.(48) of BZ model or eq.(113) of GBZ model is the Lorentz force equation. It is well-known that this equation is only approximate, since it does not take into account the interaction of charged particle with its self-field. If the effects of the self-interaction are taken into account, then Lorentz equation has to be modified to include the radiation reaction force Γ,

$$\Gamma = \frac{2e^2}{3}(\ddot{v} + (\dot{v})^2 v), \tag{164}$$

and the equation becomes the Lorentz-Dirac one [29]:

$$\dot{P} = eF \cdot v + \frac{2e^2}{3}(\ddot{v} + (\dot{v})^2 v). \tag{165}$$

There is no need to comment about the importance of constructing a spinning particle model which includes the radiation reaction effects, and our objective in this paper is just to generalize the BZ model in order to include those effects.

Our model is based on the following modification of the BZ lagrangian ($\hbar = c = 1$):

$$L = \left\langle \tilde{\psi}\dot{\psi}\gamma_2\gamma_1 + p(\dot{x} - \psi\gamma_0\tilde{\psi}) + eA\psi\gamma_0\tilde{\psi} + K(\Pi - \ddot{x} \wedge \dot{x}) + \lambda(\Pi \cdot \dot{x})(\psi\gamma_0\tilde{\psi}) \right\rangle_0, \tag{166}$$

where the bivector K is introduced as a Lagrange multiplier and the bivector Π is a new variable (which will be interpreted later). The equations of motion that follows from this lagrangian are ($v = \dot{x}$):

$$\dot{\psi}\gamma_2\gamma_1 = (p - eA - \lambda\Pi \cdot v)\psi\gamma_0, \tag{167}$$

$$\dot{x} = v = \psi\gamma_0\tilde{\psi}, \tag{168}$$

$$\Pi = \dot{v} \wedge v, \tag{169}$$

$$\frac{d}{d\tau}\left[(p - eA - \lambda(\Pi \cdot v)) + 2K \cdot v + \dot{K} \cdot v\right] = eF \cdot v, \tag{170}$$

$$K = -\lambda v \wedge (\psi\gamma_0\tilde{\psi}). \tag{171}$$

We observe that in deriving eq.(170) one must add a term $\frac{d^2}{d\tau^2}\left(\frac{\partial L}{\partial \ddot{x}}\right)$ to the usual Euler-Lagrange equation because our lagrangian involves $\ddot{x}$ [30]. Now, due to eq.(168), we have from eq.(171) that

$$K = 0, \tag{172}$$

and our system of equations become

$$\dot{\psi}\gamma_2\gamma_1 = \pi\psi\gamma_0, \tag{173}$$

$$\dot{x} = v = \psi\gamma_0\tilde{\psi}, \tag{174}$$

218

$$\Pi = \dot{v} \wedge v, \tag{175}$$

$$\dot{\pi} = eF \cdot v, \tag{176}$$

where we denoted

$$\pi = P - \lambda(\Pi \cdot v) = p - eA - \lambda(\Pi \cdot v), \tag{177}$$

P being the kinetic momentum. In what follows we shall identify (as in the BZ model) τ with the center of mass proper time, in such a way that $v^2 = 1$.

Let us write

$$P = mv, \tag{178}$$

so that

$$m = P \cdot v = \pi \cdot v, \tag{179}$$

as in the BZ model. Let us calculate $\dot{m}$; we have

$$\dot{m} = \dot{\pi} \cdot v + \pi \cdot \dot{v} = mv \cdot \dot{v} - \lambda(\Pi \cdot v) \cdot \dot{v} = -\lambda \Pi \cdot (v \wedge \dot{v}), \tag{180}$$

where we used $v \cdot \dot{v} = 0$ (which follows from $v^2 = 1$); and, by eq.(175), it follows that

$$\dot{m} = -\lambda(\dot{v})^2. \tag{181}$$

Using this result we have:

$$
\begin{aligned}
\dot{\pi} &= \dot{P} - \lambda(\dot{\Pi} \cdot v) - \lambda(\Pi \cdot \dot{v}) = \\
&= \dot{m}v + m\dot{v} - \lambda\ddot{v} + \lambda v(v \cdot \ddot{v}) + \lambda v(\dot{v})^2 = \\
&= m\dot{v} - \lambda(\dot{v})^2 v - \lambda\ddot{v},
\end{aligned}
\tag{182}
$$

where we used $v \cdot \ddot{v} = -\dot{v}^2$, and now eq.(176) becomes

$$m\dot{v} = eF \cdot v + \lambda(\ddot{v} + (\dot{v})^2 v), \tag{183}$$

which is the Lorentz-Dirac equation once we identify

$$\lambda = \frac{2e^2}{3}. \tag{184}$$

Now, from the definition of π, we have that

$$\pi \wedge v = -\lambda(\Pi \cdot v) \wedge v. \tag{185}$$

But from the equation for Π we have

$$\Pi \cdot v = \dot{v}, \tag{186}$$

and this result gives

$$\pi \wedge v = -\lambda\Pi. \tag{187}$$

Let us define now a bivector S such that

$$\dot{S} = -\lambda\Pi. \tag{188}$$

From eq.(187) we have that

$$\frac{d}{d\tau}(S + x \wedge \pi) = x \wedge f, \tag{189}$$

where $f = eF \cdot v$ is the Lorentz force. Quantity S plays the role of internal angular momentum. We must observe now the intriguing association between zitterbewegung and radiation reaction. In fact, if zitterbewegung is interpreted as a local circulatory motion, then the momentum π and the velocity v cannot be collinear, and this implies that $\pi \wedge v \neq 0$, which from eq.(187) implies $\lambda \neq 0$ since in this case $\Pi \neq 0$. Another way to see this is to use eq.(178) and eq.(186) in order to write π as

$$\pi = mv - \lambda \dot{v}, \qquad (190)$$

which implies $\lambda \neq 0$. Indeed, it is physically reasonable that *the zitterbewegung (understood as a local circulatory motion) be associated with self-interactions instead of the spin* (in fact, our previous analysis showed that the BZ model has an intrinsic spin not related to zitterbewegung).

Another fact that deserves further comments is eq.(181). From eq.(184) it gives $\dot{m} = -\frac{2e^2}{3}\dot{v}^2$, and since mass m in our model comes from energy, that is, $m = P \cdot v$, then we must have associated with $\dot{m}$ a variation in time of the energy-momentum. If we write $\dot{m} = \dot{Q} \cdot v$, it follows that the quantity $\dot{Q}$ is given by $\dot{Q} = -\frac{2e^2}{3}\dot{v}^2 v$, which is just [29] the radiation energy momentum.

Now consider eq.(173). If we multiply it on the right by $\tilde{\psi}$ and subtract from the result the reverse of eq.(173) multiplied on the left by ψ we obtain that

$$\frac{d}{d\tau}\left(\frac{1}{2}\psi\gamma_2\gamma_1\tilde{\psi}\right) + v \wedge \pi = 0, \qquad (191)$$

and comparing this equation with eq.(187) we get:

$$S = \frac{1}{2}\psi\gamma_2\gamma_1\tilde{\psi}. \qquad (192)$$

On the other hand, from eq.(174) we can calculate $\dot{v}$ by using eq.(173). We obtain that

$$\dot{v} = 4S \cdot \pi, \qquad (193)$$

where we used eq.(192). But using eq.(190) we can write eq.(193) as

$$\dot{v} = 4mS \cdot v + 4\lambda S \cdot \dot{v}. \qquad (194)$$

In addition to eq.(194) for $\dot{v}$, we have the Lorentz-Dirac equation, which can be rewritten as

$$\dot{v} = \left(\frac{e}{m}F + \frac{\lambda}{m}\dot{\Pi}\right) \cdot v, \qquad (195)$$

and also, from eq.(175):

$$\dot{v} = \Pi \cdot v = -\frac{1}{\lambda}\dot{S} \cdot v, \qquad (196)$$

where we used eq.(188). Comparing eq.(194) and eq.(196) we obtain

$$\dot{S} \cdot v = -4m\lambda S \cdot v + (2\lambda)^2 S \cdot \dot{v}. \qquad (197)$$

This equation shows that we cannot have $S \cdot v = 0$ unless $\lambda = 0$; in fact, if $S \cdot v = 0$ then $\dot{S} \cdot v = -S \cdot \dot{v}$, and eq.(197) would give $(2\lambda)^2 = -1$. We must have therefore $S \cdot v \neq 0$ when $\lambda \neq 0$, that is, when radiation reaction is included; but $S \cdot v \neq 0$ means that we have an *electric dipole moment*, and according to our model, this electric dipole moment is a radiation reaction effect. It is important to observe that some experimental data

[31] give an upper limit for the electron's electric dipole moment – $(-0.3 \pm 0.8) \times 10^{-26}$ e-cm [32] (see also [33]).

It is natural now to study the present model along the same lines we studied in the preceeding sections BZ model and generalizations. These studies, however, will be postponed for another occasion.

Acknowledgments: We are grateful to W. Seixas for helpful discussions and for checking some of our calculations using Reduce, and also to G. Cabrera, E. Recami and Q. A. G. Souza, and at NATO-ASI to D. Bambusi, A. Barut, E. Hynds, A. Laufer, A. Orlowski, J. Ralph, J. Reignier, F. Stumpf, D. Taylor, Z. Turakulov, N. Unal, T. Waite and A. Weis. We are also grateful to CNPq for finnancial support.

References

[1] Pavšic, M., Recami, E., Rodrigues, Jr., W. A. Maccarrone, G. D., Raciti, F. and Salesi, G., Phys. Lett. **B 318**, 481 (1993).

[2] Barut, A. O. and Zanghi, N., Phys. Rev. Lett. **52**, 2009 (1984).

[3] Berezin, F. A. and Marinov, M. S., Ann. Physics **104**, 336 (1977).

[4] DeWitt, B., *Supermanifolds*, Cambridge University Press (1984).

[5] Oziewicz, Z., in *Clifford Algebras and Their Applications in Mathematical Physics*, Chisholm, J. S. R. and Common, A. K. (eds.), pg. 245, D. Reidel (1985).

[6] Rodrigues, Jr., W. A., Vaz, Jr., J., Recami, E. and Salesi, G., Phys. Lett. **B 318**, 623 (1993).

[7] Rodrigues, Jr., J. and Vaz, Jr., J., in *Clifford Algebras and Their Applications in Mathematical Physics*, Brackx, F., Delanghe, D. and Serras, H. (eds.), pg. 397, Kluwer (1993).

[8] Barut, A. O. and Duru, I. H., Phys. Rev. Lett. **53**, 2355 (1984).

[9] Heisenberg, W., *Introduction to the Unified Field Theory of Elementary Particles*, Interscience (1966).

[10] Lounesto, P., Found. Phys. **23**, 1203 (1993).

[11] Figueiredo, V. L., Oliveira, E. C. and Rodrigues, Jr., W. A., Int. J. Theor. Phys. **29**, 371 (1990).

[12] Rodrigues, Jr., W. A., Souza, Q. A. G., Vaz, Jr., J. and Lounesto, P., "Dirac-Hestenes Spinor Fields, Their Covariant Derivatives, and Their Formulation on Riemann-Cartan Manifolds", preprint RP 64/93 IMECC-UNICAMP, submitted for publication.

[13] Hestenes, D., Found. Phys. **20**, 1213 (1991).

[14] Faria-Rosa, M. A. and Rodrigues, Jr., W. A., Found. Phys. **19**, 705 (1989).

[15] Sachs and Wu, *General Relativity for Mathematicians*, Springer-Verlag (1977).

[16] Salam, A. and Strathdee, J., Nucl. Phys. **76** 477 (1974).

[17] Schönberg, M., An. Acad. Brasileira Ciências **28**, 11 (1956).

[18] Witten, E., Mod. Phys. Lett. **A5**, 487 (1990).

[19] Rodrigues, Jr., W. A., Vaz, Jr., J. and Pavšič, M., "The Clifford Bundle and the Dynamics of the Superparticle", preprint IMECC-UNICAMP, submitted for publication.

[20] Rodrigues, Jr., W. A., Souza, Q. A. G. and Vaz, Jr., J., in *Gravitation: The Spacetime Structure*, Letelier, P. S. and Rodrigues, Jr., W. A. (eds.), pg. 534, World Scientific (1994).

[21] Lasenby, A., Doran, C. and Gull, S., Found. Phys. **23**, 1295 (1993).

[22] Gull, S., in *The Electron*, Hestenes, D. and Weingartshof, A. (eds.), pg. 37, Kluwer (1991).

[23] Rawnsley, J., Lett. Math. Phys. **24**, 331 (1992).

[24] Barut, A.O. and Pavšič, M., *Class. Quantum Gravity*, **4**, L41 (1987); **4** L131 (1987); **5** 707 (1988).

[25] Barut, A.O. and Unal, N., *Found. Phys.*, **23**, 1423 (1993).

[26] Recami, E., Found. Phys. **17**, 239 (1987).

[27] de Broglie, L., *Ondes Électromagnétiques et Photons*, Gauthiers-Villars (1967).

[28] Hestenes, D., J. Math. Phys. **15**, 1768 (1974).

[29] Barut, A.O., *Electrodynamics and Classical Theory of Fields and Particles*, Dover Publ. (1980).

[30] Lovelock, D. and Rund, H., *Tensors, Differential Forms and Variational Principles*, John Wiley & Sons (1975).

[31] MacGregor, M., *The Enigmatic Electron*, Kluwer (1992).

[32] Particle Data Group, "Review of Particle Properties", *Phys. Rev.* **D45**, Part II (1992).

[33] Weis, A., in these proceedings.

THE PURELY ELECTROMAGNETIC ELECTRON RE-VISITED

Tom Waite[1], Asim O. Barut[2], and José R. Zeni[3]

[1]13778 Shablow Ave.
Sylmar, CA 91342
USA

[2]Dept. of Physics
Colorado University
Boulder, CO 80309
USA

[3]Depto. Ciencas Naturai
FUNREI
Sao Joao Del Rei
MG 36300, Brazil

1. INTRODUCTION AND SUMMARY

Shortly after Thompson established the existence of the electron as a particle with charge (-e) and mass (m_o), Abraham[1] in 1903 and Lorentz[2] in 1904 proposed that the electron was a Purely Electromagnetic Particle (PEP). Their electron model, when at rest, was simply a uniform sphere of negative charge. They each noted that, when such a sphere of charge moves, it generates a magnetic field such that the Poynting vector or inertial momentum density is not zero. Therefore, if the sphere of charge is accelerated, there is a time rate of change of inertial momentum in its self fields. By Newton's Law, this time rate of change in momentum can be generated only by imposing forces (external fields) on the sphere of charge.

Abraham and Lorentz each calculated the change in inertial momentum due to changing velocity. The coefficient in the first derivative of velocity with respect to time, corresponding to Newtonian mass (m_o), was found to be $[4/(3c^2)]$ times the energy required to bring the charge density from infinity into the sphere at rest. The equivalence of mass and energy $[E = mc^2]$ was not recognized until a year later by Einstein, at which time the (4/3) numerical factor indicated a problem. Lorentz's 1904 electron model displayed Lorentz/Fitzgerald contraction, but lacked perfect Lorentz invariance.

Poincaré[3] in 1905 and 1906 brought attention to the fact that the sphere of charge is unstable, and that the electric forces will drive it apart. He analyzed the problem in sufficient depth to conclude that a PEP with internal charge/current density $[(\rho,\mathbf{j}) \neq 0]$ cannot exist without imposing other non-electromagnetic forces. Ehrenfest[4] in 1907 generalized this PEP Non-Existence Theorem, and it has been accepted as physics dogma for eighty years.

There is an exception to that PEP non-existence theorem which justifies a re-examination of the PEP concept. That exception is based on the equation given by Einstein[5] in 1919 and by Pauli[6] in 1921 as necessary and sufficient for the existence of a purely electromagnetic electron with finite and continuous internal charge/current density. Examples of exact analytic solutions and a general existence theorem for solutions exploiting that exception are presented here.

Continuous solutions to that Einstein/Pauli PEP equation have global properties with discrete topological structures. Each discrete PEP structure is centered on a localized charge/current density droplet with vortex-like internal re-circulation. Waite[7] has derived from that Einstein/Pauli PEP equation a relativistic Helmholtz-like flux conservation theorem which assures particle-like integrity and permanence for these vortex-like PEP droplets, called *vortons*. All PEP vorton droplets are pairwise disjoint by virtue of the Einstein/Pauli PEP equation.

Electron Theory and Quantum Electrodynamics: 100 Years Later
Edited by Dowling, Plenum Press, New York, 1997

The PEP internal vortex-like recirculation has a superluminal component that produces no net charge/current displacements, so that causality is not violated. Orthogonal to this superluminal flow is a subluminal component of charge/current density which accounts for any and all PEP displacements and for allowed PEP deformations which preserve the PEP topology. The superluminal component of pure re-circulation provides magnetic force which eliminates the need for Poincaré (non-electromagnetic) stresses. This is the basis for the exception to the Poincaré/Ehrenfest PEP non-existence theorem, which assumes that $(\rho, \mathbf{j})$ is a time-like 4-D vector field. Since the superluminal component of recirculation produces no time variation in $(\rho, \mathbf{j})$, it produces no radiation and no radiation damping, so that the superluminal recirculation is a perpetual flow.

Real world experiments have never been able to detect a 3-D droplet of charge/current density at the center of each fundamental particle. But that is not inconsistent with the PEP model. The relativistic Helmholtz flux conservation theorem[7] establishes, topologically, that each such PEP droplet of charge/current density is pairwise disjoint with respect to every other PEP droplet. Assume for a moment that, within a system described by the Einstein/Pauli PEP equation, there exist Intra-System Observers (ISO). The ISO measuring instruments and probe particles are all comprised of PEP droplets obeying the Einstein/Pauli PEP equation. Since every ISO/probe PEP droplet is pairwise disjoint from the measured PEP droplet, the ISO cannot penetrate, probe or measure any fields inside of the measured PEP droplet. To the ISO, the measured PEP droplets are structureless objects with inexplicable topological integrity and permanence and an extended PEP self field. This is not unlike the experimental situation in the real world.

In the PEP system, the only PEP mass (energy) and momentum are $(1/2)\ (E^2 + B^2)$ and $(\mathbf{E} \times \mathbf{B})$ integrated over the PEP's extended wave-like self fields, which fall to zero only at infinity. This extended inertial mass and momentum cause each PEP charge/current density droplet to have wave-like propagation, in spite of its particle-like integrity. To the ISO, the PEP droplet is a structureless point particle with wave-like propagation.

The ISO can measure only PEP properties integrated or averaged over a volume greater than the PEP droplet. These properties might include PEP total charge, PEP total mass, PEP total linear momentum, average position of the PEP center of mass, and PEP total integrated spin angular momentum. Accurate measurement of total mass (energy) and momentum requires an experiment which integrates over the extended self fields well outside the $(\rho, \mathbf{j})$ droplet, also.

Since the ISO cannot measure the initial fields inside the PEP charge/current density, they cannot time integrate the deterministic Einstein/Pauli PEP equation to predict future PEP properties. The ISO must consider probability statistics for the ensemble of all PEP droplets whose unmeasureable internal fields are consistent with the few integrated or averaged properties of the PEP droplet which the ISO can measure.

These unusual conclusions by Intra-System Observers (ISO) regarding vorton solutions to the Einstein/Pauli PEP equation, arise as a result of the unusual topological properties of all solutions to that PEP equation. The existence of PEP solutions to the Einstein/Pauli PEP equation, and these unusual properties of that PEP system, as seen from the inside by any ISO, provide the motivation for this re-examination of the purely electromagnetic electron concept.

2. THE POINCARÉ/EINSTEIN/PAULI PEP EQUATION

Proof of the results summarized above, requires only 3-D vector analysis with Gauss' Theorem and Stoke's Theorem on 3-D space. However, the most meaningful derivation of the PEP equation is that of Einstein using 4-D tensor analysis. That proof is stated briefly before returning to 3-D vector analysis for the remainder of this paper.

Einstein[5] considered the PEP in the curved space time of general relativity in an effort to determine whether the gravitational force could provide the non-electromagnetic "Poincaré Stresses". His analysis is valid on flat space time, also. He noted that, in principle, the 4-D Stress Energy Momentum (SEM) tensor $[S^{\alpha\beta}]$ of a system can be written as the sum of a purely electromagnetic SEM tensor:

$$T^{\alpha}{}_{\beta} = F^{\alpha\gamma}F_{\gamma\beta} - (1/4)(F^{\gamma\delta}F_{\delta\gamma})\delta^{\alpha}{}_{\beta} \qquad (1)$$

and a non-electromagnetic SEM tensor $W^\alpha{}_\beta$ such that $S^\alpha{}_\beta = T^\alpha{}_\beta + W^\alpha{}_\beta$. Here, $F_{\alpha\beta}$ is the electromagnetic field. The relativistic conservation of energy and momentum requires that the divergence of the total system SEM tensor $S^\alpha{}_\beta$ vanishes everywhere:

$$S^\alpha{}_{\beta,\alpha} = T^\alpha{}_{\beta,\alpha} + W^\alpha{}_{\beta,\alpha} = 0 \tag{2}$$

Here Greek indices run from 0 to 3, terms are summed on repeated indices, and indices following a comma indicate covariant derivatives.

Einstein defined a PEP as a 3-D particle having no non-electromagnetic (Poincaré) stresses and no non-electromagnetic mass or momentum density. Therefore $[W^\alpha{}_\beta = 0]$ and Eq. (2) becomes:

$$S^\alpha{}_{\beta,\alpha} = T^\alpha{}_{\beta,\alpha} = J^\alpha F_{\alpha\beta} = (A^{\alpha,\gamma}{}_\gamma - A^{\gamma,\alpha}{}_\gamma)(A_{\beta,\alpha} - A_{\alpha,\beta}) = 0 \tag{3}$$

where:

$$F_{\alpha\beta} \stackrel{\text{def}}{=} A_{\beta,\alpha} - A_{\alpha,\beta} \tag{4}$$

$$J^\alpha \stackrel{\text{def}}{=} F^{\alpha\beta}{}_{,\beta} \tag{5}$$

This defines the electromagnetic fields ($F_{\alpha\beta}$) and the charge/current density (J^α) in terms of the potential (A^α). This assures that Maxwell's equations are obeyed and leaves only the four component Eq. (4) to determine the four components of the potential (A_α). Immediately after presenting Eq. (4), Einstein[5] remarked that no solutions can exist with $[J^\alpha \neq 0]$, apparently overlooking or ignoring the solutions with $\det |F_{\alpha\beta}| = (\mathbf{E} \cdot \mathbf{B})^2 = 0$.

Pauli[6] in 1921, reviewed the earlier work on the PEP electron as part of his classic review of relativity theory. He considered the behavior of a localized, isolated charge current density ($\rho, \mathbf{j}$) droplet subjected to its own self generated fields $[\mathbf{E}_{\text{self}}, \mathbf{B}_{\text{self}}]$ and to externally applied fields $[\mathbf{E}_{\text{ext}}, \mathbf{B}_{\text{ext}}]$ generated by remote sources. By virtue of $[E = mc^2]$, that PEP's self mass M is the part of T^{00} generated by the self fields:

$$M = \iiint (1/2)(E^2_{\text{self}} + B^2_{\text{self}})dV \tag{6}$$

and the electromagnetic self momentum $\mathbf{P}$ is the part of T^{0k} (with $k = 1,2,3$) generated by the self fields:

$$\mathbf{P} = \iiint (\mathbf{E}_{\text{self}} \times \mathbf{B}_{\text{self}})dV \tag{7}$$

Taking the time derivative of Eq. (7) and applying vector identities (cf. Jackson[8]), one may prove that:

$$\dot{\mathbf{P}} = -\iiint (\mathbf{j} \times \mathbf{B}_{\text{self}} + \rho\mathbf{E}_{\text{self}}) \, dV + \text{surface integrals} \tag{8}$$

where the surface integrals are zero or represent physical self radiation. But Newton's law and the relativistic Lorentz force require that, in external fields:

$$\dot{\mathbf{P}} = \iiint (\mathbf{j} \times \mathbf{B}_{\text{ext}} + \rho\mathbf{E}_{\text{ext}}) \, dV + \text{radiation terms} \tag{9}$$

Equations (9) and (10) are self consistent for any and all external fields, if and only if:

$$\mathbf{j} \times (\mathbf{B}_{\text{self}} + \mathbf{B}_{\text{ext}}) + \rho(\mathbf{E}_{\text{self}} + \mathbf{E}_{\text{ext}}) = 0 \tag{10}$$

A similar analysis of $[\dot{M}]$ gives the fourth equation:

$$\mathbf{j} \cdot (\mathbf{E}_{\text{self}} + \mathbf{E}_{\text{ext}}) = 0 \qquad (11)$$

which is not independent of Eq. (10). In every Lorentz frame on flat space time, Einstein's PEP Eq. (3) is equivalent to Pauli's PEP Eqs. (10), with or without Eq. (11). Poincaré[3] had used Eq. (10) based on D'Alambert's principle.

Pauli, within sentences of presenting Eq. (10), concurred in the conclusions of Poincaré and Ehrenfest, noting that they had considered only the charge density and electric terms in the PEP Eq. (10). That limited analysis by Poincaré and Ehrenfest is valid if, and only if, the charge/current density $(\rho,\mathbf{j})$ is a time-like 4-D vector field satisfying:

$$j^2 - \rho^2 \leqq 0 \qquad (12)$$

where the velocity of light is one. Then, for each point, there exists a Lorentz frame such that: $\mathbf{j} = 0$, $\rho \neq 0$ in that frame at that point. There also exists a gauge such that: $\nabla^2 \phi = -\rho$, where ϕ is the electric potential. Therefore ϕ cannot have an extremum, and, in that frame at that point, $[\nabla \phi \neq 0]$. With the pathological exception that: $[\dot{\mathbf{A}} = -\nabla \phi]$ in each such frame, at each such point, it follows that:

$$\mathbf{j} = 0 \ , \quad \rho \neq 0 \ , \quad \mathbf{E} \neq 0 \qquad (13)$$

at every point in at least one Lorentz frame. Therefore, there are no solutions to Eq. (10) with $[(\rho,\mathbf{j}) \neq 0]$, where Eq. (12) is satisfied.

The exception to the Poincaré/Ehrenfest PEP non-existence proof arises for space-like and null charge/current density fields:

$$j^2 - \rho^2 \geqq 0 \qquad (14)$$

Because of the customary interpretation of charge/current density as moving point charges, Eq. (14) immediately implies a violation of causality to most physicists. But here $(\rho,\mathbf{j})$ is not a system of moving point particles. As defined by Maxwell:

$$\rho \equiv \nabla \cdot \mathbf{E}$$

$$\mathbf{j} \equiv \nabla \times \mathbf{B} - \dot{\mathbf{E}} \qquad (15)$$

charge/current density may be interpreted as a derivative of the finite continuous electromagnetic field existing on the space/time vacuum, so that $(\rho,\mathbf{j})$ itself is a finite and continuous field on the vacuum. A space-like charge current density $(j^2 > \rho^2)$ satisfying the Einstein Pauli PEP Eq. (3) does not violate causality. Heuristically, transmission of information or signals occurs only by transport of signal energy as defined by the electromagnetic SEM tensor. But that transport velocity for the signal energy in the self fields of a superluminal charge/current density is still:

$$v = \mathbf{E} \times \mathbf{B} / ([1/2] [E^2 + B^2]) \leqq 1 \qquad (16)$$

With greater mathematical precision, it is proven in Section (4) that, if the PEP Eq. (3) or (11) is obeyed, and if the scalar invariants satisfy:

$$j^2 - \rho^2 \geqq 0 \ ; \quad B^2 - E^2 \geqq 0 \qquad (17)$$

then no disturbance can propagate through the electromagnetic field with velocity greater than one, the speed of light. One may expect that, in regions where $[j^2 \geqq \rho^2]$, the self magnetic fields will be greater than the self electric fields. Then the resulting magnetic forces satisfy $[B^2 \geqq E^2]$ and may eliminate the need for Poincaré stresses (non-electromagnetic forces).

226

There exists PEP charge density but no PEP magnetic monopole density. Outside the regions of charge/current density, **B** falls off more rapidly than **E**. Therefore: $B^2 \geqq E^2$ need not apply, nor is it required for causality, where $(\rho, \mathbf{j}) = 0$.

In all of the PEP solutions with charge/current density, the superluminal component of internal current flow is a steady state recirculation everywhere parallel to closed or endless magnetic field lines, with no net displacement of charge/current density. Displacement of the charge/current density and propagation of disturbances through the charge/current density occur only perpendicular to **B** at velocity **E** x **B**/B^2, which is never faster than the speed of light, so long as $[B^2 - E^2 \geqq 0]$ inside that charge/current density. This is proven in Section (4) below.

There exist such solutions to the Einstein/Pauli PEP Eq. (3) with recirculating superluminal charge/current density which does not violate causality. Some examples of such exact analytic solutions are presented in Section (3) and an Existence Theorem for more general solutions is stated and proved in Section (6). Because the PEP Eq. (3) is non-linear, the solutions of greatest interest must be generated by computer analysis.

3. EXAMPLE ANALYTIC PEP SOLUTIONS

The Einstein/Pauli PEP Eq. (4) is an intractable non-linear equation. Exact analytic solutions may be constructed by general methods only if one imposes severe constraints on the solutions corresponding to a high degree of symmetry. Although one can visualize and give a physical interpretation to the fields and charge/current densities occurring in such constrained solutions, they rarely correspond to anything which one can identify in the real world. In spite of this limitation, one may use such constrained, exact solutions to infer things about more general exact solutions. In particular, they illustrate the chiral, vortex-like $(\rho, \mathbf{j})$ droplets typical of solutions to the PEP Eq.(3) or (10).

This section considers the limited class of constrained solutions to the PEP Eq. (3) which, in one singular Lorentz frame, are time independent with the electric field **E** everywhere zero. In any other Lorentz frame, moving at even a millimeter per century, the transformed solutions are time dependent and the electric field is not zero. But in that singular frame:

$$\mathbf{j} \times \mathbf{B} + \rho \mathbf{E} = (\nabla \times \mathbf{B}) \times \mathbf{B} = 0 \tag{18}$$

with [**E** • **B**] and [**E** • **j**] identically zero. This Eq. (18) is satisfied if the magnetic field is everywhere parallel to its curl, the current field: $\mathbf{j} = \nabla \times \mathbf{B} = f\mathbf{B}$, where: $\mathbf{B} \cdot \nabla f = 0$]. The solutions to this equation with [f] constant have been studied in relation to other physical systems by Chandrasekhar and Kendall[9,10] and others[11].

When f is constant:

$$\nabla \times \mathbf{B} = \chi \kappa \mathbf{B} \tag{19}$$

where χ is chirality and κ is the soliton wave number. Taking the curl of Eq. (19) one obtains the sufficient, but not necessary equation:

$$\nabla \times \nabla \times \mathbf{B} = \kappa^2 \mathbf{B} \tag{20}$$

This is the spatial part of the Helmholtz vector wave equation. It's well known solutions are the vector multipole fields. But those solutions occur in pairs of opposite parity, $\mathcal{E}_{lm}$ and $\mathcal{B}_{lm}$ such that: $\nabla \times \mathcal{B} = -i\kappa\mathcal{E}$ and $\nabla \times \mathcal{E} = i\kappa\mathcal{B}$ which do not satisfy Eq. (21). However, the chiral, mixed parity fields:

$$\mathbf{B}_{lm}^{\chi} (\kappa r, \phi, \theta) = \mathcal{B}_{lm} (\kappa r, \phi, \theta) - i\chi\mathcal{E}_{lm} (\kappa r, \phi, \theta) \tag{21}$$

satisfy Eq. (19) where χ is the screw sense or chirality of the fields, and κ is the wave number of the solitons in each solution. If one uses the spatial parts of the standing wave

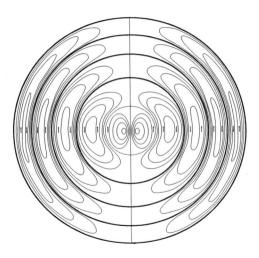

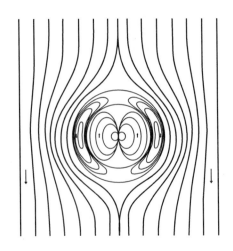

Figure 1. Schematic diagram of the meridional plane components of some of the magnetic and current flow field lines in the simplest array of PEP vortons as given by Eq. (22).

Figure 2. Schematic diagram of the meridional plane components of some of the magnetic field lines for an isolated PEP spherical shell vorton, bound to a point dipole in a uniform external field as given by Eqs. (22)–(24).

multipole fields, the solutions are finite, single valued and continuous at each point, and vanish at infinity.

The simplest solution corresponds to the magnetic dipole field. It is of the form:

$$B_r = 2bB^* \{[\kappa r]^{-1} j_1(\kappa r)\} \cos\theta$$

$$B_\theta = bB^* \{(\kappa r)^{-1} j_1(\kappa r) - j_0(\kappa r)\} \sin\theta$$

$$B_\phi = \chi bB^* \{j_1(\kappa r)\} \sin\theta \qquad (22)$$

with $\mathbf{j} = \nabla \times \mathbf{B} = \chi\kappa\mathbf{B}$, $\mathbf{E} = 0$, and $\rho = \rho_0$. Here, $j_n(\kappa r)$ is the spherical Bessel function of order n. The uniform background charge density $[\rho_0]$ is set equal to zero in the only case considered here. Waite[7] has described the geometry of this solution in detail because it occurs as a solution in a variety of relativistic and non-relativistic systems with flux conservation analogous to perfect vorticity conservation.

Figure (1) illustrates the topology of the magnetic and parallel current field lines in the meridional plane for the magnetic and current field lines of Eq. (22). The similarity with the familiar dipole radiation field picture requires that one distinguish the magnetic fields of Eq. (22) as time independent, as possessing an azimuthal (ϕ) component not shown in Fig. (1), and as possessing a non-zero current density parallel to the magnetic field throughout space.

When Figure (1) is rotated about its polar axis, the dark circles form 2-D spheres which separate 3-D space into a central 3-D sphere and concentric 3-D spherical shells. Since the normal components of $[\mathbf{B}]$ and $[\mathbf{j}]$ vanish on each of these 2-D spheres separating the 3-D shells, the magnetic field lines in this continuous field are separated into discrete sets, one set in each 3-D spherical shell. The charge/current density, similarly, segregates into discrete re-circulating flow fields in each 3-D shell, with no flow across the boundaries between the discrete 3-D shells.

When Fig. (1) is rotated about its polar axis, each set of nested 1D ellipses in Fig. (1) forms a set of nested 2-D toruses of elliptic cross section. The magnetic field lines (and current flow lines) all lie on the nested 2-D toruses, just as they lay on the nested ellipses of Fig. (1). However, as in Eq. (22), the magnetic and current fields possess an azimuthal (ϕ) component which causes them to wind around the 2-D toruses with screw sense or chirality, χ. The winding number must be constant on each 2-D torus so that the field lines never intersect, as required by the single valued fields of Eq. (22). The winding number, although constant on each 2-D torus, varies continuously from infinity on the innermost nested 2-D torus (which is degenerate as a 1-D circle), down to zero on the outermost 2-D torus (which is degenerate with its polar axis as part of the degenerate torus hole). Note that each magnetic field line in any one set of nested toruses links with every other magnetic field line in that same set of nested toruses. The linking of any two field lines defines a chirality (χ) with the same screw sense as the winding number of the field lines on the 2-D toruses. This same topology of field lines winding on nested 2-D toruses is repeated in each of the 3-D spherical shells formed by rotating Fig. (1) about its polar axis.

Each 3-D spherical shell forms a discrete topological structure in a continuous field. Each has its own discrete vortex-like recirculating charge/current density, independent of, but continuous with, the vortex-like recirculation in each of the other 3-D spherical shells. These are the discrete *vortons* which occur, when the requirement that solutions be finite, single valued and continuous, causes the unique local properties of partial differential equations, such as the PEP Eq. (3), to impose global properties on their solutions. These vortons are studied from a global geometric viewpoint in Sections (4) and (5) below.

One may create a solution for a single PEP vorton by discarding all other vortons in the array, and requiring that the fields outside the retained PEP vorton satisfy Maxwell's free space equations [$(\rho, \mathbf{j}) = 0$] with continuous fields at the boundaries of the retained PEP vorton, as shown in Fig. (2). If one retains only the vorton in the spherical shell between $r = R_n$ and $r = R_{n+1}$ where the R_n are the zeroes of the Bessel function $j_1(\kappa R_n) = 0$, then the fields for $r \geq R_{n+1}$ are:

$$B_r = B^* \, [1 - (R_{n+1}/r)^3] \cos\theta$$

$$B_\theta = B^* \, [1 + (1/2) \, (R_{n+1}/r)^3] \sin\theta$$

$$B_\phi = 0 \tag{23}$$

with $\mathbf{j} = 0$, $\mathbf{E} = 0$ and $\rho = \rho_0 = 0$. The solution given in Eq. (22) for the array now applies only for $R_n \leq r \leq R_{n+1}$ with b chosen to give a continuous magnetic field at $r = R_{n+1}$.

The fields for $r \leq R_n$ are:

$$B_r = B_0 \, [1 - (R_n/r)^3] \cos\theta$$

$$B_\theta = B_0 \, [1 + (1/2) \, (R_n/r)^3] \sin\theta$$

$$B_\phi = 0 \tag{24}$$

with $\mathbf{j} = 0$, $\mathbf{E} = 0$ and $\rho = \rho_0 = 0$. Here B_0 is chosen so that the magnetic field is continuous at $r = R_n$. A schematic diagram of the magnetic field line topology is given in Fig. (2). The externally applied fields for the PEP Eq. (3), are $\mathbf{E}_{ext} = 0$, with $\mathbf{B}_{ext}$ a uniform magnetic field of magnitude B^* due to sources at infinity, and a point dipole field: $B_r = -B_0 \, (R_n/r)^3 \cos\theta$, $B_\theta = (1/2)B_0 \, (R_n/r)^3 \sin\theta$, $B_\phi = 0$, with its source at the origin. The self field is the difference between the fields given by Eqs. (22)-(24) and the sum of these two externally applied fields.

Physically, one might interpret this steady state solution as a 3-D PEP vorton in a non-radiating bound state, bound to a point magnetic dipole of infinite mass in a uniform external field. Such bound PEP vorton solutions are suggestive of Schroedinger's[12] attempts to describe his atomic wave function as physically real charge/current density with [$\rho = \psi^*\psi$] and [$\mathbf{j} = \psi^*\nabla\psi - \psi\nabla\psi^*$]. Using this interpretation and time dependent perturbation theory, Schroedinger derived the atomic radiation frequencies, the selection rules of atomic spectra and computed the radiating atomic dipole moment magnitudes from this oscillating charge/current density. He attempted to explain each stationary bound state as a steady state charge/current density with no time dependent multipole moments, analogous to this PEP spherical shell charge/current density bound to a point dipole. However, Schroedinger's linear equation failed to give to the electron, the integrity and permanence of a soliton, which the non-linear PEP equation gives to its soliton solutions.

4.   PEP FLUX CONSERVATION AND CAUSALITY

The vortex-like steady state solutions to the Einstein/Pauli PEP Eq. (3) suggests that there exists a vorticity like conserved flux which gives topological integrity and permanence to the structure in these solutions. That is indeed the case. Waite[7] has stated and proved a Lorentz invariant generalization of the Helmholtz flux conservation theorem applicable to a broad class of partial differential equations. That class includes the PEP Eq. (3) as well as the equations of fluid mechanics. That theorem may be stated and proved more simply for the specific PEP system:

The PEP Magnetic Flux Theorem: GIVEN an electromagnetic field which satisfies:

$$\nabla \cdot \mathbf{B} = 0 ; \quad \nabla \times \mathbf{E} + \dot{\mathbf{B}} = 0 \tag{25}$$

$$\nabla \cdot \mathbf{E} = \rho ; \quad \nabla \times \mathbf{B} - \dot{\mathbf{E}} = \mathbf{j} \tag{26}$$

$$\mathbf{j} \cdot \mathbf{E} = 0 ; \quad \mathbf{j} \times \mathbf{B} + \rho \mathbf{E} = 0 \tag{27}$$

in a region where:

$$\rho^2 \leqq \mathbf{j}^2 \neq 0 , \quad E^2 \leqq B^2 \neq 0 \tag{28}$$

THEN, in that region, the magnetic field lines are:
  (P1) pairwise disjoint,
  (P2) closed or endless,
  (P3) conserved, and
  (P4) moving point by point with the conserved charge/current density at velocity:

$$\mathbf{v} = \mathbf{j}/\rho = (\mathbf{v} \cdot \mathbf{B}/B^2) \mathbf{B} + \mathbf{E} \times \mathbf{B}/B^2 \tag{29}$$

PROOF: Property (P1) follows because the magnetic field can not have two directions at any one point as would occur if two field lines intersect, even momentarily to pass through one another. Exceptions as field line bifurcation points occur where the magnitude of the field is zero. These are excluded by Eq. (28) of the theorem, but they do occur; and these zero valued topological singularities are responsible for the topological structure in the field.
        Property (P2) follows from $\nabla \cdot \mathbf{B} = 0$ of Eq. (25). If a magnetic field line loose end occurs inside of any infinitesimal sphere, then by Gauss' theorem, $\nabla \cdot \mathbf{B} \neq 0$ inside that sphere.
        Properties (P3) and (P4) are proven by the construction illustrated in Fig. (3). Consider any arbitrary 2-D surface increment Σ bounded by any arbitrary closed curve L, in a region of 3-D space where there exist two vector fields $\mathbf{B}(x,y,z;t)$ and $\mathbf{j}/\rho = \mathbf{v}(x,y,z;t)$. Let the arbitrary Σ and L move point by point at velocity $\mathbf{v}$, so as to move from Σ_1 to Σ_2 during time (dt), as illustrated in Fig. (3). The number of field lines through Σ at any time is $N = \iint B_n d\Sigma$ where the subscript n indicates the component normal to (dΣ) at each point on Σ. The change in N during the time interval (dt) is:

$$\dot{N} \, dt = \iint (\dot{\mathbf{B}}_n \, dt) \, d\Sigma + \oint \mathbf{B} \cdot [(\mathbf{v}dt) \times d\mathbf{L}]$$

plus higher order terms in (dt). The second integral extends around the closed curve L in Fig. (3). Rotating the vector triple product and applying Stoke's Theorem to the second integral, one obtains: $\dot{N} = \iint [\dot{\mathbf{B}} - \nabla \times (\mathbf{v} \times \mathbf{B})]_n \, d\Sigma$. Substituting for $\dot{\mathbf{B}}$ from Eq. (5), one obtains:

$$\dot{N} = \iint [-\nabla \times (\mathbf{E} + \mathbf{v} \times \mathbf{B})]_n d\Sigma = 0$$

which vanishes for all arbitrary Σ by Eq. (27) since $\mathbf{v} = \mathbf{j}/\rho$.

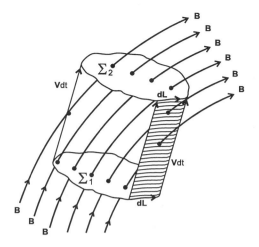
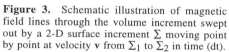
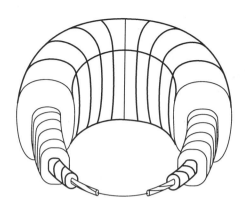

Figure 3. Schematic illustration of magnetic field lines through the volume increment swept out by a 2-D surface increment Σ moving point by point at velocity $\mathbf{v}$ from Σ_1 to Σ_2 in time (dt).

Figure 4. Schematic illustration of some conserved magnetic field lines in an elementary chiral PEP vorton, shown in 3-D perspective with some outer field lines foliating 2-D toruses cut away to expose those inside.

But the number of magnetic field lines penetrating each and every arbitrary Σ surface increment moving at velocity $\mathbf{v} = \mathbf{j}/\rho$ can remain unchanged, if and only if, each and every magnetic field line is conserved and moves point by point at velocity $\mathbf{v} = \mathbf{j}/\rho$ with each and every such Σ it penetrates. This proves the theorem except for showing that $\mathbf{v} = \mathbf{j}/\rho$ satisfies Eq. (29).

From Eq. (27), it is obvious that $\mathbf{E} \cdot \mathbf{B} = 0$, so that $\mathbf{v} = \mathbf{j}/\rho$ can be resolved onto the moving orthogonal triad ($\mathbf{E}$, $\mathbf{B}$, $\mathbf{E} \times \mathbf{B}$). By Eq. (27), $\mathbf{j} \cdot \mathbf{E} = 0$. The coefficient of $\mathbf{B}$ in Eq. (29) is an identity. To obtain $\mathbf{j} \cdot (\mathbf{E} \times \mathbf{B})$, form the scalar product of Eq. (27) with ($\mathbf{E} \cdot$) and rotate the triple product to obtain the $\mathbf{E} \times \mathbf{B}/B^2$ term in Eq. (29). The Theorem is proved.

This proof is similar to one used by Sommerfeld[13] to establish conditions under which fluid vorticity conservation is perfect. Here, the theorem applies in every Lorentz frame, because the given Eqs. (25)-(28) are Lorentz invariant.

Causality is preserved in all solutions satisfying the PEP Magnetic Flux Theorem, in spite of the fact that $\rho^2 \lesseqgtr j^2 \neq 0$. This follows from the fact that $|\mathbf{E} \times \mathbf{B}/B^2| \lesseqgtr 1$ by Eq. (28). Since $\mathbf{v} = \mathbf{j}/\rho$ is superluminal, and $\mathbf{E} \times \mathbf{B}/B^2$ is subluminal, the superluminal component must be the component everywhere parallel to the magnetic field $\mathbf{B}$. But $\mathbf{v}$ is the velocity of the magnetic field lines as well as the velocity of charge/current flow. However, the component of $\mathbf{B}$ field line motion everywhere parallel to each closed or endless field line, does not displace any field line and does not produce any change in $\mathbf{B}$. This is true in every Lorentz frame because Eqs. (25)–(28) are Lorentz invariant. Therefore, the component of $\mathbf{v} = \mathbf{j}/\rho$ parallel to $\mathbf{B}$ produces no change in $\mathbf{E}$ in any Lorentz frame, either. Since this superluminal component of charge/current flow produces no change in $\mathbf{E}$ or $\mathbf{B}$, it produces no change in ρ or $\mathbf{j}$. Therefore it can not violate causality.

The superluminal component of $\mathbf{v} = \mathbf{j}/\rho$ parallel to $\mathbf{B}$ produces no time dependent multipole moments. Therefore it does not radiate. It is undamped and perpetual as a steady state recirculation.

The subluminal component $\mathbf{E} \times \mathbf{B}/B^2$ of $\mathbf{v} = \mathbf{j}/\rho$ perpendicular to $\mathbf{B}$, does displace magnetic field lines and does change the $\mathbf{B}$ and $\mathbf{E}$ fields. To do this, it must change ρ and $\mathbf{j}$. This subluminal component is responsible for the translation, rotation, expansion, contraction and deformation of the ($\rho,\mathbf{j}$) droplet in each soliton.

231

5. PEP SOLITON/VORTON TOPOLOGY

Because the conserved fluid-like charge density ρ moves point by point with the conserved magnetic field lines of **B**, the fluid-like ρ behaves as if comprised of long, thin, unbreakable strings with properties (P1)-(P4). This gives structure to the flow in that fluid-like entity, and gives topological integrity and permanence to certain vorton structures. The charge/current density has even greater structure because field lines with infinitesimal separation at one point have infinitesimal separation, and are nearly congruent, over their entire closed or endless lengths. That structure even takes on a degree of rigidity in chiral vortons with mutually interlinked field lines, as described below, and in Section (3) above. This is analogous to the integrity and permanence of vortex structures in fluids. However, systems satisfying the Theorem have no viscosity or other dissipation mechanisms as in real fluids. Therefore, the integrity and permanence of vortices in real fluids, provide a mere suggestion of the absolute integrity and absolute permanence of the vortex-like structures in a PEP field.

It is essential that one consider only well behaved fields which, at each point, are finite, single valued and continuous functions. It is these properties which assure that the local properties imposed by the partial differential equations extend to global properties in the vortex-like field structures.

A PEP *vorton* is defined as a 3-D region bounded by a closed oriented 2-D surface inside of which there exists a magnetic field which is zero at most on a set of points of measure zero, and satisfies Eqs. (25)-(29) of the Magnetic Flux Theorem, and whose field lines on the 2-D boundary surface have no component normal to that surface.

The only closed oriented 2-D surfaces are spheres with 2 p-holes connected by p-handles, where p is the genus number of the closed 2-D surface. When [p = 0], the surface is a sphere; when [p = 1] is a torus; etc.

By virtue of the vorton definition, no magnetic field lines extend through the vorton's 2-D boundary surface. As a result magnetic field lines inside are distinct and separate from those outside. Figure (1) illustrating Eq. (22) shows PEP vortons in concentric 3-D spherical shells separated by 2-D spherical surfaces on which the normal component of the magnetic field is zero. Each vorton in Figure (1) is a *chiral vorton*.

A *chiral PEP vorton* is defined as a PEP vorton for which all magnetic field lines inside are linked directly or in chains with all other magnetic field lines inside.

The chirality of a single magnetic field line may be defined as the sign of its average torsion, averaged over the closed or endless length of the field line. If a field line has non-zero average torsion, it twists around its nearest neighbor field lines, which are nearly congruent to itself, and links to them. Any two linked field lines define a screw sense or chirality. That linking has the same chirality as the single field line chirality defined from its torsion. Since the field line torsion is a well behaved function, the 3-D regions of positive average torsion are separated from the 3-D regions of negative, average torsion by 2-D surfaces on which field lines have zero average torsion. Consequently, chirality of all magnetic field lines in a single elementary vorton tend to have the same sign. Since linked magnetic field lines cannot unlink, the chirality of the vorton is conserved.

An *elementary PEP vorton* is defined as a PEP vorton for which the magnetic field is nowhere zero inside or on the surface.

The only closed, oriented 2-D surface which can be foliated by closed or endless, conserved magnetic field lines, with no zeroes in the field on that surface, is the 2-D torus[14]. Therefore, every elementary PEP vorton is bounded by a 2-D torus. Fig. (4) illustrates an elementary chiral vorton. It's field lines have the same topology as in each vorton in the vorton array in Fig. (1), except that there are zeroes in the magnetic fields of Fig. (1) where the polar axis penetrates each 2-D sphere boundary of the 3-D spherical shells. Those zeroes in the field permit the boundary surfaces to be degenerate toruses. Also, the zeroes are field line bifurcation points which permit the field lines on the boundary surfaces to bifurcate, some turning into each of the different vortons which share that boundary surface. It is correct to say that zero valued topological singularities in vorton theory replace the infinite valued topological singularities in point particle theory.

From the Magnetic Flux Theorem and the definitions of vortons, and chirality given above, the following theorem is obvious:

The PEP Vorton Conservation Theorem: A PEP vorton has the following properties:

(V1) All magnetic field lines initially inside the vorton remain forever inside

(V2) All magnetic field lines initially outside the vorton remain forever outside

(V3) All increments of the conserved charge ρ initially inside the vorton remain forever inside, and those increments of ρ initially outside remain forever outside, so that the total integrated charge inside is invariant.

(V4) The topology of the knotting and linking of the magnetic field lines inside the vorton is preserved

(V5) The chirality of elementary vortons is conserved

(V6) All vortons are pairwise disjoint

All of the solutions to the PEP Eq. (10) given by Eq. (21), and their Lorentz transformation are segregated into discrete chiral vortons, but their fields remain everywhere continuous. A great variety of other vorton solutions to the PEP Eq. (10) have been generated. The existence of such PEP vorton solutions is thoroughly established. The particle-like integrity and permanence of these PEP vortons is established by the PEP Magnetic Flux and PEP Vorton Conservation Theorems.

6. ELECTRON-LIKE PEP TOPOLOGY

A solution to the Einstein/Pauli Eq. (3) or (10) for a PEP vorton analogous to the electron, should be the simplest chiral PEP vorton with conserved magnetic field lines embedded in a 3-D toroidal conserved charge/current distribution, similar, topologically, to Fig. (4). That 3-D toroid is foliated by nested 2-D tori, the innermost of which is degenerate forming the core 1-D circle on which all other tori are nested. The conserved magnetic field lines each lie on one of these 2-D tori. The field line winding number on each 2-D torus is constant so that field lines do not intersect. However, the winding number can vary continuously from one nested 2-D torus in the vorton to the next. The winding number on the innermost torus, the 1-D circle, is required, by its topology, to be infinite (if the field is not zero there). The winding number on the outermost 2-D torus bounding the 3-D toroid must be zero in order to fit continuously onto the external magnetic dipole field when the vorton is stationary. Therefore, the winding number must vary continuously from infinity on the innermost 2-D torus to zero on the outermost.

Since the PEP vorton must always have a superluminal component of current flow everywhere parallel to the embedded magnetic field lines, the dominant (superluminal) components of **j** will lie on the nested 2-D tori, also. Since the PEP Eq. (11) requires [**E** • **B** = 0] and [**E** • **j** = 0], the dominant component of **E** will be perpendicular to the nested 2-D tori on which the field lines of **B** lie. The nested 2-D tori will be equipotential surfaces in time independent solutions, but not for moving, deforming PEP vortons. Any component of **E** lying in one of the nested 2-D torus surfaces must be perpendicular to **B** on that torus. Since [**E** • **j** = 0] everywhere, any contribution to [**E** • **j**] by the component of **E** lying on that 2-D torus must be negated by a contribution to [**E** • **j**] of opposite sign due to a component of **j** perpendicular to the 2-D torus surface. Such subluminal perpendicular components of **j** are not recirculation. They are responsible for changes in size, shape, and orientation of the nested 2-D toruses inside the PEP vorton, and for PEP vorton translation, as given by Eqs. (25)-(29) of the Permaflux Theorem. These subluminal components of flow may be (or may not be) subject to radiation damping.

It is only necessary to find solutions to the PEP Eq. (3) which give quantitative agreement with the experimental values of the electron parameters. That is:

$$-e = \iiint \rho dV$$

$$m_0 = \iiint (1/2)\,(E^2 + B^2)dV$$

$$\hbar/2 = \iiint \mathbf{r} \times (\mathbf{E} \times \mathbf{B})dV \qquad (30)$$

with the gyromagnetic ratio equal to 2.

One may generate an approximate solution to the PEP Eq. (3) or (10) with the above topology using an exact analytic solution for an infinitely long, straight, spiraling charge/current density of radius a. Such a time independent exact solution in cylindrical coordinates [R,Φ,Z] is, for [R ≤ a] given by:

$$j_R = 0$$

$$j_\Phi = \chi b \sqrt{1 - u^2} \; J_1 (\alpha R)$$

$$j_Z = b \; J_0 (\alpha R)$$

$$\rho = bu \; J_0 (\alpha R) \tag{31}$$

where $[\chi = \pm 1]$ is the field chirality and J_n is the cylindrical Bessel function of order n. For $[R > a]$, the charge/current density is zero. The charge/current density of Eq. (31), in its self generated fields, is an exact solution to the PEP Eq. (10) for all values of the parameters a, b and u. One may "cut" a length $[L \sim (137)^2 a]$ and "bend" this long thin charge/current density into a circle of circumference L, with "adjustments" in functional form to assure charge/current conservation.

One may then adjust the four parameters [a,b,u and L] so that Eq. (30) is satisfied. The resulting toroidal charge/current density and its self generated fields are only an approximate solution to the Einstein/Pauli PEP equation. However, the unbalanced force $[\mathbf{j} \times \mathbf{B} - \rho \mathbf{E}]$ is orders of magnitude smaller than either $[\mathbf{j} \times \mathbf{B}]$ or $[\rho \mathbf{E}]$ everywhere in this Lorentz frame. In this sense, this electron-like PEP vorton is a good approximate solution.

There are some fundamental difficulties in generating an exact solution to the PEP Eq. (3), for a localized time independent "free" vorton. Those subtle problems seem less serious if one seeks time independent, localized solutions bound to an attractive potential fixed in space by sources of infinite mass, such as that for a magnetic vorton bound to a magnetic point dipole, as in Section (3) above, or a charged vorton bound in a coulomb or Dehmelt[15] trap potential. As another difficulty, the PEP theory is invariant to the choice of field and charge scale, admitting solutions of arbitrary charge. However, that choice does not change dimensionless parameters of the theory, such as $[e^2/(\hbar c)]$, which may prove to be topological invariants of solutions.

7. PEP EXISTENCE THEOREM

Because the PEP Eq. (10) is non-linear, general methods of finding exact analytic solutions are not yet available. Only highly symmetric and/or highly constrained analytic solutions have been constructed.

A very powerful and general existence theorem for solutions to the PEP Eq. (10) can be proven by solving the PEP Eq. (10) for the time derivatives of the fields. Then the Cauchy Kovaleska Theorem[16] may be applied to prove the existence of time integrated solutions with a great variety of conserved PEP vortons which may be constructed into the initial fields.

The time derivatives also provide a powerful tool for computer generation of explicit solutions to the PEP Eq. (3) with specific PEP vortons in specific external fields. One simply constructs the desired PEP vortons and desired external fields into the initial fields imposed at time $[t = t_0]$. One then applies the finite element method to iteratively compute the time derivatives and new fields after each small time increment $[\Delta t]$.

In regions where $[(\rho, \mathbf{j}) = 0]$, the PEP Eq. (10) is satisfied identically by solutions to Maxwell's free space equations:

$$\dot{\mathbf{B}} = -\nabla \times \mathbf{E} \tag{32}$$

$$\dot{\mathbf{E}} = -\nabla \times \mathbf{B} \tag{33}$$

with initial fields which satisfy:

$$\nabla \cdot \mathbf{B} \; (x,y,z,t = t_0) = 0 \tag{34}$$

$$\nabla \cdot \mathbf{E} \ (x,y,z,t = t_0) = 0 \tag{35}$$

The time derivatives of $[\nabla \cdot \mathbf{B}]$ and $[\nabla \cdot \mathbf{E}]$ are zero by virtue of Eqs. (32) and (33), so that the initial condition of Eqs. (34) and (35) are preserved on time integration. The Cauchy Kovaleska theorem assures the existence of solutions around any point where $[\dot{\mathbf{E}}]$ and $[\dot{\mathbf{B}}]$ in Eqs. (32) and (33) are real analytic functions.

In regions where $[(\rho,\mathbf{j}) \neq 0]$, one must replace Eq. (33) and the initial condition $[\nabla \cdot \mathbf{E} = 0]$ from Eq. (35), by:

$$\dot{\mathbf{E}} = -\left[\frac{\mathbf{E} \cdot (\nabla \times \mathbf{B})}{E^2}\right] \mathbf{E} + \left[\frac{\mathbf{E} \cdot (\nabla \times \mathbf{E})}{B^2}\right] \mathbf{B} + \left[\frac{\mathbf{E} \cdot [\mathbf{B} \times (\nabla \times \mathbf{B})] - E^2 \ (\nabla \cdot \mathbf{E})}{E^2 B^2}\right] (\mathbf{E} \times \mathbf{B}) \tag{36}$$

and by the initial condition:

$$\mathbf{E} \ (x,y,z,t = t_0) \cdot \mathbf{B} \ (x,y,z,t = t_0) = 0 \tag{37}$$

The initial condition of Eq. (37) is preserved on the time integration of Eq. (36). With these initial conditions and time derivatives of the fields, and with Maxwell's Eq. (15) in the presence of charge/current density, the PEP Eq. (10) is satisfied. This may be proven by direct substitution into the PEP Eq. (10) from Eqs. (32), (34), (36) and (37) above. Again, the Cauchy Kovaleska theorem applies and time integrated solutions exist around any point where the time derivatives given by Eqs. (32) and (36) are real analytic function. Again, Eqs. (32) and (36) provide a powerful tool for computer generation of solutions to the PEP Eq. (10) by iterative time integration using the finite element method and rather arbitrary initial conditions.

Special problems arise at vorton boundaries between regions where $[(\rho,\mathbf{j}) = 0]$ and regions where $[(\rho,\mathbf{j}) \neq 0]$, because the vorton boundary is, in general moving at velocity $[\mathbf{v} = \mathbf{j}/\rho]$, where $[\rho,\mathbf{j}]$ are given just inside the vorton boundary. Furthermore, analytic solutions with $[(\rho,\mathbf{j}) \neq 0]$ cannot exist if there are extended 3-D regions where $[(\rho,\mathbf{j}) = 0]$, unless $(\rho,\mathbf{j})$ is pure imaginary with zero real part outside the PEP vorton boundary (which may require a zero valued singularity at every point on that vorton boundary surface). However, C^n functions with continuous derivatives to order n, and even C^∞ functions, are possible with extended 3-D regions where $[(\rho,\mathbf{j}) = 0]$. In spite of these difficulties, the applicable existence theorems are as powerful as are generally available in practical problems. These difficulties are emphasized primarily to identify difficulties which arise in computer generation of solutions.

If one chooses initial fields which satisfy the initial conditions of Eq. (34) and either Eq. (35) or (37) and, which include PEP vortons constructed into the initial fields where $[(\rho,\mathbf{j}) \neq 0]$, then those vortons are conserved by the Vorton Conservation Theorem in the time integrated solutions which exist by virtue of the Cauchy Kovalevska Theorem. This provides a very powerful proof of the existence of solutions to the Einstein/Pauli PEP Eq. (3) which include PEP vortons with $[(\rho,\mathbf{j}) \neq 0]$. Again, using the finite element method and iterative time integration, it provides a powerful tool for explicitly generating such solutions.

8. CONSTRAINTS ON MEASUREMENT OF PEP VORTONS

At the beginning of the twentieth century, physicists adopted a new philosophy:

If man cannot measure it, it does not exist; at least not as a
variable or parameter in any theoretical physics model.

Einstein initiated that philosophy by eliminating the unmeasureable ether from electromagnetic theory.

Heisenberg[17] attempted to extend that philosophy to quantum physics, by eliminating the unmeasureable electron orbits. When Schroedinger and de Broglie attempted to interpret wave functions (and off-diagonal matrix elements) as real physical standing wave/particles, their interpretation was rejected as giving physical reality to the unmeasureable.

Schroedinger's charge and current densities, given by $[\rho = \Psi * \Psi]$ and $[j = \Psi^*\nabla\Psi - \Psi\nabla\Psi^*]$, like all of Schroedinger's results, were adopted by the Copenhagen school as their own, and were then re-interpreted as probabilities, lacking in directly measurable physical reality.

For the above reasons, twentieth century physics, which has never been able to directly measure a 3-D charge/current distribution inside a particle, denies the existence of any such 3-D charge/current density. The Einstein/Pauli PEP Eq. (3) or (10) is, by any direct physical interpretation, a violation of this twentieth century physics philosophy. The PEP Eq. (10), written with both self fields and external fields, emphasizes that this unmeasureable PEP charge/current density changes its size, shape and density distribution whenever the externally applied field changes. Only the topology of the embedded magnetic field lines is invariant. Yet the only structure ever measured is the indirect impact of the Compton wave length or the de Broglie wave length in scattering experiments, and the excluded volume associated with generalizations of the Bohr radius in atoms, molecules and crystals. Twentieth century physics assumes that the only charges are point charges, and that the only currents are moving point charges.

If one is to give physical meaning to the charge/current density in the PEP Eq. (3) or (10), one must adopt the philosophy expressed by Einstein in his later years, which might be paraphrased as:

> *God may have created things which man cannot measure. If man guesses their existence and the equations which these "unmeasureables" obey, and includes them in his theoretical physics models, that theory must be applied to both the measured entity and the measurement devices in the model, so that*[18]*:"the theory itself can determine what is, and what is not, measurable."*

The final test of the theoretical model is whether or not the things which it predicts to be measurable, are experimentally measurable with the predicted values or probabilities.

The PEP system is remarkable in that the mere existence of Intra System Observers (ISO) leads to the logical conclusion that there are entities in the solutions to the PEP Eq. (3) which the ISO cannot measure. The ISO measuring instruments and their probe particles are all comprised of PEP vorton droplets of charge/current density. By the Vorton Conservation Theorem, vorton droplets are topologically pairwise disjoint. Therefore, the PEP vorton droplets in the ISO measuring instruments and probe particles can <u>not penetrate</u>, can <u>not probe</u>, and can <u>not measure</u> the fields inside the measured PEP vorton droplet. To the ISO, the PEP vorton droplets appear to be structureless objects with topological integrity and permanence which propagate as waves.

Furthermore, the only PEP properties which the ISO can measure are integrals or averages over the PEP vorton droplet. These include total PEP charge, total PEP mass, average position of the PEP mass or charge centroid, and the spin angular momentum about that centroid. Measurements of total energy and total momentum must be done by experiments which integrate over the extended self field of the vorton, which falls to zero only at infinity. Since the point by point fields, and the distribution of mass and charge cannot be measured inside the vorton, there are fundamental uncertainties in these integrals and averages.

Because the initial fields inside the PEP vorton droplet cannot be measured, it is impossible for the ISO to time integrate the PEP field Eq. (10) to deterministically predict future values of the ISO measurable quantities from past measured values. The ISO must time integrate the PEP field Eq. (10) for each and every different initial field consistent with the initial ISO measured PEP observables, and make statistical predictions. The ISO live in an uncertain and indeterminant world, and can only make probabilistic predictions based on the statistics of such ensembles of solutions.

The wave equations of quantum theory appear to describe particle wave functions whose self interference or diffraction give square amplitudes approximating particle ensemble probabilities, in the same sense that the squared amplitudes of light waves give approximate photon ensemble statistics. This leads to the conjecture that the empirical rules and recipes which the Copenhagen school have adopted to generate empirical probabilities, quite generally, agree with the correct soliton ensemble statistics. Barut[19] has given further credence to that conjecture. He has computed the statistics for an ensemble of particle/waves which obey a deterministic wave equation as applied to the double slit and Stern Gerlock experiments. Using an ensemble of different localized initial wavepackets, each consistent with the initial uncertainty and experimental error, Barut[19] obtained results in agreement with experiment and the quantum theory predictions. In this sense, quantum theory might be a Bohr correspondence principle limiting approximation to ensemble statistics for solitons with uncertain fields obeying deterministic field equations.

9.   CONCLUSIONS

The Poincaré/Ehrenfest Purely Electromagnetic Particle (PEP) Non-Existence Theorem has been accepted physics dogma for eighty years. But there is a broad exception to that PEP Non-Existence Theorem. That Theorem applies only to charge/current densities which are time-like 4-D vector fields.

Einstein[5] and Pauli[6] each derived the same equation as the necessary and sufficient condition for the existence of PEP with non-zero 3-D charge/current density. Contrary to physics dogma, there exists a plethora of soliton solutions to that Einstein/Pauli PEP Eq. (10). Each PEP soliton in such solutions is centered on a discrete, localized 3-D droplet of charge/current density with vortex-like internal recirculation. The recirculation has a superluminal component which does not displace or deform the droplet. It is a space-like 4-D vector field without violating causality.

The component of the magnetic field normal to the closed 2-D surface of each PEP droplet is zero on that 2-D surface, so that magnetic field lines inside each PEP droplet are completely separate from those outside. In regions where the charge/current density is not zero, solutions to the Einstein/Pauli PEP equation satisfy the conditions of the Relativistic Helmholtz Flux Conservation Theorem derived by Waite[7]. The magnetic field lines embedded in the PEP droplet are (P1) pairwise disjoint, (P2) closed or endless, (P3) conserved, and (P4) moving point by point with the conserved charge/current density at velocity $[\mathbf{v} = \mathbf{j}/\rho]$. The PEP droplets are called *vortons* (vortex-like particles), because the conserved magnetic flux gives each vortex-like droplet a particle-like integrity and permanence analogous to fluid vortices, but without any dissipative terms analogous to viscosity in classical fluids.

The vortex-like recirculation velocity in each PEP charge/current density droplet has the functional form $[\mathbf{v} = (\mathbf{j}/\rho) = \alpha\mathbf{B} + \mathbf{E} \times \mathbf{B}/B^2]$, as in Eq. (29). It has a *superluminal* component $[\alpha\mathbf{B}]$ everywhere *parallel* to the embedded closed or endless magnetic field lines. That *superluminal* component produces only time independent recirculation with *no net* displacement of the charge current density. And, it does not produce displacement of the embedded magnetic field lines, because $[\mathbf{v} = \mathbf{j}/\rho]$ for that superluminal component is everywhere parallel to the closed or endless magnetic field lines. There may exist a second *subluminal* component $[\mathbf{E} \times \mathbf{B}/B^2]$ of charge/current density flow which is everywhere *perpendicular* to the embedded magnetic field lines. That *subluminal* component is responsible for all net displacements, deformations and orientation changes of the PEP charge/current density droplet and its embedded magnetic field lines. Causality is not violated by solutions to the Einstein/Pauli PEP equation with $[j^2 \gtreqqless \rho^2]$ which satisfy $[B^2 \gtreqqless E^2]$ of Eq. (28) in the PEP Magnetic Flux Theorem.

The superluminal charge/current density recirculation inside the PEP droplet provides magnetic forces which make Poincaré (non-electromagnetic) stresses unnecessary. Because the integrated mass (energy) in the self field of the PEP charge/current density droplet: $\iiint(1/2) (E^2 + B^2)dV$, is finite, there is no need to ignore it as in classical point particle physics, nor to provide an infinite negative "bare" (non-electromagnetic) mass as in quantum field theory. However, Waite[7] had shown that it is possible to generalize the PEP model to include other forces and other sources of finite mass inside the vorton charge/current density droplet, while retaining conserved flux vortons which obey the laws of electromagnetic theory and the conservation laws of relativistic mechanics. One may use these generalizations, if necessary, to account for finite non-electromagnetic forces or masses (energy) which may exist inside the charge/current density of a PEP vorton, but which cannot be directly measured by Intra-System Observers (ISO).

Each PEP vorton droplet has particle-like integrity and permanence. But by itself it has no inertial mass or momentum. The only inertia possessed by the PEP vorton droplet exists as the integrated mass (energy) equal to $(1/2)(E^2 + B^2)$ in its self field, and the integrated momentum ($E \times B$) in its self field. The extended self fields are wave-like, falling to zero only at infinity. The PEP vorton droplet gives particle-like integrity and permanence to its soliton-like self field; and the soliton-like self field gives wave-like inertia and propagation to the PEP vorton droplet.

Each PEP vorton droplet is pairwise disjoint with respect to every other PEP vorton droplet, by virtue of the topology and global symmetry imposed by the Einstein/Pauli PEP equation on its finite, single valued and continuous solutions.

Consider the postulate that Intra-System Observers (ISO) exist within a system defined by the Einstein/Pauli PEP equation. The ISO measuring instruments and their probe particles are all comprised of discrete PEP vorton droplets which obey that PEP equation. Because all PEP vorton droplets are pairwise disjoint, the ISO cannot penetrate, probe and measure the field inside of any single PEP vorton droplet. The measuring probe droplet is pairwise disjoint from the measured PEP droplet.

Consequently, any ISO existing in a PEP system experimentally observes each PEP droplet as a structureless point particle with wave-like propagation. The ISO can measure only PEP properties which are integrated or averaged over the PEP vorton droplet, giving a limited set of measurable PEP observables which have fundamental uncertainties. The ISO, even if they know that the deterministic PEP equation controls their world, cannot use it to make deterministic predictions of future PEP properties based on past PEP measurements. This follows because the ISO can never measure and know the initial values of the fields inside of the PEP vorton droplet, as required for the deterministic integration with respect to time. The ISO may compute only probability statistics for an entire ensemble, which includes all PEP vorton droplets whose internal fields are consistent with the few integrated or average PEP properties which the ISO can measure directly. These integrals or averages over the PEP droplet have fundamental uncertainties, so that the ensembles are large.

The above conclusions all follow from the global symmetries and topological properties of solutions to the Einstein/Pauli PEP equation. The global symmetry and topology apply to all PEP solutions which, at each point, are finite, single valued and continuous functions. It is this condition of regularity imposed on the solutions, which assures that the topology and global symmetry will follow from the local differential equations.

In most cases, those global solutions for PEP vorton droplets have natural limiting approximations, outside the charge/current density droplets, as linear fields with infinite valued singularities on point particles or closed strings. That is, there is a Bohr correspondence principle limiting approximation in terms of classical point particles or strings with mass, charge and multipole moments, with an independent but coupled electromagnetic field. But coupling of the point charges or point multipole moments to the independent electromagnetic fields, leads to infinite energy in the system. This must be ignored or it may be "re-normalized" by the addition of non-real infinite negative bare (non-electromagnetic) mass to each point particle or string.

It is interesting to conjecture that the process of quantizing the point particle or string, which is an approximation to a PEP vorton, is a Bohr correspondence principle limiting approximation to quantization of the PEP vortons/solitons, also. Perhaps the empirical quantization process generates the ensemble probability statistics which the ISO must use to make predictions of PEP properties from the limited set of past PEP properties which the ISO are able to measure. However, that conjecture should not be allowed to obscure and detract from the very concrete examples of exact vorton solutions to the Einstein/Pauli PEP equation; the existence theorem for more general vorton solutions, the PEP Magnetic Flux Theorem and the general theorems which establish the unusual global properties of vorton solutions to that Einstein/Pauli PEP equation. Those concrete results more than justify a re-examination of the purely electromagnetic electron concept, independent of any more elusive conjectures which they may suggest.

REFERENCES

1. M. Abraham, *Ann. Phys.*, Lpz. 10, 105, (1903).
2. H.A. Lorentz, *Proc. Acad. Sci.* Amst. 6, 809, (1904).
3. H. Poincaré, *C.R. Acad. Sci.*, Paris, 140, 1504 (1905) and R.C. Circ. mat. Palermo, 21 129 (1906).
4. P. Ehrenfest, *Ann. Phys.*, Lpz. 23, 204, (1907).

5. A. Einstein, *Sitzungsberichte der Preussischen Akad. d. Wissenschaften* (1919), as translated in *The Principle of Relativity*, H. A. Lorentz, A. Einstein, H. Minkowski and H. Weyl, Dover, NY (1952) p. 192.
6. W. Pauli, *Relativitätstheorie*, Encyklopädie der mathematischem Wissenschaften, 19 B. G. Teubner, Lupzig (1921), as translated in *Theory of Relativity*, Dover, NY, (1958) p. 184.
7. T. Waite, to be published in *Physics Essays*, March (1995).
8. G.D. Jackson, *Classical Electrodynamics*, John Wiley, NY (1975), p. 238.
9. S. Chandrasekhar, *Proc. Nat. Acad. Sc.*, 42, 1 (1956).
10. S. Chandrasekhar and P.C. Kendall, *Astrophys.*, J., 126, 2544 (12969).
11. For a recent review and bibliography, see H. Zaghloul and O. Barajas, *Am. J. Phys.*, 58, 783 (1990).
12. E. Schroedinger, Quantization as a problem in proper values IV, *Ann. d. Physik*, 4-81 (1926) as translated in *Collected Papers on Wave Mechanics*, Chelsea Publishing, New York, NY (1982) p. 123.
13. A. Sommerfeld, *The Mechanics of Deformable Bodies*, Academic Press NY (1950), p. 132.
14. V. Guillemin and A. Pollack,*Differential Topology*, p. 146, Prentice Hall Inc., Englewood Cliffs, N.J. (1974).
15. H. Dehmelt, *Rev. Mod. Phys.* 62, 525, (1990).
16. Michael Spivak, *A Comprehensive Introduction to Differential Geometry*, vol. 5, p. 57, Publish or Perish, Inc., Berkeley (1979).
17. W. Heinsenberg, *The Physical Principles of Quantum Theory*, English translation by University of Chicago Press (1930) reprinted by Dover Publishing Co., NY, NY, (1949).
18. A. Einstein as quoted in J.A. Wheeler and W.H. Zurek, editors, *Quantum Theory and Measurement*, Princeton University Press, Princeton, NJ (1983) p. 58.
19. A.O. Barut, *Found of Phys. Lett*, 1, 47 (1988).

FIELD THEORY OF THE SPINNING ELECTRON:
I – INTERNAL MOTIONS [(*)]

Giovanni SALESI[1] and Erasmo RECAMI[2]

[1]Dipart. di Fisica, Università Statale di Catania,
57 Corsitalia, 95129–Catania, Italy.
[2]Facoltà di Ingegneria, Università Statale di Bergamo,
24044–Dalmine (BG), Italy;
INFN, Sezione di Milano, Milan, Italy; and
Dept. of Applied Math., State University at Campinas,
Campinas, S.P., Brazil.

> "If a spinning particle is not quite a point particle, nor
> a solid three dimensional top, what can it be?"
>
> Asim O. Barut

ABSTRACT and INTRODUCTION

This paper is dedicated to the memory of Asim O. Barut, who so much contributed
to clarifying very many fundamental issues of physics, and whose work constitutes a
starting point of these articles.

We present here a field theory of the spinning electron, by writing down a new
equation for the 4-velocity field v^μ (different from that of Dirac theory), which allows
a classically intelligible description of the electron. Moreover, we make explicit the
noticeable kinematical properties of such velocity field (which also result different from
the ordinary ones). At last, we analyze the internal *zitterbewegung* (zbw) motions, for
both time-like and light-like speeds. We adopt in this paper the ordinary tensorial
language. Our starting point is the Barut–Zanghi classical theory for the relativistic
electron, which related spin with zbw.

[(*)] Work supported in part by INFN, CNR, MURST, and by FAPESP, CNPq.

A NEW MOTION EQUATION FOR THE
SPINNING (FREE) ELECTRON

Attempts to put forth classical models for the spinning electron are known since more than seventy years [1]. In the Barut–Zanghi (BZ) theory,[2] the classical electron was actually characterized, besides by the usual pair of conjugate variables (x^μ, p^μ), by a second pair of conjugate classical *spinorial* variables $(\psi, \overline{\psi})$, representing internal degrees of freedom, which were functions of the (proper) time τ measured in the electron global center-of-mass (CM) system; the CM frame (CMF) being the one in which $\boldsymbol{p} = 0$ at every instant of time. Barut and Zanghi, then, introduced a classical lagrangian that in the free case (i.e., when the *external* electromagnetic potential is $A^\mu = 0$) writes $[c = 1]$

$$\mathcal{L} = \frac{1}{2}i\lambda(\overline{\psi}\dot{\psi} - \dot{\overline{\psi}}\psi) + p_\mu(\dot{x}^\mu - \overline{\psi}\gamma^\mu\psi) , \tag{1}$$

where λ has the dimension of an action and ψ and $\overline{\psi} \equiv \psi^\dagger\gamma^0$ are ordinary $\mathbb{C}^4$–bispinors, the dot meaning derivation with respect to τ. The four Euler–Lagrange equations, with $-\lambda = \hbar = 1$, yield the following motion equations:

$$\begin{cases} \dot{\psi} + ip_\mu\gamma^\mu\psi = 0 & \text{(2a)} \\ \dot{x}^\mu = \overline{\psi}\gamma^\mu\psi & \text{(2b)} \\ \dot{p}^\mu = 0 , & \text{(2c)} \end{cases}$$

besides the hermitian adjoint of eq.(2a), holding for $\overline{\psi}$. From eq.(1) one can also see that

$$H \equiv p_\mu v^\mu = p_\mu\overline{\psi}\gamma^\mu\psi \tag{3}$$

is a constant of the motion [and precisely is the energy in the CMF].[2-4] Since H is the BZ hamiltonian in the CMF, we can suitably set $H = m$, quantity m being the particle rest-mass. The general solution of the equations of motion (2) can be shown to be:

$$\psi(\tau) = [\cos(m\tau) - i\frac{p_\mu\gamma^\mu}{m}\sin(m\tau)]\psi(0) , \tag{4a}$$

$$\overline{\psi}(\tau) = \overline{\psi}(0)[\cos(m\tau) + i\frac{p_\mu\gamma^\mu}{m}\sin(m\tau)] , \tag{4b}$$

with $p^\mu = \text{constant}$; $p^2 = m^2$; and finally:

$$\dot{x}^\mu \equiv v^\mu = \frac{p^\mu}{m} + [\dot{x}^\mu(0) - \frac{p^\mu}{m}]\cos(2m\tau) + \frac{\ddot{x}^\mu}{2m}(0)\sin(2m\tau) . \tag{4c}$$

This general solution exhibits a classical analogue of the phenomenon known as zitterbewegung: in fact, the velocity v^μ contains the (expected) term p^μ/m plus a term describing an oscillating motion with the characteristic zbw frequency $\omega = 2m$. The velocity of the CM will be given by p^μ/m. Let us explicitly observe that the general solution (4c) represents a helical motion in the ordinary 3-space of a "constituent" Q: a result that has been met also by means of other, alternative approaches.[5,6]

Before studying the time evolution of our electron, we want to write down its motion equation in a "kinematical" form, suitable a priori for describing a point-like object;

i.e., at variance with eqs.(2), expressed not in terms of ψ and $\overline{\psi}$, but on the contrary in terms of quantities related to the particle trajectory (such as p^μ and v^μ). To this aim, we can introduce the spin variables, and adopt the set of dynamical variables

$$x^\mu, \ p^\mu; \ v^\mu, S^{\mu\nu},$$

where

$$S^{\mu\nu} \equiv \frac{i}{4}\overline{\psi}[\gamma^\mu, \gamma^\nu]\psi; \tag{5a}$$

then, we get the following motion equations:

$$\dot{p}^\mu = 0; \quad v^\mu = \dot{x}^\mu; \quad \dot{v}^\mu = 4S^{\mu\nu}p_\nu; \quad \dot{S}^{\mu\nu} = v^\nu p^\mu - v^\mu p^\nu. \tag{5b}$$

[By varying the action corresponding to $\mathcal{L}$, one finds as generator of space-time rotations the conserved quantity $J^{\mu\nu} = L^{\mu\nu} + S^{\mu\nu}$, where $L^{\mu\nu} \equiv x^\mu p^\nu - x^\nu p^\mu$ is the orbital angular momentum tensor, and $S^{\mu\nu}$ is just the particle spin tensor: so that $\dot{J}^{\mu\nu} = 0$ implies $\dot{L}^{\mu\nu} = -\dot{S}^{\mu\nu}$].

By deriving the third one, and using the first one, of eqs.(5b), we obtain

$$\ddot{v}^\mu = 4\dot{S}^{\mu\nu}p_\nu; \tag{6}$$

by substituting now the fourth one of eqs.(5b) into eq.(6), and imposing the previous constraints $p_\mu p^\mu = m^2$ and $p_\mu v^\mu = m$, we end with the time evolution[3] of the *field four-velocity*:

$$v^\mu = \frac{p^\mu}{m} - \frac{\ddot{v}^\mu}{4m^2}, \tag{7}$$

such *a new motion equation* corresponding to the whole system of eqs.(2). Let us recall, for comparison, that the analogous equation *for the standard Dirac case*:[1]

$$v^\mu = \frac{p^\mu}{m} - \frac{i}{2m}\dot{v}^\mu \tag{7'}$$

was totally devoid of a classical, intuitive meaning, because of the known appearance of an imaginary unit i in front of the acceleration (connected with the well-known fact that the position operator is not hermitian therein).

Let us observe that, by differentiating the relation $p_\mu v^\mu = m = $ constant, one immediately finds that the (internal) acceleration $\dot{v}^\mu \equiv \ddot{x}^\mu$ is orthogonal to the electron impulse p^μ, since $p_\mu \dot{v}^\mu = 0$ at any instant. To conclude, let us stress that, while the Dirac electron has no classically meaningful internal structure, our electron on the contrary (an *extended–type* particle) does possess an internal structure, and internal motions, which are all endowed with a "realistic" meaning, from both the geometrical and kinematical points of view: as we are going to see in the next section.

SPIN AND INTERNAL KINEMATICS

We wish first of all to make explicit the kinematical definition of v^μ, *rather different from the ordinary one* valid for scalar particles.[7]. In fact, from the very definition of

v^μ, we get

$$v^\mu \equiv dx^\mu/d\tau \equiv (dt/d\tau; d\boldsymbol{x}/d\tau) \equiv (\frac{dt}{d\tau}; \frac{d\boldsymbol{x}}{dt}\frac{dt}{d\tau})$$

$$= (1/\sqrt{1 - \boldsymbol{w}^2}; \ \boldsymbol{u}/\sqrt{1 - \boldsymbol{w}^2}), \qquad [\boldsymbol{u} \equiv d\boldsymbol{x}/dt] \qquad (8)$$

where $\boldsymbol{w} = \boldsymbol{p}/m$ is the velocity of the CM in the chosen reference frame (i.e., in the frame in which quantities x^μ are measured). Below, it will be convenient to choose as reference frame the CMF (even if quantities as $v^2 \equiv v_\mu v^\mu$ are frame invariant); so that

$$v_{\mathrm{CM}}^\mu = V^\mu \equiv (1; \boldsymbol{V}), \qquad (9)$$

wherefrom one deduces for the speed $|\boldsymbol{V}|$ of the internal motion (i.e., for the zbw speed) the em new conditions:

$$0 < V^2(\tau) < 1 \quad \Leftrightarrow \quad 0 < \boldsymbol{V}^2(\tau) < 1 \qquad \text{(time-like)}$$

$$V^2(\tau) = 0 \quad \Leftrightarrow \quad \boldsymbol{V}^2(\tau) = 1 \qquad \text{(light-like)} \qquad (10)$$

$$V^2(\tau) < 0 \quad \Leftrightarrow \quad \boldsymbol{V}^2(\tau) > 1 \qquad \text{(space-like)},$$

where $V^2 = v^2$. Notice that, in general, the value of V^2 does vary with τ; except in special cases (e.g., the case of polarized particles: as we shall see). Coming back to the expression of the 4-velocity, eq.(4c), it is possible after some algebra to recast this equation in a "spinorial" form, i.e., to write it as a function of the initial spinor $\psi(0)$:

$$v^\mu = p^\mu/m + E^\mu \cos(2m\tau) + H^\mu \sin(2m\tau), \qquad (11)$$

where $[\alpha^\mu \equiv \gamma^0 \gamma^\mu]$

$$E^\mu = \frac{1}{2}\overline{\psi}(0)[\frac{\not{p}}{m}, \alpha^\mu]\psi(0) \qquad (12a)$$

$$H^\mu = \frac{i}{2}\overline{\psi}(0)(\alpha^\mu - \frac{\not{p}}{m}\alpha^\mu\frac{\not{p}}{m})\psi(0). \qquad (12b)$$

In the chosen CM frame, eqs.(12) read:

$$E^\mu = \overline{\psi}(0)\gamma^\mu\psi(0) - \frac{p^\mu}{m} \qquad (13a)$$

$$H^\mu = i\overline{\psi}(0)(\alpha^\mu - g^{0\mu})\psi(0), \qquad (13b)$$

where $g^{\mu\nu}$ is the metric tensor. Bearing in mind that (in the CMF) it holds $v^0 = 1$ [cf. eq.(9)], and therefore $\overline{\psi}\gamma^0\psi = 1$ (which, incidentally, implies the normalization $\psi^\dagger\psi = 1$ in the CMF), one obtains

$$E^\mu = (0; \ \overline{\psi}(0)\vec{\gamma}\psi(0)) \qquad (14a)$$

$$H^\mu = (0; \ i\overline{\psi}(0)\vec{\alpha}\psi(0)). \qquad (14b)$$

By eq.(4), for V^2 we can write:

$$V^2 = 1 + E^2 \cos^2(2m\tau) + H^2 \sin^2(2m\tau) + 2E_\mu H^\mu \sin(2m\tau)\cos(2m\tau). \qquad (15).$$

Now, let us single out the solutions ψ of eq.(2) corresponding to *constant* V^2 and A^2, where $A^\mu \equiv dV^\mu/d\tau \equiv (0; \boldsymbol{A})$, quantity $V^\mu \equiv (1; \boldsymbol{V})$ being the zbw velocity. In the present frame, therefore, we shall suppose quantities

$$V^2 = 1 - \boldsymbol{V}^2 \quad ; \quad A^2 = -\boldsymbol{A}^2$$

to be constant in time:

$$V^2 = \text{constant} \; ; \quad A^2 = \text{constant} , \tag{16}$$

so that $\boldsymbol{V}^2$ and $\boldsymbol{A}^2$ are constant in time too. (Let us recall that we are dealing with the internal motion only, in the CMF; thus, our results are independent of the global 3-impulse $\boldsymbol{p}$ and hold both in the relativistic and in the non-relativistic case). Requirements (16), inserted into eq.(15), yield the following interesting constraints:[7]

$$\begin{cases} E^2 = H^2 & \text{(17a)} \\ E_\mu H^\mu = 0 . & \text{(17b)} \end{cases}$$

Constraints (17) are necessary and sufficient (initial) conditions to get a circular *uniform* motion (the only finite, uniform motion possible in the CMF). Since both E and H do not depend on τ, also eqs.(17) hold at any time. In the euclidean 3-dimensional space, and at any time, constraints (17) may read:

$$\begin{cases} \boldsymbol{A}^2 = 4m^2 \boldsymbol{V}^2 & \text{(18a)} \\ \boldsymbol{V} \cdot \boldsymbol{A} = 0 & \text{(18b)} \end{cases}$$

which explicitly correspond to a uniform circular motion with radius

$$R = |\boldsymbol{V}|/2m . \tag{19}$$

Quantity R is the "zitterbewegung radius"; the zbw frequency was already found to be $\Omega = 2m$. By means of eqs.(14), conditions (17) or (18) can be written in spinorial form (still for any time instant τ) as follows:

$$\begin{cases} (\overline{\psi}\vec{\gamma}\psi)^2 = -(\overline{\psi}\vec{\alpha}\psi)^2 & \text{(20a)} \\ (\overline{\psi}\vec{\gamma}\psi) \cdot (\overline{\psi}\vec{\alpha}\psi) = 0 . & \text{(20b)} \end{cases}$$

At this point, let us show that this classical uniform circular motion, around the z-axis (which in the CMF can be chosen arbitrarily, while in a generic frame is parallel to the global three-impulse $\boldsymbol{p}$, as we shall see below), does just correspond to the case of *polarized* particles with $s_z = \pm\frac{1}{2}$. It may be interesting to notice that in this case the *classical* requirements (17) or (18) —namely, the uniform motion conditions— play the role of the ordinary *quantization* conditions $s_z = \pm\frac{1}{2}$.

It is straightforward to realize also that the most general spinors $\psi(0)$ satisfying the conditions

$$s_x = s_y = 0 \tag{21a}$$

$$s_z = \frac{1}{2}\overline{\psi}(0)\Sigma_z\psi(0) = \pm\frac{1}{2} \tag{21b}$$

($\vec{\Sigma}$ being the spin operator) possess in the standard representation the form

$$\psi^T_{(+)}(0) = (\,a\;0\mid 0\;d\,) \tag{22a}$$

$$\psi^T_{(-)}(0) = (\,0\;b\mid c\;0\,)\,, \tag{22b}$$

and obey in the CMF the normalization constraint $\psi^\dagger\psi = 1$. [It could be easily shown that, for generic initial conditions, it is always $-\frac{1}{2} \le s_z \le \frac{1}{2}$]. Notice that the set of our spinors $\psi_\pm$ include the Dirac spinors, but is an ensemble larger than Dirac's. In eqs.(22) we separated the first two from the second two components, bearing in mind that in the standard Dirac theory (and in the CMF) they correspond to the positive and negative frequencies, respectively. With regard to this point, let us observe that the "negative-frequency" components c and d do *not* vanish at the non-relativistic limit (since, let us repeat, in the CMF it is $\boldsymbol{p} = 0$); but the field hamiltonian H is *nevertheless* positive and equal to m, as already stressed. Now, from relation (22a) we are able to deduce that (with $* \equiv$ complex conjugation):

$$< \vec{\gamma} > \;\equiv\; \overline{\psi}\vec{\gamma}\psi = 2(\mathrm{Re}[a^*d], +\mathrm{Im}[a^*d], 0)$$
$$< \vec{\alpha} > \;\equiv\; \overline{\psi}\vec{\alpha}\psi = 2i(\mathrm{Im}[a^*d], -\mathrm{Re}[a^*d], 0)$$

and analogously, from eq.(22b), that

$$< \vec{\gamma} > \;\equiv\; \overline{\psi}\vec{\gamma}\psi = 2(\mathrm{Re}[b^*c], -\mathrm{Im}[b^*c], 0)$$
$$< \vec{\alpha} > \;\equiv\; \overline{\psi}\vec{\alpha}\psi = 2i(\mathrm{Im}[b^*c], +\mathrm{Re}[b^*c], 0)\,,$$

which just imply relations (20):

$$\begin{cases} < \vec{\gamma} >^2 = - < \vec{\alpha} >^2 \\[2mm] < \vec{\gamma} > \cdot < \vec{\alpha} > = 0 \; . \end{cases}$$

In conclusion, the (circular) polarization conditions, eqs.(21), do imply the internal zbw motion to be uniform and circular ($V^2 =$ constant; $A^2 =$ constant); equations (21), in other words, imply simultaneously that s_z be conserved and quantized.[7].

When passing from the CMF to a generic frame, eqs.(21) transform into

$$\lambda \equiv \frac{1}{2}\overline{\psi}(x)\frac{\vec{\Sigma}\cdot\boldsymbol{p}}{|\boldsymbol{p}|}\psi(x) \;=\; \pm\frac{1}{2} \;=\; \text{constant}\,. \tag{23}$$

Therefore, to get a uniform motion around the $\boldsymbol{p}$-direction [cf. eq.(4c)], we have to require that the field helicity λ be constant (in space and in time), and quantized in the ordinary way: $\lambda = \frac{1}{2}$.

It may be interesting also to calculate $|\boldsymbol{V}|$ as a function of the spinor components a and d. With reference to eq.(22a), since $\psi^\dagger\psi \equiv |a|^2 + |d|^2 = 1$, we obtain (for the $s_z = +\frac{1}{2}$ case):

$$\boldsymbol{V}^2 \equiv < \vec{\gamma} >^2 = 4|a^*d|^2 = 4|a|^2\,(1 - |a|^2) \tag{24a}$$

$$\boldsymbol{A}^2 \equiv (2im < \vec{\alpha} >)^2 = 4m^2\boldsymbol{V}^2 = 16m^2|a|^2\,(1 - |a|^2)\,, \tag{24b}$$

246

and therefore the normalization value (valid now in any frame, at any time):

$$\overline{\psi}\psi = \sqrt{1 - \boldsymbol{V}^2} \, , \qquad (24c)$$

showing that to the same speed and acceleration there correspond two spinors $\psi(0)$, related by an interchange of a and d. From eq.(24a) we derive also that, as $0 \leq |a| \leq 1$, it is:

$$0 \leq \boldsymbol{V}^2 \leq 1 \, ; \quad 0 \leq \overline{\psi}\psi \leq 1 \, . \qquad (24d)$$

Correspondingly, from eq.(19c) we would obtain for the zbw radius that $0 \leq R \leq \frac{1}{2}m$.

The second of eqs.(24d) is a *new*, rather interesting (normalization) boundary condition. From eq.(24c) one can easily see that: (i) for $\boldsymbol{V}^2 = 0$ (no zbw) we have $\overline{\psi}\psi = 1$ and ψ is a "Dirac spinor"; (ii) for $\boldsymbol{V}^2 = 1$ (light-like zbw) we have $\overline{\psi}\psi = 0$ and ψ is a "Majorana" spinor"; (iii) for $0 < \boldsymbol{V}^2 < 1$ we meet, instead, spinors with characteristics "intermediate" between the Dirac and Majorana ones.

The "Dirac" case, corresponding to $\boldsymbol{V}^2 = \boldsymbol{A}^2 = 0$, i.e., to *no* zbw internal motion, is trivially represented (apart from phase factors) by the spinors:

$$\psi^{\mathrm{T}}(0) \equiv (1 \quad 0 \mid 0 \quad 0) \qquad (25)$$

and (interchanging a and d):

$$\psi^{\mathrm{T}}(0) \equiv (0 \quad 0 \mid 0 \quad 1) \, . \qquad (25')$$

This is the unique case (together with the analogous one for $s_z = -\frac{1}{2}$) in which the zbw disappears, while the field spin is still present! In fact, even in terms of eqs.(25)–(25') one still gets that $\frac{1}{2}\overline{\psi}\Sigma_z\psi = +\frac{1}{2}$.

Since we have been discussing a classical theory of the relativistic electron, let us finally notice that even the well-known change in sign of the fermion wave function, under 360°-rotations around the z-axis, gets in our theory a natural classical interpretation. In fact, a 360°-rotation of the coordinate frame around the z-axis (passive point of view) is indeed equivalent to a 360°-rotation of the constituent $\mathcal{Q}$ around the z-axis (active point of view). On the other hand, as a consequence of the latter transformation, the zbw angle $2m\tau$ does suffer a variation of 360°, the proper time τ does increase of a zbw period $T = \pi/m$, and the pointlike constituent does describe a complete circular orbit around the z-axis. It appears then obvious that, since the period $T = 2\pi/m$ of spinor $\psi(\tau)$ in eq.(4c) is *twice* as large as the zbw orbital period, the wave function of eq.(4c) does suffer a phase–variation of 180° only, and then does change its sign: as it occurs in the standard theory.

SPECIAL CASES: LIGHT-LIKE MOTIONS
AND LINEAR MOTIONS

Let us first fix our attention on the special case of *light-like* motions.[7,6] The spinor fields $\psi(0)$, corresponding to $V^2 = 0; V^2 = 1$, are given by eqs.(22) with $|a| = $

$|d|$ for the $s_z = +\frac{1}{2}$ case, or $|b| = |c|$ for the $s_z = -\frac{1}{2}$ case; as it follows from eqs.(24) for $s_z = +\frac{1}{2}$, as well as from the analogous equations

$$\boldsymbol{V}^2 = 4|b^*c| = 4|b|^2(1 - |b|^2) \tag{26a}$$

$$\boldsymbol{A}^2 = 4m^2\boldsymbol{V}^2 = 16m^2|b|^2(1 - |b|^2) , \tag{26b}$$

for the case $s_z = -\frac{1}{2}$. It can be easily seen that a difference in the phase factors of a and d (or of b and c, respectively) does *not* change the motion kinematics, nor its rotation direction; but it merely shifts the zbw phase angle at $\tau = 0$. Thus, one is entitled to choose *purely real* spinor components (as we did above). As a consequence, the *simplest* spinors may be written:

$$\psi^{\mathrm{T}}_{(+)} = \frac{1}{\sqrt{2}}(1 \ \ 0 \ | \ 0 \ \ 1) \tag{27a}$$

$$\psi^{\mathrm{T}}_{(-)} = \frac{1}{\sqrt{2}}(0 \ \ 1 \ | \ 1 \ \ 0) ; \tag{27b}$$

and then

$$< \vec{\gamma} >_{(+)} = (1, 0, 0) ; \ \ < \vec{\alpha} >_{(+)} = (0, -i, 0)$$

$$< \vec{\gamma} >_{(-)} = (1, 0, 0) ; \ \ < \vec{\alpha} >_{(-)} = (0, i, 0)$$

which, inserted into eqs.(14), yield

$$E^\mu_{(+)} = (0; 1, 0, 0) ; \ \ \ \ H^\mu_{(+)} = (0; 0, 1, 0) .$$

$$E^\mu_{(-)} = (0; 1, 0, 0) ; \ \ \ \ H^\mu_{(-)} = (0; 0, -1, 0) .$$

Because of eq.(11), we meet now for $s_z = +\frac{1}{2}$ an *anti-clockwise* internal motion, with respect to the chosen z-axis:

$$V_x = \cos(2m\tau); \ \ V_y = \sin(2m\tau); \ \ V_z = 0 ; \tag{28}$$

and a *chockwise* internal motion for $s_z = -\frac{1}{2}$:

$$V_x = \cos(2m\tau); \ \ V_y = -\sin(2m\tau); \ \ V_z = 0 . \tag{29}$$

Let us explicitly observe that spinor (27a), associated with $s_z = +\frac{1}{2}$ (i.e., with an anti-clockwise internal rotation), gets contributions of equal magnitude from the positive–frequency spin-up component and from the negative–frequency spin-down component: in full agreement with our "reinterpretation" in terms of particles and antiparticles, given in refs.[8]. Analogously, spinor (27b), associated with $s_z = -\frac{1}{2}$ (i.e., with a clockwise internal rotation), gets contributions of equal magnitude from the positive-frequency spin-down component and the negative-frequency spin-up component.[8]

Let us observe also that, having recourse to the light-like solutions, one is actually entitled to regard the electron spin as totally arising from the zbw motion, since the *intrinsic* term $\Delta^{\mu\nu}$ entering the BZ theory[2] does *vanish* when v^μ tends to c.

As we have seen above [cf. eq.(23)], in a *generic* reference frame the polarized states are characterized by a helical uniform motion around the $\boldsymbol{p}$-direction; therefore, the

$\lambda = +\frac{1}{2}$ $[\lambda = -\frac{1}{2}]$ spinor will correspond to an anti-clockwise [a clockwise] helical motion with respect to the $\boldsymbol{p}$-direction.

Going back to the CMF, we have to remark that eq.(19) yields in this case for the zbw radius R the traditional result:

$$R = \frac{|\boldsymbol{V}|}{2m} \equiv \frac{1}{2m} \equiv \frac{\lambda}{2} , \tag{30}$$

where λ is the Compton wave-length. Of course, $R = \frac{1}{2}m$ represents the *maximum* size (in the CMF) of the electron, among all the uniform motion (A^2 =const.; V^2 =const.) solutions. The minimum, $R = 0$, corresponding to the limiting Dirac case with no zbw ($V = A = 0$), represented by eqs.(25), (25'): so that the Dirac free electron is a pointlike, extensionless object.

Before concluding this Section, let us shortly consider what happens when *releasing* the conditions (22)–(25) (and therefore abandoning the assumption of circular uniform motion), so to obtain an internal oscillating motion along a constant straight line. For instance, one may choose either

$$\psi^{\mathrm{T}}(0) \equiv \frac{1}{\sqrt{2}}(1 \;\; 0 \;|\; 1 \;\; 0) , \tag{31}$$

or $\psi^{\mathrm{T}}(0) \equiv \frac{1}{\sqrt{2}}(1 \;\; 0 \;|\; i \;\; 0)$, or $\psi^{\mathrm{T}}(0) \equiv \frac{1}{2}(1 \;\; -1 \;|\; -1 \;\; 1)$, or $\psi^{\mathrm{T}}(0) \equiv \frac{1}{\sqrt{2}}(0 \;\; 1 \;|\; 0 \;\; 1)$, and so on.

In case (31), for example, one actually gets

$$< \vec{\gamma} > \equiv (0,0,1) ; \; < \vec{\alpha} > \equiv (0,0,0)$$

which, inserted into eqs.(14), yield

$$E^\mu = (0; 0, 0, 1) ; \; H^\mu = (0; 0, 0, 0).$$

Therefore, because of eq.(21a), we have now a *linear, oscillating* motion [for which equations (22), (23), (24) and (25) do *not* hold: here $V^2(\tau)$ does vary from 0 to 1!] along the z-axis:

$$V_x(\tau) = 0; \quad V_y(\tau) = 0; \quad V_z(\tau) = \cos(2m\tau) .$$

All the spinors written above could describe an unpolarized, mixed state, since it holds

$$\boldsymbol{s} \equiv \frac{1}{2}\overline{\psi}\vec{\Sigma}\psi = (0,0,0) ,$$

in agreement with the existence of a linear oscillating motion. Furthermore for such new spinors it holds $\overline{\psi}\psi = \overline{\psi}\gamma^5\psi = 0$, but $\overline{\psi}\gamma^5\gamma^\mu\psi \neq 0$ and $\overline{\psi}S^{\mu\nu}\psi \neq 0$. This *new*

class of spinors has been very recently proposed and extensively studied by Lounesto,[9)] by employing a new concept, called "boomerang", within the framework of Clifford algebras. A physical realization of those new spinors[9] seems now to be provided by our electron, in the present case.

ACKNOWLEDGEMENTS

The authors are grateful to J.P. Dowling for having extended to them the permission to contribute to this Volume of Proceedings in memory of Professor Barut. They acknowledge continuous, stimulating discussions with M. Pavšič, S. Sambataro, D. Wisnivesky, J. Vaz and particularly W.A. Rodrigues Jr. Thanks for useful discussions and kind collaboration are also due G. Andronico, G.G.N. Angilella, M. Baldo, M. Borrometi, A. Buonasera, A. Bugini, F. Catara, A. Del Popolo, C. Dipietro, M. Di Toro, G. Giuffrida, A.A. Logunov, J. Keller, C. Kiihl, G.D. Maccarrone, J.E. Maiorino, R. Maltese, G. Marchesini, R. Milana, R.L. Monaco, E.C. de Oliveira, M. Pignanelli, P.I. Pronin, G.M. Prosperi, M. Sambataro, J.P. dos Santos, P.A. Saponov, G.A. Sardanashvily, E. Tonti, P. Tucci, R, Turrisi, M.T. Vasconselos and J.R. Zeni.

References

[1] A.H. Compton: Phys. Rev. **14** (1919) 20, 247; E. Schrödinger: Sitzunger. Preuss. Akad. Wiss. Phys. Math. Kl. **24** (1930) 418. See also P.A.M. Dirac: *The Principles of Quantum Mechanics*, 4th edition (Claredon; Oxford, 1958), p.262; J. Frenkel: Z. Phys. **37** (1926) 243; M. Mathisson: Acta Phys. Pol. **6** (1937) 163; H. Hönl and A. Papapetrou: Z. Phys. **112** (1939) 512; **116** (1940) 153; M.J. Bhabha and H.C. Corben: Proc. Roy. Soc. (London) A**178** (1941) 273; K. Huang: Am. J. Phys. **20** (1952) 479; H. Hönl: Ergeb. Exacten Naturwiss. **26** (1952) 29; A. Proca: J. Phys. Radium **15** (1954) 5; M. Bunge: Nuovo Cimento **1** (1955) 977; F. Gursey: Nuovo Cimento **5** (1957) 784; W.H. Bostick: "Hydromagnetic model of an elementary particle", in *Gravity Res. Found. Essay Contest* (1958 and 1961), and refs. therein; H.C. Corben: Phys. Rev. **121** (1961) 1833; T.F. Jordan and M. Mukunda: Phys. Rev. **132** (1963) 1842; B. Liebowitz: Nuovo Cimento A**63** (1969) 1235; H. Jehle: Phys. Rev. D**3** (1971) 306; F. Riewe: Lett. Nuovo Cim. **1** (1971) 807; G.A. Perkins: Found. Phys. **6** (1976) 237; D. Gutkowski, M. Moles and J.P. Vigier: Nuovo Cim. B**39** (1977) 193; A.O. Barut: Z. Naturforsch. A**33** (1978) 993; J.A. Lock: Am. J. Phys. **47** (1979) 797; M. Pauri: in *Lecture Notes in Physics, vol.135* (Springer; Berlin, 1980), p.615; J. Maddox: "Where Zitterbewegung may lead", Nature **325** (1987) 306; M. H. McGregor: *The enigmatic electron* (Kluwer; Dordrecht, 1992); W.A. Rodrigues, J. Vaz and E. Recami: Found. Phys. **23** (1993) 459.

[2] A.O. Barut and N. Zanghi: Phys. Rev. Lett. **52** (1984) 2009. See also A.O. Barut and A.J. Bracken: Phys. Rev. D**23** (1981) 2454; D**24** (1981) 3333; A.O. Barut and M. Pavsic: Class. Quantum Grav. **4** (1987) L131; A.O. Barut: Phys. Lett. B**237** (1990) 436.

[3] E. Recami and G. Salesi: "Field theory of the electron: Spin and zitterbewegung", in *Particles, Gravity and Space-Time*, ed. by P.I. Pronin and G.A. Sardanashvily (World Scient.; Singapore, 1996), pp.345-368.

[4] M. Pavsic, E. Recami, W.A. Rodrigues, G.D. Maccarrone, F. Raciti and G. Salesi: Phys. Lett. **B318** (1993) 481.

[5] M. Pavsic: Phys. Lett. **B205** (1988) 231; **B221** (1989) 264; Class. Quant. Grav. **7** (1990) L187.

[6] A.O. Barut and M. Pavsic: Phys. Lett. **B216** (1989) 297; F.A. Ikemori: Phys. Lett. **B199** (1987) 239. See also D. Hestenes: Found. Phys. **20** (1990) 1213; S. Gull, A. Lasenby and C. Doran: "Electron paths, tunneling and diffraction in the space-time algebra", to appear in Found. Phys. (1993); D. Hestenes and A. Weingartshofer (eds.): *The electron* (Kluwer; Dordrecht, 1971), in particular the contributions by H. Krüger, by R. Boudet, and by S. Gull; A. Campolattaro: Int. J. Theor. Phys. **29** (1990) 141; D. Hestenes: Found. Phys. **15** (1985) 63.

[7] W.A. Rodrigues Jr., J. Vaz, E. Recami and G. Salesi: Phys. Lett. **B318** (1993) 623.

[8] For the physical interpretation of the negative frequency waves, without any recourse to a "Dirac sea", see e.g. E. Recami: Found. Phys. **8** (1978) 329; E. Recami and W.A. Rodrigues: Found. Phys. **12** (1982) 709; **13** (1983) 533; M. Pavsic and E. Recami: Lett. Nuovo Cim. **34** (1982) 357. See also R. Mignani and E. Recami: Lett. Nuovo Cim. **18** (1977) 5; A. Garuccio *et al.*: Lett. Nuovo Cim. **27** (1980) 60.

[9] P. Lounesto: "Clifford algebras, relativity and quantum mechanics", in *Gravitation: The Space-Time Structure — Proceedings of Silarg-VIII*, ed. by W.A. Rodrigues et al. (World Scient.; Singapore, 1994).

FIELD THEORY OF THE SPINNING ELECTRON:
II – THE NEW NON-LINEAR FIELD EQUATIONS [*]

Erasmo RECAMI[1] and Giovanni SALESI[2]

[1]Facoltà di Ingegneria, Università Statale di Bergamo,
24044–Dalmine (BG), Italy;
INFN, Sezione di Milano, Milan, Italy; and
Dept. of Applied Math., State University at Campinas,
Campinas, S.P., Brazil.
[2]Dipart. di Fisica, Università Statale di Catania,
57 Corsitalia, 95129–Catania, Italy.

ABSTRACT and INTRODUCTION

One of the most satisfactory picture of spinning particles is the Barut–Zanghi (BZ) classical theory for the relativistic electron, that relates the electron spin to the so-called *zitterbewegung* (zbw). The BZ motion equations constituted the starting point for two recent works about spin and electron structure, co-authored by us, which adopted the Clifford algebra language. Here, employing on the contrary the tensorial language, more common in the (first quantization) field theories, we "quantize" the BZ theory and derive for the electron field a non-linear Dirac equation (NDE), of which the ordinary Dirac equation represents a particular case.

We then find out the general solution of the NDE. Our NDE does imply *a new probability current* J^μ, that is shown to be a *conserved* quantity, endowed (in the center-of-mass frame) with the zbw frequency $\omega = 2m$, where m is the electron mass. Because of the conservation of J^μ, we are able to adopt the ordinary *probabilistic interpretation* for the fields entering the NDE.

At last we propose a natural generalization of our approach, for the case in which an external electromagnetic potential A^μ is present; it happens to be based on a new system of *five* first–order differential field equations.

---

[*] Work partially supported by INFN, CNR, MURST, and by CNPq, FAPESP.

ON BARUT–ZANGHI THEORY
FOR THE SPINNING (FREE) ELECTRON

Classical models of spin and classical electron theories have been investigated for about seventy years[1]. For instance, Schrödinger's suggestion[2] that the electron spin was related to *zitterbewegung* did originate a large amount of subsequent work, including Pauli's. Let us quote, among the others, ref.[3], where one meets even the proposal of models with clockwise and anti-clockwise "internal motions", as classical analogues of quantum relativistic spinning particles and antiparticles, respectively. The use of Grassmann variables in a classical lagrangian formulation for spinning particles was proposed by Berezin and Marinov[4] and by Casalbuoni[4]. A recent approach, based on a generalization of Dirac non-linear electrodynamics, where the electron spin is identified with the momentum in such a theory, can be found in ref.[5]. In the Barut–Zanghi (BZ) theory,[6] the classical electron was actually characterized, besides by the usual pair of conjugate variables (x^μ, p^μ), by a second pair of conjugate classical *spinorial* variables $(\psi, \overline{\psi})$, representing internal degrees of freedom, which were functions of the (proper) time τ measured in the electron global center-of-mass (CM) system; the CM frame (CMF) being the one in which $\boldsymbol{p} = 0$ at any instant of time. Barut and Zanghi, then, introduced a classical lagrangian that in the free case (i.e., when the *external* electromagnetic potential is $A^\mu = 0$) writes [$c = 1$]

$$\mathcal{L} = \frac{1}{2} i\lambda(\dot{\overline{\psi}}\psi - \overline{\psi}\dot{\psi}) + p_\mu(\dot{x}^\mu - \overline{\psi}\gamma^\mu\psi) \,, \tag{1}$$

where λ has the dimension of an action, and ψ and $\overline{\psi} \equiv \psi^\dagger\gamma^0$ are ordinary $\mathbb{C}^4$–bispinors, the dot meaning derivation with respect to τ. The four Euler–Lagrange equations, with $-\lambda = \hbar = 1$, yield the following motion equations:

$$\begin{cases} \dot{\psi} + ip_\mu\gamma^\mu\psi = 0 & \text{(2a)} \\ \dot{x}^\mu = \overline{\psi}\gamma^\mu\psi & \text{(2b)} \\ \dot{p}^\mu = 0 \,, & \text{(2c)} \end{cases}$$

besides the hermitian adjoint of eq.(2a), holding for $\overline{\psi}$. From eq.(1) one can also see that

$$H \equiv p_\mu v^\mu = p_\mu \overline{\psi}\gamma^\mu\psi \tag{3}$$

is a constant of the motion [and precisely is the energy in the CMF].[7,8] Since H is the BZ hamiltonian in the CMF, we can suitably set $H = m$, where m is the particle rest-mass. The general solution of the equations of motion (2) can be shown to be:

$$\psi(\tau) = [\cos(m\tau) - i\frac{p_\mu\gamma^\mu}{m}\sin(m\tau)]\psi(0) \,, \tag{4a}$$

$$\overline{\psi}(\tau) = \overline{\psi}(0)[\cos(m\tau) + i\frac{p_\mu\gamma^\mu}{m}\sin(m\tau)] \,, \tag{4b}$$

with $p^\mu = $ constant; $p^2 = m^2$; and finally:

$$\dot{x}^\mu \equiv v^\mu = \frac{p^\mu}{m} + [\dot{x}^\mu(0) - \frac{p^\mu}{m}]\cos(2m\tau) + \frac{\ddot{x}^\mu}{2m}(0)\sin(2m\tau) \,. \tag{4c}$$

This general solution exhibits the classical analogue of the phenomenon known as zit-terbewegung (zbw): in fact, the velocity v^μ contains the (expected) term p^μ/m plus a term describing an oscillating motion with the characteristic zbw frequency $\omega = 2m$. The velocity of the CM will be given by p^μ/m. Let us explicitly observe that the general solution (4c) represents a helical motion in the ordinary 3-space: a result that has been met also by means of other, alternative approaches.[9,10] The BZ theory has been recently studied also in the lagrangian and hamiltonian simplectic formulations, both in flat and in curved space-times.[6]

ABOUT A NEW NON-LINEAR DIRAC–LIKE EQUATION (NDE) FOR THE FREE ELECTRON

At this point we want to introduce the spinorial fields $\psi(x), \overline{\psi}(x)$ and the velocity 4-vector field $V(x)$, starting from the spinor variables $\psi(\tau), \overline{\psi}(\tau)$ and from the 4-velocity $v(\tau)$ along the helical paths. Let us indeed consider a spinorial field $\psi(x)$ such that its *restriction* $\psi(x)|_\sigma$ to the world-line σ (along which the "particle" moves) coincides with $\psi(\tau)$. Consider at the same time a velocity field $V(x)$ together with its integral lines (or stream-lines). Then the velocity distribution $V(x)$ is required to be such that its restriction $V(x)|_\sigma$ to the world-line σ coincide with the ordinary 4-velocity $v(\tau)$ of the considered "particle". Therefore, for the tangent vector along any line σ the relevant relation holds:

$$\frac{d}{d\tau} \equiv \frac{dx^\mu}{d\tau}\frac{\partial}{\partial x^\mu} \equiv \dot{x}^\mu \partial_\mu \; . \tag{5}$$

Inserting eq.(5) into eq.(2a) one gets the non-linear equation:

$$i\dot{x}^\mu \partial_\mu \psi = p_\mu \gamma^\mu \psi \; ,$$

and, since $\dot{x}^\mu = \overline{\psi}\gamma^\mu\psi$ because of eq.(2b), one arrives at the following interesting equation:[7,8,11]

$$i\overline{\psi}\gamma^\mu\psi\partial_\mu\psi = p_\mu\gamma^\mu\psi \; . \tag{6}$$

Let us notice that, differently from eqs.(1)–(2), equation (6) can be valid a priori even for massless spin $\frac{1}{2}$ particles, since the CMF proper time does not enter it any longer.

The *non-linear* equation (6) corresponds to the whole system of eqs.(2): quantizing the BZ theory, therefore, does *not* lead to the Dirac equation, but rather to our non-linear, Dirac–like equation (6), that we shall call the NDE. Let us add, furthermore, that the analogous of eq.(3), now holding for our field $\psi(x)$, is the following noticeable normalization constraint:[#1]

$$p_\mu\overline{\psi}\gamma^\mu\psi = m \; . \tag{7}$$

[#1] In refs.[9] it was moreover assumed $p_\mu\dot{x}^\mu = m$, which actually does imply the very general relation $p_\mu\overline{\psi}\gamma^\mu\psi = m$, but *not* the Dirac equation $p_\mu\gamma^\mu\psi = m\psi$, as claimed therein; these two equations in general are not equivalent.

This non-linear equation is very probably the simplest[12] non-linear *Dirac–like* equation.

In a generic frame, the general solution of NDE can be easily shown to be the following $[\not{p} \equiv p_\mu \gamma^\mu]$:

$$\psi(x) = [\frac{m - \not{p}}{2m} \, e^{ip_\mu x^\mu} + \frac{m + \not{p}}{2m} \, e^{-ip_\mu x^\mu}] \, \psi(0) ; \qquad (8a)$$

which, in the CMF, reduces to

$$\psi(\tau) = [\frac{1 - \gamma^0}{2} \, e^{im\tau} + \frac{1 + \gamma^0}{2} \, e^{-im\tau}] \, \psi(0) \qquad (8b)$$

(or, in simpler form, to eq.(4a)).

Let us explicitly observe that, by inserting eq.(8a) or eq.(8b) into eq.(7), one obtains that every solution of the NDE does correspond to the CMF field hamiltonian $H = m$, which is always *positive* even if it appears (as expected) to be a suitable superposition of plane waves endowed with *positive and negative* frequencies.[13]

It can be also noticed that superposition (8a) is a solution of eq.(6), due to its non-linearity, *only* for suitable pairs of plane waves with weights

$$\frac{m \pm \not{p}}{2m} = \Lambda_\pm ,$$

respectively; which are nothing but the usual projectors Λ_+ (Λ_-) onto the positive (negative) energy states of the standard Dirac equation. In other words, the plane wave solution (for a fixed value of p) of the Dirac eigenvalue equation $\not{p}\psi = m\psi$ is a particular case of the general solution of eq.(6): corresponding, namely, to

$$\text{either} \qquad \Lambda_+\psi(0) = 0 \qquad \text{or} \qquad \Lambda_-\psi(0) = 0 . \qquad (9)$$

Therefore, the solutions of the Dirac eigenvalue equation are a *subset* of the set of solutions of our NDE. It is worthwhile to repeat that, for each fixed p, the wave function $\psi(x)$ describes both particles and antiparticles: all corresponding however to positive *energies*, in agreement with the reinterpretation forwarded in refs.[13], as well as with the already mentioned fact that we can always choose $H = m > 0$.

THE CONSERVED CURRENT J^μ

We want now to study the probability current J^μ corresponding to the wave functions (8a,b) and (4a). Let us define it as follows:[7]

$$J^\mu \equiv \frac{m}{\mathcal{E}} \, \overline{\psi}\gamma^\mu\psi , \qquad (10)$$

where the normalization factor $m/\mathcal{E}$ (the 3-volume V being assumed to be equal to 1, as usual; so that $\mathcal{E}V \equiv \mathcal{E}$) is required to be such that the classical limit of J^μ, that is $(m/\mathcal{E})\,v^\mu$, equals $(1; \boldsymbol{v})$, like for the ordinary quantum probability currents. Notice also that $J^0 \equiv 1$, which means that we have one particle inside the unitary 3-volume

256

$V = 1$. This normalization allows us to recover, in particular, the Dirac current $J^\mu_D = p^\mu/\mathcal{E}$ when considering the (trivial) solutions, without zbw, corresponding to relations (9). Actually, if we insert quantity $\psi(x)$ given by eq.(8a) into eq.(10), we get

$$J^\mu = \frac{p^\mu}{\mathcal{E}} + E^\mu \cos(2p_\mu x^\mu) + H^\mu \sin(2p_\mu x^\mu) , \qquad (10')$$

where

$$E^\mu \equiv J^\mu(0) - p^\mu/\mathcal{E} ; \qquad H^\mu \equiv \dot{J}(0)/2m . \qquad (10'')$$

If we now impose conditions (9), we have $E^\mu = H^\mu = 0$ and get therefore the Dirac current $J^\mu = J^\mu_D = p^\mu/\mathcal{E} = $ constant. Let us observe that the normalization factor $\sqrt{m/\mathcal{E}}$ cannot be included into the expressions of ψ and $\overline{\psi}$, as it would seem convenient, because of the non-linearity of eq.(6) and/or of constraint (7).

From the fact that $p_\mu E^\mu \equiv p_\mu J^\mu(0) - p_\mu p^\mu/\mathcal{E} = m^2/\mathcal{E} - m^2/\mathcal{E} = 0$ (where we used eq.(7) for $x = 0$), that $p_\mu H^\mu \equiv p_\mu \dot{J}^\mu(0)/2m = 0$, obtained deriving both members of eq.(7), and that both E^μ and H^μ are orthogonal to p^μ, it follows that

$$\partial_\mu J^\mu = 2p_\mu H^\mu \cos(2px) - 2p_\mu E^\mu \sin(2px) = 0 . \qquad (10''')$$

We may conclude, with reference to the NDE, that our current J^μ is *conserved*: we are therefore allowed to adopt the ordinary probabilistic interpretation for the fields $\psi, \overline{\psi}$. Equation (10') does clearly show that our conserved current J^μ, as well as its classical limit $mv^\mu/\mathcal{E}$ [see eq.(4c)], are endowed with a zitterbewegung–type motion: precisely, with an oscillating motion having the CMF frequency $\Omega = 2m \simeq 10^{21}$ s^{-1} and period $T = \pi/m \simeq 10^{-20}$ s (we may call Ω and T the zbw frequency and period, respectively).

From eq.(10') one can immediately verify that in general

$$J^\mu \neq p^\mu/\mathcal{E} , \qquad J^\mu \equiv J^\mu(x) ;$$

whilst the Dirac current J^μ_D for the free electron with fixed p, as already mentioned, is constant:

$$J^\mu_D = p^\mu/\mathcal{E} = \text{ constant} ,$$

which correspond to *no* zbw. In other words, our current behaves differently from Dirac's, even if both of them obey[#2] the constraint [cf. eq.(7)]:

$$p_\mu J^\mu = p_\mu J^\mu_D = m^2/\mathcal{E} .$$

It is noticeable, moreover, that our current J^μ goes into the Dirac one, not only in the (no-zbw) case of eq.(9), but also when considering its time–average over a zbw period:

$$< J^\mu >_{\text{zbw}} = \frac{p^\mu}{\mathcal{E}} \equiv J^\mu_D . \qquad (11)$$

[#2] In the Dirac case, this is obtained by getting, from the ordinary Dirac equation $p_\mu \gamma^\mu \psi_D = m\psi_D$, the non-linear constraint $p_\mu \overline{\psi}_D \gamma^\mu \psi_D = m\overline{\psi}_D \psi_D$, and therefore replacing $\overline{\psi}_D \psi_D$ by $m/\mathcal{E}$, consistently with the ordinary normalization $\psi_D = e^{-ipx} u_p/\sqrt{2\mathcal{E}}$, where $\overline{u}_p u_p = 2m$.

GENERALIZATION OF THE NDE FOR THE NON-FREE CASES

Let us now pass to consider the presence of *external* electromagnetic fields: $A^\mu \neq 0$. For the non-free case, Barut and Zanghi[6] proposed the lagrangian

$$\mathcal{L} = \frac{1}{2} i (\dot{\overline{\psi}} \psi - \overline{\psi} \dot{\psi}) + p_\mu (\dot{x}^\mu - \overline{\psi} \gamma^\mu \psi) + e A_\mu \overline{\psi} \gamma^\mu \psi \tag{12}$$

which in our opinion should be better rewritten in the following form, obtained directly from the free lagrangian *via* the minimal prescription procedure:

$$\mathcal{L} = \frac{1}{2} i (\dot{\overline{\psi}} \psi - \overline{\psi} \dot{\psi}) + (p_\mu - e A_\mu)(\dot{x}^\mu - \overline{\psi} \gamma^\mu \psi), \tag{13}$$

all quantities being expressed as functions of the (CMF) proper time τ, the generalized impulse being now $p^\mu - eA^\mu$. We shall call as usual $F^{\mu\nu} \equiv \partial^\mu A^\nu - \partial^\nu A^\mu$.

Lagrangian (13) does yield, in this case, the following system of differential equations:

$$\begin{cases} \dot{\psi} + i(p_\mu - e A_\mu)\gamma^\mu \psi = 0 & (14a) \\ \dot{x}^\mu = \overline{\psi} \gamma^\mu \psi & (14b) \\ \dot{p}^\mu - e \dot{A}^\mu = e F^{\mu\nu} \dot{x}_\nu \, . & (14c) \end{cases}$$

As performed at the beginning of Sect.2, we can insert the identity (5) into eqs.(14a), (14b), and exploit the definition of the velocity field, eq.(14c). We easily get the following *five* first–order *differential equations* (one scalar plus one vector equation) containing the five (independent) unknown functions $\psi(x)$ and $p^\mu(x)$:

$$\begin{cases} i(\overline{\psi} \gamma^\mu \psi) \partial_\mu \psi = (p_\mu - e A_\mu) \gamma^\mu \psi & (15a) \\ (\overline{\psi} \gamma^\mu \psi) \partial_\mu (p^\nu - e A^\nu) = e F^{\mu\nu} \overline{\psi} \gamma_\mu \psi \, , & (15b) \end{cases}$$

which are now *field* equations (quantities $\psi, \overline{\psi}, p$ and A being all functions of x^μ).

The solutions $\psi(x)$ of system (15) may be now regarded as the classical spinorial fields for relativistic spin-$\frac{1}{2}$ fermions, in presence of an electromagnetic potential $A^\mu \neq 0$. We can obtain from eqs.(15) well–defined time evolutions, both for the velocity p^μ/m of the CMF and for the "particle" velocity v^μ. It is rather likely that, by imposing the condition of finite motions, i.e., that $v(\tau)$ and $p(\tau)$ be periodic in time (and ψ vanish at spatial infinity), one will be able to find a discrete spectrum, out from the continuum set of solutions of eqs.(15). Therefore, without solving any eigenvalue equations, within our field theory we may *expect* to be able to single out discrete energy level spectra for the stationary states, in analogy with what we already found[7] in the free case (in which the *uniform* motion condition *implied* the z-component s_z of spin s to be discrete).[7] We shall expand on this point elsewhere: paying attention to applications, especially to "classical" problems so as hydrogen atom, Zeeman effect, and tunnelling through a barrier.

ACKNOWLEDGEMENTS

This paper is dedicated to the memory of Asim O. Barut, who so much contributed to clarifying very many issues of physics, and whose work is a starting point of the present articles. The authors are grateful to J.P. Dowling for having extended to them the permission to contribute to this Volume of Proceedings in memory of Professor Barut. They wish to acknowledge stimulating discussions with M. Pavšič, S. Sambataro, D. Wisnivesky, and particularly W.A. Rodrigues Jr. and J. Vaz Jr. Thanks for useful discussions and kind collaboration are also due to M. Baldo, A. Bonasera, A. Bugini, F. Catara, L. D'Amico, C. Dipietro, G. Dimartino, M. Di Toro, P. Falsaperla, M. Gambera, G. Giuffrida, A. Lamagna. A.A. Logunov, J. Keller, C. Kiihl, J.E. Maiorino, G. Marchesini, R.L. Monaco, E.C. Oliveiras, M. Pignanelli, P.I. Pronin, G.M. Prosperi, M. Sambataro, J.P. dos Santos, P.A. Saponov, G.A. Sardanashvily, Q.A.G. Souza, E. Tonti, P. Tucci, M.T. Vasconselos and J.R. Zeni.

References

[1] A.H. Compton: Phys. Rev. **14** (1919) 20, 247, and refs. therein. See also W.H. Bostick: "Hydromagnetic model of an elementary particle", in *Gravity Res. Found. Essay Contest* (1958 and 1961); J. Frenkel: Z. Phys. **37** (1926) 243; M. Mathisson: Acta Phys. Pol. **6** (1937) 163; H. Hönl and A. Papapetrou: Z. Phys. **112** (1939) 512; **116** (1940) 153; M.J. Bhabha and H.C. Corben: Proc. Roy. Soc. (London) A**178** (1941) 273; K. Huang: Am. J. Phys. **20** (1952) 479; H. Hönl: Ergeb. Exacten Naturwiss. **26** (1952) 29; A. Proca: J. Phys. Radium **15** (1954) 5; M. Bunge: Nuovo Cimento **1** (1955) 977; F. Gursey: Nuovo Cimento **5** (1957) 784; B. Liebowitz: Nuovo Cimento A**63** (1969) 1235; H. Jehle: Phys. Rev. D**3** (1971) 306; F. Riewe: Lett. Nuovo Cim. **1** (1971) 807; G.A. Perkins: Found. Phys. **6** (1976) 237; D. Gutkowski, M. Moles and J.P. Vigier: Nuovo Cim. B**39** (1977) 193; A.O. Barut: Z. Naturforsch. A**33** (1978) 993; J.A. Lock: Am. J. Phys. **47** (1979) 797; M.H. McGregor: *The enigmatic electron* (Kluwer; Dordrecht, 1992); W.A. Rodrigues, J. Vaz and E. Recami: Found. Phys. **23** (1993) 459.

[2] E. Schrödinger: Sitzunger. Preuss. Akad. Wiss. Phys. Math. Kl. **24** (1930) 418. See also P.A.M. Dirac: *The principles of quantum mechanics* (Claredon; Oxford, 1958), 4th edition, p. 262; J. Maddox: "Where Zitterbewegung may lead", Nature **325** (1987) 306.

[3] H.C. Corben: Phys. Rev. **121** (1961) 1833; Nuovo Cim. **20** (1961) 529; Phys. Rev. D**30** (1984) 2683; Am. J. Phys. **61** (1993) 551; "Primitive quantization of Zitterbewegung", preprint (June 1994).

[4] F.A. Berezin and M.S. Marinov: J.E.T.P. Lett. **21** (1975) 320; R. Casalbuoni: Nuovo Cimento A**33** (1976) 389.

[5] W.A. Rodrigues Jr., J. Vaz and E. Recami: Found. Phys. 23 (1993) 469.

[6] A.O. Barut and N. Zanghi: Phys. Rev. Lett. **52** (1984) 2009. See also A.O. Barut and A.J. Bracken: Phys. Rev. D**23** (1981) 2454; D**24** (1981) 3333; A.O. Barut and M. Pavsic: Class. Quantum Grav. **4** (1987) L131; A.O. Barut: Phys. Lett. B**237** (1990) 436.

[7] G. Salesi and E. Recami: Phys. Lett. A**190** (1994) 137; A**195** (1994) E389; E. Recami and G. Salesi: "Field theory of the electron: Spin and Zitterbewegung", in *Particles, Gravity and Space-Time*, ed. by P.I. Pronin and G.A. Sardanashvily (World Scient.; Singapore, 1996), pp.345-368.

[8] M. Pavsic, E. Recami, W.A. Rodrigues, G.D. Maccarrone, F. Raciti and G. Salesi: Phys. Lett. B**318** (1993) 481.

[9] M. Pavsic: Phys. Lett. B**205** (1988) 231; B**221** (1989) 264; Class. Quant. Grav. **7** (1990) L187.

[10] A.O. Barut and M. Pavsic: Phys. Lett. B**216** (1989) 297; F.A. Ikemori: Phys. Lett. B**199** (1987) 239. See also D. Hestenes: Found. Phys. **20** (1990) 1213; S. Gull, A. Lasenby and C. Doran: "Electron paths, tunneling and diffraction in the space-time algebra", to appear in Found. Phys. (1993); D. Hestenes and A. Weingartshofer (eds.): *The electron* (Kluwer; Dordrecht, 1991), in particular the contributions by H. Krüger, by R. Boudet, and by S. Gull; A. Campolattaro: Int. J. Theor. Phys. **29** (1990) 141; D. Hestenes: Found. Phys. **15** (1985) 63.

[11] W.A. Rodrigues Jr., J. Vaz, E. Recami and G. Salesi: Phys. Lett. B**318** (1993) 623.

[12] V.I. Fushchich and R.Z. Zhdanov: Sov. J. Part. Nucl. **19** (1988) 498.

[13] For the physical interpretation of the negative frequency waves, without any recourse to a "Dirac sea", see e.g. E. Recami: Found. Phys. 8 (1978) 329; E. Recami and W.A. Rodrigues: Found. Phys. **12** (1982) 709; **13** (1983) 533; M. Pavsic and E. Recami: Lett. Nuovo Cim. **34** (1982) 357. See also R. Mignani and E. Recami: Lett. Nuovo Cim. **18** (1977) 5; A. Garuccio *et al.*: Lett. Nuovo Cim. **27** (1980) 60.

QUANTUM THEORY OF SELF-ORGANIZING ELECTRICALLY CHARGED PARTICLES: SOLITON MODEL OF THE ELECTRON

V. P. Oleinik

Department of General and Theoretical Physics
Kiev Polytechnic Institute
prospect Pobedy, 37, Kiev, 252056, Ukraine

INTRODUCTION

The investigation of the physical nature of the electron and the study of its internal structure is one of the key problems of present-day physics. Though the electron became the first elementary particle discovered experimentally (J. Thomson, 1897), the description of its unique properties on the basis of an uncontradictory model remains the most important scientific problem, which was formulated by W. Thomson as follows: "Tell me what the electron is and I'll explain to you everything else." Progress in solving this problem will undoubtedly open up great possibilities for producing radically new electronic devices, materials, and technologies that are based on using intra-electron processes.

In the conventional formulation of quantum electrodynamics (QED), one proceeds from the assumption that the electron is a structureless point particle which does not experience the Coulomb self-action. This assumption seems to be the main cause of serious mathematical and logical difficulties of QED[1-3]. One of the most significant of them is the divergence of the electron self-energy and the other is the failure of the theory to explain the stability of the point electron. Numerous attempts to overcome these difficulties and to work out a consistent quantum theory of internal structure of electron have failed. The main reason consists in using the standard scheme of quantum mechanics, the framework of which proved to be too narrow to take into consideration the Coulomb self-action of the electron and to describe its internal structure.

As is known, one of the most bold ideas concerning the physical nature of the electron belongs to E. Schrödinger who believed that the dimensions of the electron are the same as those of the atom[4,5]. According to Schrödinger's interpretation of quantum mechanics the quantity $e|\psi|^2$ is the density of spatial distribution of the electron's charge (e is the charge and ψ is the wave function of the electron).

The interest in Schrödinger's hypothesis has been revived during the last few decades in connection with the new approaches to calculation of radiative corrections. E. Jaynes and co-workers[3,6] have developed a semiclassical theory of radiative phenomena which is based on the Schrödinger interpretation and does not use the second quantization of the electromagnetic field. J. Ackerhalt and J. Eberly[7] have worked out a source-field approach to the quantum theory of radiation which relies completely on the time evolution of dynamical variables, in accordance with the Heisenberg picture, without resorting to vacuum fluctuations and second quantization of the electromagnetic field.

An important stride towards the true understanding of the physical nature of the electron was made by A. Barut and his collaborators[8–12]. They formulated and developed the quantum theory of electromagnetic processes based entirely on the self-energy picture (the Self-Field QED). Using the expression for the total self-energy of the electron, they managed to derive the formulae for the Lamb shift and other radiative corrections and to show that the radiative corrections may be calculated in terms of the action without treating the second quantization. As is pointed out in Reference 9, the correct equation of motion for the radiating electron is not the Dirac or Schrödinger equation for a bare electron, but an equation with an additional nonlinear self-energy term.

In References 13–24, new lines of approach to the problem are proposed which change QED into a theory of self-organization of electrically charged matter. Mechanism of self-organization consists in the back action of the Coulomb field created by the particle upon the same particle and is described by the model of an open system with the wave function belonging to the indefinite metric space.

Basic to the approach are the following physical ideas:

1. The electron is a quantum (an elementary excitation) of the charged matter field localized in a bounded region of space and subject to the Coulomb self-action. This means that the ability of the electron to produce the Coulomb forces and Coulomb self-action are the physical properties intrinsically inherent in the charged matter and should be included from the very beginning in the definition of the particle. Mathematically, from this it follows that the behavior of the electron should be governed by the nonlinear dynamical equation. Physically, the electron becomes a *self-organizing system*, whose geometric shape and linear dimensions are determined in a self-consistent way from the solution of a dynamical equation.

2. Since the electron is a clump of the charged matter producing the long-range Coulomb forces in the surroundings, its environment becomes a medium which can have a determining effect on the physical properties of the particle. Thus, the electron turns to *an open system* inseparable from the surrounding medium. In a sense the whole universe takes part in the formation of the electron.

Obviously, to properly describe the electron being treated as an open self-organizing system, with the electric charge distributed in space, one needs some essentially new mathematical tools. In this article we use the functional space characterized by a quadratic form with indefinite metric.

It should be noted that in the standard approach, when studying the interaction of the electron with the electromagnetic field, the bare electron isolated from its own Coulomb field is taken as an initial approximation to the real particle, with the Coulomb field being considered as a self-dependent degree of freedom of the electromagnetic field. The Coulomb field and the transverse electromagnetic waves are described by the single 4-vector potential whose components are assumed to be the independent dynamical variables of the electromagnetic field. Such an approach leads to simple, symmetric, and clearly covariant, equations of motion for the electron and the electromagnetic field. However, this simplicity doesn't seem to reflect the true physical nature of electromagnetic interaction because there is a fundamental difference between the Coulomb field and the transverse electromagnetic waves. Unlike the waves, the Coulomb field is produced only by the electrically charged particles; it cannot exist without electric charges and is inseparable from them. Since the scalar photons do not exist in nature, it is intolerable to at first introduce these photons into consideration and then to direct all the power of theoreticians at removing them from the theory. As can be seen from the results obtained, it is the Coulomb self-action rather than the transverse electromagnetic waves that is primarily responsible for the formation of clots of the charged matter; the self-action is not small and thus cannot be taken into account by perturbation theory.

One of the most unexpected, from the point of view of the standard approach, conclusions of the theory being developed concerns the dimensions characterizing the internal structure of the electron. By the internal structure are meant here the spatial inhomogeneities in the distribution of charge of the particle. The simple estimation based on using the fundamental equation for the self-acting electron shows the size of a free electron in the ground state to be of the order of the Bohr radius a_0 ($\sim 10^{-10}$m). At first sight, this conclusion is at variance with experimental data according to which the electron structure does not manifest itself in experiments on particles scattering up to the distances of the order of $(10^{-16}/10^{-17})$m; (see, for example, Reference 25). At least two reasons may be pointed out why here there is no contradiction with experiment. First, since the details of the internal structure were looked for at distances much smaller than a_0, it is quite natural that the results proved to be negative. Second, according to our results, the self-acting electron is a soliton, i.e., such a cloud of the electrically charged substance which tends to maintain its size and shape when interacting with other particles. The scattering theory of this kind of particle has yet to be constructed and so at present one has no idea of how the internal structure of the particle may be manifested in the particle scattering experiments.

The aim of this article is to briefly outline the main results of the theory of self-organizing electrons which represents the synthesis of the standard QED and the theory of self-organization in physical systems[26-28]. The foundation of the theory is the relativistically invariant action which takes into account both the Coulomb self-action and the interaction of charged particles with the transverse electromagnetic waves and is based on the model of an open system. The fundamental dynamical equation derived from the action principle[17,20] is a generalization of the Dirac equation to the case of the self-organizing electron. The solutions to this equation are indicative of the soliton nature of the electron and allows one to determine the internal energy, dimensions, and geometric shape of the particle in different quantum states. It should be emphasized that the theory proposed fits the fundamental principles of symmetry, gives an insight into the problem of electron's stability, and does not lead to the divergence of self-energy. The calculations of the hydrogen atom's dimensions, Balmer's spectrum and total angular momentum of the electron, which have been made to date, are in agreement with experimental data.

Section 1 deals with the equation of motion for the self-acting electron in the nonrelativistic approximation. In Section 2 the relativistic generalization of the fundamental dynamical equation is obtained. In Section 3 the energy characteristics of the system of interacting fields are considered. Section 4 is concerned with the quantum model of the self-acting electron. Solutions of the fundamental dynamical equations are discussed in Section 5. The main results of the paper are summarized in the Conclusions section.

FUNDAMENTAL DYNAMICAL EQUATION FOR THE SELF-ORGANIZING ELECTRON: NONRELATIVISTIC APPROXIMATION

In order to represent the Coulomb field produced by the electron as one of its physical properties we need to derive the dynamical equation which allows for the Coulomb self-action. One of the hints as to how to do this can be obtained from Maxwell's equations for the electric and magnetic fields. From them it follows that the total energy W of the Coulomb field can be written by

$$W = \frac{1}{2} \int dr_1 \int dr_2 \left| r_1 - r_2 \right|^{-1} \rho\left(r_1, t\right) \rho\left(r_2, t\right), \tag{1}$$

$\rho = \rho(r,t)$ being the charge density of the particles. Quantity (1) is the potential energy of the Coulomb interaction between the charges including the self-energy of each particle. Obviously, when deriving the equation of motion for the self-acting electron from the action principle, we have to include the additional term $-W$ in the Lagrangian L of the electron field, i.e.,

$$L = L_0 - W, \tag{2}$$

where L_0 is the Lagrangian for a free particle in the absence of the Coulomb field. Making use of the nonrelativistic approximation,

$$L_0 = \int dr \left[\frac{i}{2} \Psi^* \partial_t \Psi - \frac{1}{2m} \left(\nabla \Psi^* \right) \cdot \left(\nabla \Psi \right) \right], \tag{3}$$

and putting

$$\rho = e \Psi^* \Psi, \tag{4}$$

we arrive from the action principle at the following equation for the wave function $\Psi = \psi(r,t)$ of the nonrelativistic electron,

$$i \frac{\partial \Psi}{\partial t} = \left(-\frac{1}{2m} \nabla^2 + U \right) \Psi, \tag{5}$$

$$U(r,t) = e^2 \int dr_1 \left| r - r_1 \right|^{-1} \left| \Psi\left(r_1, t\right) \right|^2 \equiv U. \tag{6}$$

An inspection shows, however, that Eq. (5) with the potential energy function (6), has no solutions satisfying the reasonable physical requirements. From the physical point of view, this is due to the fact that the Coulomb forces of repulsion are trying to tear the electron to pieces. Formally, the potential energy function U, Eq. (6), is a potential hump rather than the potential well and so Eq. (5) cannot have solutions that would describe the stable states of the particle.

Thus, the negative result is obtained: we had a try, remaining within the framework of the standard theoretical scheme, to take into account the self-action of the electron and arrived at the equation that has no reasonable physical solutions at all. This result seems to mean that it is impossible to construct, without resorting to essentially new physical ideas, a consistent quantum model of the electron.

264

As was noted in the Introduction, the self-acting electron differs essentially from the bare electron in its physical properties. The main difference is that the real electron, as distinct from the bare one, produces the long-range Coulomb field in the surroundings and as a result cannot be treated as an isolated system.

To take into account that the real electron is inseparably linked with the surrounding medium, we should first of all increase the number of dynamical variables describing it. Here we adopt the simplest version of the theory in which the number of variables is doubled as compared with the isolated system, namely, to each dynamical variable of the bare electron, Ψ, there correspond two dynamical variables which are denoted by Ψ and $\widetilde{\Psi}$. These quantities are considered as components of the wave functions describing the quantum state of particle. One of them, say, Ψ, corresponds in a sense to the particle alone (to the bare electron) and the other, $\widetilde{\Psi}$, to the surrounding medium in which the particle moves.

The fundamental quadratic form defining the metric of the wave function space is assumed to be given by

$$\widetilde{\Psi}^{*}\Psi + \Psi^{*}\widetilde{\Psi}. \tag{7}$$

This quantity is used instead of the positively defined quadratic form $\Psi^{*}\Psi$ underlying the conventional formulation of quantum mechanics. As the electric charge density we take the quantity

$$\rho(x) = e\left(\widetilde{\Psi}^{*}(x)\,\Psi(x) + \Psi^{*}(x)\,\widetilde{\Psi}(x)\right), \quad x = (t,r) \tag{8}$$

and as the Lagrangian of the free electron field we use the function [cf (3)]

$$L_0 = \int dr\left\{\frac{i}{2}\left[\widetilde{\Psi}^{*}\partial_t\Psi + \Psi^{*}\partial_t\widetilde{\Psi}\right] - \frac{1}{2m}\left[\left(\nabla\widetilde{\Psi}^{*}\right)\cdot\left(\nabla\Psi\right) + \left(\nabla\Psi^{*}\right)\cdot\left(\nabla\widetilde{\Psi}\right)\right]\right\}. \tag{9}$$

The Lagrangian L of the nonrelativistic self-acting electron is given by Eq. (2) where W and ρ are defined by Eqs. (1) and (8), respectively. The action principle with the Lagrangian L, Eq. (2), gives rise to the following nonlinear equations of motion

$$\left(i\frac{\partial}{\partial t} + \frac{\nabla^2}{2m} - U(x)\right)\begin{pmatrix}\Psi(x)\\[1mm]\widetilde{\Psi}(x)\end{pmatrix} = 0, \tag{10}$$

$$U(x) = e\int dr_1\left|r-r_1\right|^{-1}\rho\left(r_1,t\right) \tag{11}$$

the quantity $\rho(r,t)$ being given by Eq. (8).

An analysis shows that Eq. (10) has the solutions describing the stationary states of the electron at $N = -1$ where N is the normalization constant

$$N = \int dr \left(\widetilde{\Psi}^* \Psi + \Psi^* \widetilde{\Psi} \right). \tag{12}$$

As is seen from Eq. (7), the wave functions of the self-acting electron belong to indefinite metric space. The presence of two components of the nonrelativistic electron wave function without regard for the spin variables, Ψ and $\widetilde{\Psi}$, implies that the particle has an additional degree of freedom. In the theory under consideration, this degree of freedom is characterized by the sign of the normalizing factor N (where $N = \pm 1$) which acts as a quantum number, taking into account the Coulomb self-action of the particle. It is of interest that the states of the free electron and of the atomic one differ from one another by the sign of N: $N = +1$ for the atomic electron, and $N = -1$ for the free one[23,29].

In connection with the quadratic form, Eq. (7), it should be stressed that there is no way of describing the electron as an open system without using the indefinite metric space. This point is worthy of special attention because the quadratic form determines the properties of the wave function space and thus the physical behavior of the system. At first sight, the quadratic form corresponding to the functional space of two dynamical variables, Ψ and $\widetilde{\Psi}$, should be given by

$$\Psi^* \Psi + \widetilde{\Psi}^* \widetilde{\Psi}, \tag{13}$$

instead of Eq. (7). Accordingly, the Lagrangian $\widetilde{L} = \widetilde{L}(\Psi, \widetilde{\Psi})$ of the real electron should be constructed in the standard manner:

$$\widetilde{L}\left(\Psi,\widetilde{\Psi}\right) = L_1 (\Psi) + L_2\left(\widetilde{\Psi}\right) + L_{int}\left(\Psi,\widetilde{\Psi}\right) \tag{14}$$

where $L_1(\Psi)$ is the Lagrangian of the bare electron (that is, of the electron isolated from the medium), $L_2(\widetilde{\Psi})$ is the Lagrangian of the medium created by the particle and estranged from it, and L_{int} is the Lagrangian describing the interaction of the bare electron with the medium, with the equalities

$$L_{int}\left(0,\widetilde{\Psi}\right) = L_{int}\left(\Psi,0\right) = 0$$

being fulfilled. If we now neglect the dynamical variables of the medium, that is, if we put $\widetilde{\Psi} = 0$, we shall come to the Lagrangian of the bare electron

$$\widetilde{L}\left(\Psi,0\right) = L_1(\Psi) \neq 0 \tag{15}$$

being considered as the zeroth approximation for the real particle. The other limiting case, $\Psi = 0$, leads to the Lagrangian of the medium alone

$$\tilde{L}\left(0,\tilde{\Psi}\right) = L_2\left(\tilde{\Psi}\right) \neq 0. \tag{16}$$

We should take into account, however, that the real electron is indissolubly related to the surrounding medium. The two objects, the bare electron and the bare medium created by it, taken separately, do not exist in nature. Therefore, the use of the Langrangian, Eq. (14), subject to the conditions, Eqs. (15) and (16), as a basis of the theory, is intolerable. At the same time, using the quadratic form, Eq. (7), and accordingly the Langrangian $L = L(\Psi, \dot\Psi)$, Eq. (2), where L_0, W and ρ are defined by Eqs. (9), (1), and (8), we allow for the inseparability of the particle from the medium. In particular, the equalities,

$$L(\Psi,0) = L\left(0,\tilde{\Psi}\right) = 0$$

are fulfilled, which means that the bare electron approximation has no physical meaning; in either case, with no electron or with no medium, we have no physical system.

RELATIVISTIC FUNDAMENTAL DYNAMICAL EQUATIONS FOR THE SELF-ACTING PARTICLES

Let us introduce some designations which are necessary for deriving the relativistic fundamental equation of motion. As is known, any vector field, say $E = E(r)$, can be split into potential $(E_{||})$ and vortex $(E_\perp)$ components, $E = E_{||} + E_\perp$, which are defined by

$$\nabla \times E_{||} = 0 \quad , \quad \nabla \cdot E_{||} \neq 0 \ (or = 0)$$
$$\nabla \times E_{||} \neq 0 \quad , \quad \nabla \cdot E_\perp = 0. \tag{17}$$

Analogously, any four-vector field $A^\mu = (A^0, A)$ can be represented as a sum of potential $\left(A^\mu_{||}\right)$ and vortex $\left(A^\mu_\perp\right)$ components, $A^\mu = A^\mu_{||} + A^\mu_\perp$, with

$$A^\mu_{||} = \left(A^0, A_{||}\right) \quad , \quad A^\mu_\perp = \left(0, A_\perp\right). \tag{18}$$

Splitting the vector fields entering into Maxwell's equations for the electric (E) and magnetic (B) fields into the vortex and potential components, we arrive at the two independent subsets of equations:

$$\partial_t B = -\nabla \times E_\perp \quad , \quad \nabla \cdot B = 0 \ ,$$
$$\partial_t E_\perp = \nabla \times B - 4\pi j_\perp \quad , \quad \nabla \cdot E_\perp = 0 \tag{19}$$

and

$$\nabla \cdot E_{||} = 4\pi\rho \ , \ \partial_t E_{||} = -4\pi j_{||} \quad , \quad \nabla \times E_{||} = 0. \tag{20}$$

Each subset is seen from Eqs. (19) and (20) to involve merely either the vortex components or the potential ones. These subsets can also be represented in the four-vector form:

$$\partial_\nu \mathscr{F}_A^{\mu\nu} = -4\pi j_A^\mu, \tag{21}$$

$$\partial^\alpha \mathscr{F}_A^{\mu\nu} + \partial^\mu \mathscr{F}_A^{\nu\alpha} + \partial^\nu \mathscr{F}_A^{\alpha\mu} = 0 \quad (A = \perp, \|)$$

where $\partial_\nu = (\partial/\partial x^\nu) = (\partial_t, \nabla)$, $x^\nu = (t,r)$ ($\nu = 0,1,2,3$).

$\mathscr{F}_\perp^{\mu\nu}$ and $\mathscr{F}_\|^{\mu\nu}$ are the vortex and potential components of the field-strength tensor $\mathscr{F}^{\mu\nu}$ defined by

$$\mathscr{F}_\perp^{\mu\nu} = \begin{pmatrix} 0 & -E_{\perp x} & -E_{\perp y} & -E_{\perp z} \\ E_{\perp x} & 0 & -B_z & B_y \\ E_{\perp y} & B_z & 0 & -B_x \\ E_{\perp z} & -B_y & B_x & 0 \end{pmatrix}, \quad \mathscr{F}_\|^{\mu\nu} = \begin{pmatrix} 0 & -E_{\|x} & -E_{\|y} & -E_{\|z} \\ E_{\|x} & 0 & 0 & 0 \\ E_{\|y} & 0 & 0 & 0 \\ E_{\|z} & 0 & 0 & 0 \end{pmatrix}.$$

(22)

$j_\perp^\mu(x)$ and $j_\|^\mu(x)$ are the components of the 4-current density $j^\mu(x) = [\rho(x), j(x)]$. It should be pointed out that the vortex $(A_\perp^\mu)$ and potential $(A_\|^\mu)$ components of the 4-vector A^μ each taken separately, are not 4-vectors. Analogously, the quantities $\mathscr{F}_\perp^{\mu\nu}$ and $\mathscr{F}_\|^{\mu\nu}$ do not behave like 4-tensors. Nevertheless, one can easily be convinced of form-invariance of Maxwell's equations (21) under Lorentz transformations.

Since the potential component of the electric field, $E_{\||}$, is not an independent degree of freedom of the electromagnetic field, we include it in the definition of the electrically charged matter to obtain the self-acting field. The vortex electromagnetic field will be treated on the same grounds as the charged matter field, using the indefinite metric space. To each dynamical variable we shall put into correspondence to variables; for one of them the old designation will be retained (E, B or $\mathscr{F}_\perp^{\mu\nu}$) and the other will be labeled by the sign $\sim$ ($\tilde{E}_\perp$, B or $\tilde{\mathscr{F}}_\perp^{\mu\nu}$)....

The action of the whole system, which consists of n electrically charged fields described by Ψ_κ and $\tilde{\Psi}_\kappa$ ($\kappa = 1,2, \dots, n$) and of the vortex electromagnetic fields represented by $\mathscr{F}_\perp^{\mu\nu}$ and $\tilde{\mathscr{F}}_\perp^{\mu\nu}$, can be written as

$$S = S_\| + S_\perp + S_{int},$$

$$S_{\parallel} = \sum_{\kappa} \int d^4x \left[\overline{\widetilde{\Psi}}_\kappa \left(\frac{i}{2} \overset{\leftrightarrow}{\partial} - m_\kappa \right) \Psi_\kappa + \overline{\Psi}_\kappa \left(\frac{i}{2} \overset{\leftrightarrow}{\partial} - m_\kappa \right) \widetilde{\Psi}_\kappa \right] -$$

$$\frac{1}{2} \int d^4x_1 \int d^4x_2 \, \delta\left(\left(x_1 - x_2 \right)^2 \right) j_{\parallel\mu}(x_1) j_{\parallel}^\mu (x_2) \tag{23}$$

$$S_\perp = - \frac{1}{16\pi} \int d^4x \, \widetilde{\mathscr{F}}_\perp^{\mu\nu}(x) \mathscr{F}_{\perp\mu\nu}(x) ,$$

$$S_{int} = - \int d^4x \, j_{\perp\mu}(x) \, A_\perp^\mu(x)$$

where $S_{\parallel}$ is the action of the self-acting charged fields, Ψ_κ and $\widetilde{\Psi}_\kappa$ are the wave function components for a particle of mass m_κ and electric charge e_x ($\kappa = 1, 2, \ldots, n$); $S_\perp$ is the action of the vortex electromagnetic fields; S_{int} describes interaction of charged particles with the electromagnetic fields; $j_{\parallel}^\mu$ and $j_\perp^\mu$ are the potential and vortex components of the 4-current density

$$j^\mu(x) = \sum_\kappa e_\kappa \left(\overline{\widetilde{\Psi}}_\kappa(x) \gamma^\mu \Psi_\kappa(x) + \overline{\Psi}_\kappa(x) \gamma^\mu \widetilde{\Psi}_\kappa(x) \right) = \left(\rho(x), \mathbf{j}(x) \right). \tag{24}$$

$\widetilde{\mathscr{F}}_\perp^{\mu\nu}$ and $\mathscr{F}_\perp^{\mu\nu}$ are vortex components of the field-strength tensors $\widetilde{\mathscr{F}}_\perp^{\mu\nu}(x)$ and $\mathscr{F}^{\mu\nu}(x)$:

$$\widetilde{\mathscr{F}}^{\mu\nu}(x) = \partial^\mu \widetilde{\mathscr{A}}^\nu(x) - \partial^\nu \widetilde{\mathscr{A}}^\mu(x) , \quad \mathscr{F}^{\mu\nu}(x) = \partial^\mu \mathscr{A}^\nu(x) - \partial^\nu \mathscr{A}^\mu(x),$$

$$\mathscr{A}_\parallel^\mu (x) = \widetilde{\mathscr{A}}_\parallel^\mu (x) = \int d^4x_1 \, \delta\left(\left(x - x_1 \right)^2 \right) j_\parallel^\mu (x_1),$$

$$\mathscr{A}_\perp^\mu(x) = - \frac{1}{4\pi} \int d^4x_1 \, \delta\left(\left(x - x_1 \right)^2 \right) \partial_{1\nu} \mathscr{F}_\perp^{\mu\nu}(x_1),$$

$$\widetilde{\mathscr{A}}_\perp^\mu(x) = - \frac{1}{4\pi} \int d^4x_1 \, \delta\left(\left(x - x_1 \right)^2 \right) \partial_{1\nu} \widetilde{\mathscr{F}}_\perp^{\mu\nu}(x_1),$$

$$A^\mu(x) = \frac{1}{2} \left(\mathscr{A}^\mu(x) + \widetilde{\mathscr{A}}^\mu(x) \right), \tag{25}$$

where $\hat{\partial} = \partial_\alpha \gamma^\alpha$, and the γ^α are Dirac's matrices.

The action principle $\delta S = 0$ gives rise to the fundamental dynamical equations for the charged particles

$$\left(i\hat{\partial} - e_\kappa \widehat{A}(x) - m_\kappa \right) \begin{pmatrix} \Psi_\kappa(x) \\ \widetilde{\Psi}_\kappa(x) \end{pmatrix} = 0 \quad (\kappa = 1, 2, \ldots, n) \tag{26}$$

and to Maxwell's equations

$$\partial_\nu \widetilde{\mathscr{F}}_\perp^{\mu\nu} = \partial_\nu \mathscr{F}_\perp^{\mu\nu} = -4\pi j_\perp^\mu,$$

$$\partial^\alpha \widetilde{\mathscr{F}}_\perp^{\mu\nu} + \partial^\mu \widetilde{\mathscr{F}}_\perp^{\nu\alpha} + \partial^\mu \widetilde{\mathscr{F}}_\perp^{\alpha\mu} = \partial^\alpha \mathscr{F}_\perp^{\mu\nu} + \partial^\mu \mathscr{F}_\perp^{\nu\mu} + \partial^\nu \mathscr{F}_\perp^{\alpha\mu} = 0. \quad (27)$$

By their appearances, the equations of motion (26) coincide with the Dirac equation for the charged particle in an external field described by the 4-potential $A^\mu = A_\parallel^\mu + A_\perp^\mu$. However, in distinction to Dirac's equation, they are nonlinear and nonlocal, with the nonlocality being of the space and time character. The quantities $A_\parallel^\mu$ and $A_\perp^\mu$ entering into Eq. (26) differ from each other by their physical nature: the first describes the Coulomb field and is expressed in terms of the wave functions of the particles, and the second describes the vortex electromagnetic fields and is uniquely determined by the field variables, $E_\parallel, E_\parallel, B, \widetilde{B}$. These two considerably different quantities are combined in our theory to form the single 4-vector.

Note that in place of A^μ and $\mathscr{A}^\mu$ and $\widetilde{\mathscr{A}}^\mu$, Eq. (25), we can equally well use the quantities $\mathscr{A}'^\mu$ and $\widetilde{\mathscr{A}}'^\mu$ defined by

$$\mathscr{A}_\parallel'^\mu = \widetilde{\mathscr{A}}_\parallel'^\mu = \left(\mathscr{A}_\parallel'^0, 0\right)$$

$$\mathscr{A}_\parallel'^0(x) = \int d\mathbf{r}_1 \left|\mathbf{r} - \mathbf{r}_1\right|^{-1} \rho\left(\mathbf{r}_1, t\right),$$

$$\mathscr{A}_\perp'^\mu = \left(0, \mathscr{A}_\perp'\right), \quad \mathscr{A}_\perp'^\mu = \left(0, \widetilde{\mathscr{A}}_\perp'\right),$$

$$\mathscr{A}_\perp'(x) = (4\pi)^{-1} \int d\mathbf{r}_1 \left|\mathbf{r} - \mathbf{r}\right|^{-1} \nabla_{\mathbf{r}_1} \times \mathbf{B}\left(\mathbf{r}_1, t\right),$$

$$\mathscr{A}_\perp'(x) = (4\pi)^{-1} \int d\mathbf{r}_1 \left|\mathbf{r} - \mathbf{r}_1\right|^{-1} \nabla_{\mathbf{r}_1} \times \widetilde{\mathbf{B}}\left(\mathbf{r}_1, t\right). \quad (28)$$

It can easily be shown that the 4-potentials $\mathscr{A}^\mu(x)$ and $\mathscr{A}'^\mu(x)$ as well as $\widetilde{\mathscr{A}}^\mu(x)$ and $\widetilde{\mathscr{A}}'^\mu(x)$ are related to each other by a gauge transformation and thus are physically equivalent.

ENERGY RELATIONS OF THE INTERACTING FIELDS

Using Expression (23) for the total action, one can derive, in the usual manner, the energy-momentum tensor of interacting fields

$$T^{\mu\nu} = T_\perp^{\mu\nu} + T_\parallel^{\mu\nu} + T_{int}^{\mu\nu} \quad (29)$$

and the law of conservation of the energy-momentum 4-vector

$$\int d\mathbf{r}\, T^{\mu 0}(x) \equiv P^\mu + P_\perp^\mu + P_\parallel^\mu + P_{int}^\mu. \quad (30)$$

Separate terms in the right-hand side of Eqs. (29) and (30) correspond, respectively, to the vortex electromagnetic field, charged particles, and interaction of the particles with the field. The quantities $P_\perp^\mu$ and $P_\parallel^\mu$ can be written as follows:

$$\left(P_\perp^0, \mathbf{P}_\perp\right) = (8\pi)^{-1}\int d\mathbf{r}\left(\widetilde{\mathbf{E}}_\perp \cdot \mathbf{E}_\perp + \widetilde{\mathbf{B}} \cdot \mathbf{B}, \ \widetilde{\mathbf{E}}_\perp \times \mathbf{B} + \mathbf{E}_\perp \times \widetilde{\mathbf{B}}\right),$$

$$P_\parallel^0 = T + V,$$

$$T = -\sum_\kappa \int d\mathbf{r}\left[\widetilde{\overline{\Psi}}_\kappa\left(i\boldsymbol{\gamma}\cdot\nabla - m_\kappa\right)\Psi_\kappa + \overline{\Psi}_\kappa\left(i\boldsymbol{\gamma}\cdot\nabla - m_\kappa\right)\widetilde{\Psi}_\kappa\right],$$

$$V = \int d\mathbf{r}\left(\frac{1}{8\pi}E_\parallel^2 - \mathbf{j}_\parallel \cdot \mathbf{A}_\parallel\right),$$

$$\mathbf{P}_\parallel = -\frac{i}{2}\sum_\kappa \int d\mathbf{r}\left[\widetilde{\overline{\Psi}}_\kappa \boldsymbol{\gamma}^0 \overset{\leftrightarrow}{\nabla}\Psi_\kappa + \overline{\Psi}_\kappa \boldsymbol{\gamma}^0 \overset{\leftrightarrow}{\nabla}\widetilde{\Psi}_\kappa\right] - \int d\mathbf{r}\rho\mathbf{A}_\parallel . \quad (31)$$

Separate terms in the expression for $P_\parallel^0$ have the following meaning: T is the sum of the internal energy of electrically charged particles and kinetic energy connected with their motion in space as a whole; V is the potential energy of particles, including the Coulomb self-action energy. Separate components of the momentum $\mathbf{P}_\parallel$, Eq. (31), can be interpreted in the following way: The first addend in the right-hand side represents the momentum of the particles due to their motion as a whole, and the second is the momentum of the Coulomb field created by the particles.

In the case of stationary states,

$$\Psi_\kappa(x) = \varphi_\kappa(\mathbf{r})\exp\left(-iE_\kappa t\right), \quad \widetilde{\Psi}_\kappa(x) = \widetilde{\varphi}_\kappa(\mathbf{r})\exp\left(-iE_\kappa t\right), \quad (32)$$

where E_κ is the energy eigenvalue of a particle associated with the field of the type k, and $\varphi_\kappa(\mathbf{r})$ and $\widetilde{\varphi}_\kappa(\mathbf{r})$ are the functions dependent only on spatial coordinates. The energy of the charged particles is given by

$$P_\parallel^0 = \sum_\kappa E_\kappa N_\kappa - V, \quad V = \frac{1}{8\pi}\int d\mathbf{r}E_\parallel^2 , \quad (33)$$

with

$$N_\kappa = \int d\mathbf{r}\left(\widetilde{\Psi}_\kappa^* \Psi_\kappa + \Psi_\kappa^* \widetilde{\Psi}_\kappa\right). \quad (34)$$

QUANTUM MODEL OF THE ELECTRON

Consider a system of n self-acting charged fields described by the action

$$\widetilde{S} = \sum_{\kappa} \int d^4x \left[\overline{\widetilde{\Psi}}_{\kappa} \left(\frac{1}{2} \frac{\overleftrightarrow{\partial}}{\partial} - m_{\kappa} \right) \Psi_{\kappa} + \overline{\Psi}_{\kappa} \left(\frac{i}{2} \frac{\overleftrightarrow{\partial}}{\partial} - m_{\kappa} \right) \widetilde{\Psi}_{\kappa} \right]$$

$$- \frac{1}{2} \int d^4x_1 \int d^4x_2 \delta \left(\left(x_1 - x_2 \right)^2 \right) j_{\alpha}(x_1) j^{\alpha}(x_2). \tag{35}$$

The action $\widetilde{S}$ can be obtained from the general expression (23) by excluding from it the vortex components $\widetilde{\mathscr{F}}_{\perp}^{\mu\nu}$ and $\mathscr{F}_{\perp}^{\mu\nu}$ of the field-strength tensors with the help of Maxwell's equations (27). If we put $\widetilde{\Psi}_{\kappa} = \Psi_{\kappa}$ in Eqs. (24) and (35), we obtain the self-field QED of A. Barut[8–12].

The action principle $\delta \widetilde{S} = 0$ gives the following equations of motion:

$$\left(i \widehat{\partial} - e_{\kappa} \widehat{A}(x) - m_{\kappa} \right) \begin{pmatrix} \Psi_{\kappa}(x) \\ \\ \widetilde{\Psi}_{\kappa}(x) \end{pmatrix} = 0,$$

$$A^{\mu}(x) = \int d^4x_1 \, \delta \left(\left(x - x_1 \right)^2 \right) j^{\mu}(x_1). \tag{36}$$

The energy-momentum tensor $\widetilde{T}^{\mu\nu}$ corresponding to our system is ($g_{\mu\nu} = 0$ at $\mu \neq \nu$, and $g_{00} = -g_{ii} = 1$, for $i = 1,2,3$):

$$\widetilde{T}^{\mu\nu} = t^{\mu\nu} + \theta^{\mu\nu} + \frac{1}{2} g^{\mu\nu} j_{\alpha} A^{\alpha} - A^{\mu} j^{\mu}, \tag{37}$$

where

$$t^{\mu\nu} = \frac{i}{2} \sum_{\kappa} \left(\overline{\widetilde{\Psi}}_{\kappa} \overleftrightarrow{\partial}^{\mu} \gamma^{\nu} \Psi_{\kappa} + \overline{\Psi}_{\kappa} \overleftrightarrow{\partial}^{\mu} \gamma^{\nu} \widetilde{\Psi}_{\kappa} \right)$$

$$- g^{\mu\nu} \left\{ \sum_{\kappa} \left[\overline{\widetilde{\Psi}}_{\kappa} \left(\frac{i}{2} \frac{\overleftrightarrow{\partial}}{\partial} - m_{\kappa} \right) \Psi_{\kappa} + \overline{\Psi}_{\kappa} \left(\frac{i}{2} \frac{\overleftrightarrow{\partial}}{\partial} - m_{\kappa} \right) \widetilde{\Psi}_{\kappa} \right] - \frac{1}{2} j_{\alpha} A^{\alpha} \right\},$$

$$\theta^{\mu\nu} = - \frac{1}{4\pi} \left(F^{\mu\alpha} F^{\nu}_{\alpha} - \frac{1}{4} g^{\mu\nu} F^{\alpha\beta} F_{\alpha\beta} \right). \tag{38}$$

The components of the 4-tensor $F^{\mu\nu} = \partial^{\mu} A^{\nu} - \partial^{\nu} A^{\mu}$ satisfying the equality

$$\partial_{\nu} F^{\mu\nu} = - 4\pi j^{\mu} \tag{39}$$

can be represented in the form $F^{\mu\nu} = F_{\parallel}^{\mu\nu} + F_{\perp}^{\mu\nu}$, with $F_{\parallel}^{\mu\nu}$ describing the Coulomb field and $F_{\perp}^{\mu\nu}$ the vortex fields. It should be stressed that in the model under study the field being described by $F^{\mu\nu}$ is not a degree of freedom independent of the charged particles. For this reason the equalities (39) are the identities, not the equations. The 4-tensor $\theta^{\mu\nu}$, Eq. (38), is the energy-momentum tensor for the field produced by the charged matter. It satisfies the equation

$$\partial_\nu \theta^{\mu\nu} = -F^\mu,$$

with $F^\mu = F^{\mu\alpha} j_\alpha$ being the force-density, 4-vector which describes the back action of the fields produced by the charged particles on the same particles.

The differential conservation law, $\partial_\nu \widetilde{T}^{\mu\nu} = 0$, results in the integral law of conservation

$$\int d\mathbf{r}\, \widetilde{T}^{\mu 0} = \widetilde{P}^\mu = \text{const.} \tag{40}$$

The energy-momentum, 4-vector components $\widetilde{P}^\mu$ can be written as follows:

$$\widetilde{P}^0 = -\sum_\kappa \int d\mathbf{r} \left[\overline{\widetilde{\Psi}}_\kappa \left(\frac{i}{2} \boldsymbol{\gamma} \cdot \overset{\leftrightarrow}{\nabla} - m_\kappa \right) \Psi_\kappa + \overline{\Psi}_\kappa \left(\frac{i}{2} \boldsymbol{\gamma} \cdot \overset{\leftrightarrow}{\nabla} - m_\kappa \right) \widetilde{\Psi}_\kappa \right]$$

$$- \int d\mathbf{r}\, \mathbf{j} \cdot \mathbf{A} + \frac{1}{8\pi} \int d\mathbf{r} \left(E_{\|}^2 + E_\perp^2 + B^2 \right),$$

$$\widetilde{\mathbf{P}} = -\frac{i}{2} \sum_\kappa \int d\mathbf{r} \left(\overline{\widetilde{\Psi}}_\kappa \gamma^0 \overset{\leftrightarrow}{\nabla} \Psi_\kappa + \overline{\Psi}_\kappa \gamma^0 \overset{\leftrightarrow}{\nabla} \widetilde{\Psi}_\kappa \right)$$

$$- \int d\mathbf{r}\, \rho \mathbf{A}_\| + \frac{1}{4\pi} \int d\mathbf{r}\, \mathbf{E}_\perp \times \mathbf{B}, \tag{41}$$

where $\mathbf{E} = -\partial_t \mathbf{A} - \nabla A^0$, $\mathbf{B} = \nabla \times \mathbf{A}$, where $\mathbf{E}$ and $\mathbf{B}$ are the electric and magnetic fields produced by the charged particles.

Consider the quantity $\int d\mathbf{r}\, \widetilde{T}^{\mu\nu} \equiv \tau^{\mu\nu}$. According to Eq. (40) the quantity $\tau^{\mu 0}$ represents a 4-vector. This is possible only when in the inertial reference frame, in which the particles are in stationary states; the component τ^{00} is the only one distinct from zero. Using the solutions of the fundamental equations corresponding to the stationary spherically symmetric states [see Eq. (32)], we have derived the following relationship:

$$\tau^{\mu\nu} = \delta_{\mu 0}\, \delta_{\nu 0} \left(\sum_\kappa E_\kappa N_\kappa - V \right), \quad V = \frac{1}{2} \int d\mathbf{r}\, \rho A^0, \tag{42}$$

where N_κ is the normalizing constant, Eq. (34). From Eq. (42) it follows that we have managed to construct an uncontradictory quantum relativistic model of the electron.

As distinct from $\tau^{\mu 0}$, the quantities $\int d\mathbf{r}\, \theta^{\mu 0} \equiv \tau_1^{\mu 0}$ and $\tau^{\mu 0} - \tau_1^{\mu 0}$ are not the 4-vectors. This means that it is impossible in principle to define correctly the notions of energy and momentum of both the particle free of its own Coulomb and vortex fields and the fields being produced by the particle and separated from it. It is only the particle being thought of as an elementary excitation of the charged matter together with the fields included in its definition that is a well-defined physical object.

Consider the inertial frames of reference K' and K moving relative to one another. Let the reference frame K' move with a velocity v relative to K along the z axis and a system of charged particles be in stationary state in reference fame K'. According to Eq. (42), in K' the energy-momentum, 4-vector of the system is of the form

$$\widetilde{P}'^{\mu} = \left(E_0, 0\right) \ , \quad E_0 = \sum_{\kappa} E_{\kappa} N_{\kappa} - V.$$

In accordance with the Lorentz transformations, in the reference frame K the energy-momentum, 4-vector will be

$$\widetilde{P}^{\mu} = \left(\gamma E_0, 0, 0, v\gamma E_0\right) \ , \quad \gamma = \left(1 - v^2\right)^{-1/2} .$$

Hence it follows that the energy E and momentum P of the system are related to each other by the equality

$$E = \pm \left[P^2 + \left(\sum_{\kappa} E_{\kappa} - V\right)^2 \right]^{1/2} .$$

For the free electron $(K = 1, m_1 = m, N_1 = -1)$

$$E_0 = -E_1 - V = -(m + \mathscr{E}_1 + V) \ , \quad |\mathscr{E}_1| \ll m, \mathscr{E}_1 + V < 0 .$$

The quantity E_0 takes on discrete values according to the fact that the electron can be in various quantum states. Denoting by n the totality of quantum numbers of the electron, we can write: $E_{0n} = -(m + \mathscr{E}_{1n} + V_n)$ where by $\mathscr{E}_{1n}$ and V_n are meant the quantities $\mathscr{E}_1$ and V pertaining to the quantum state n. Since the energy E_0 and the normalization constant N_1 are negative, it is natural to use the hole interpretation, namely, the electron can be considered as a vacancy (a hole) in a nonobservable sea of states with the energy $-E_0$ and the charge $Q = -\int d\mathbf{r}\rho = -eN_1 = e$. When adopting such an interpretation, the quantity $\mathscr{E}_{1n} + V_n < 0$ will have the meaning of the energy of the electron reckoned from n, the quantity $\mathscr{E}_{1n}$ being the binding energy (the energy necessary for the formation of the structure localized in space) and V_n the energy of the Coulomb field created by the particle.

The quantum theory of the self-acting electron given above differs qualitatively from the theory of electromagnetic mass of the electron (the Abraham-Lorentz (A-L) model). In the A-L model the stability of the electron is achieved by introducing the special attractive forces being produced by a hypothetical matter field and compensating for the Coulomb repulsive forces. In the present theory, contrastingly, there are no additional forces and no material sources creating them. It is the electrically charged matter that is the only source of both the Coulomb forces and the forces compensating for them and holding the particle stable. The stability of the electron is due to the Coulomb self-action and is achieved by the use of the functional space with indefinite metric.

As can be seen from the model discussed above, the Coulomb field plays a leading part in formation of the electron being considered as a clump of charged matter localized in some region of space. It is evident, besides, that the Coulomb self-action described by the last term in the right-hand side of Eq. (35) cannot be considered as a small perturbation. Indeed, the behavior of the electron wave function is considerably dependent on the self-action: when the self-action is absent [i.e., at $A = 0$ in Eq. (36)], the electron wave function is a plane wave, whereas at $A \neq 0$ the wave function describes the soliton — the state of particle localized in space.

SOLUTIONS TO THE FUNDAMENTAL DYNAMICAL EQUATION

The equation of motion of the self-acting electron interacting with the vortex electromagnetic field in the nonrelativistic approximation can be obtained as the nonrelativistic limit of the relativistic equation (26). This nonrelativistic equation can be written as follows (in the case of one particle):

$$
i \frac{\partial}{\partial t} \begin{pmatrix} \Psi \\ \widetilde{\Psi} \end{pmatrix} = \left\{ \frac{1}{2m} \left(-i\nabla - e\beta\mathbf{A} \right)^2 - \frac{e\beta\boldsymbol{\sigma} \cdot \mathbf{H}}{2m} + e\beta U \right\} \begin{pmatrix} \Psi \\ \widetilde{\Psi} \end{pmatrix} ,
$$

$$
\mathbf{H} = \nabla \times \mathbf{A} \ , \quad \mathbf{A} = \frac{1}{2}\left(\mathcal{A}'_\perp + \widetilde{\mathcal{A}}'_\perp \right),
$$

$$
U\left(\mathbf{r},t\right) = e \int d\mathbf{r}' \left|\mathbf{r}-\mathbf{r}'\right|^{-1} \left(\widetilde{\Psi}^+\left(\mathbf{r}',t\right)\Psi\left(\mathbf{r}',t\right) + \Psi^+\left(\mathbf{r}',t\right)\widetilde{\Psi}\left(\mathbf{r}',t\right) \right), \quad (43)
$$

where $\mathcal{A}'_\perp$ and $\widetilde{\mathcal{A}}'_\perp$ are given by Eq. (28). We have introduced above a new parameter — the self-action constant β, which is indicative of the intensity of self-action. The magnitude of β can be determined from comparing the theoretical results with experimental data[29]. In what follows we shall take the simplest version of the theory by imposing on the wave function components the constraint: $\widetilde{\Psi} = c\Psi$, $c = $ const. This is possible because the equation for $\widetilde{\Psi}$ does not differ from the equation for Ψ. Introducing the notation $\widetilde{\Psi}^\dagger \Psi + c.c. = (c+c^*) \Psi^\dagger \Psi = N\Psi^\dagger\Psi$, $N=+1$, and neglecting the vortex electromagnetic field and the spin of particle, we arrive at the Schrödinger equation (5) with the potential energy function given by

$$
U(\mathbf{r},t) = \beta e^2 N \int d\mathbf{r}' \left|\mathbf{r}-\mathbf{r}'\right|^{-1} \left|\Psi\left(\mathbf{r}',t\right)\right|^2 . \quad (44)
$$

Let us consider stationary spherically symmetric solutions corresponding to the energy eigenvalue $E = $ constant. Putting

$$
\Psi(\mathbf{r},t) = \exp\left(-i Et\right)\varphi(r) , \quad \varphi(r) = \frac{X(r)}{r} \ , \quad U(r) = U(0) + \frac{Z(r)}{r} ,
$$

we obtain the following set of equations equivalent to Eqs. (5) and (44):

$$\left(\frac{d^2}{dr^2} - \frac{Z}{r} + c\right) X = 0,$$

$$\frac{d^2}{dr^2} Z = -8\pi\beta N\frac{1}{r}X^2 \;, \quad c = E - U(0). \tag{45}$$

Here we have used as units of length and energy, respectively, the Bohr radius $a_0 = 1/(me)^2$ and the ionization energy $I_0 = me^4/2$.

The sought-for solutions of the set of Eqs. (45) have to satisfy the normalization condition $4\pi\int_0^\infty drX^2(r) = 1$, some boundary conditions at $r = 0$, and the requirement that the function x have the shape of a soliton with a fixed number of extrema. Such solutions exist and can be obtained with the aid of a computer. As an analysis shows, the self-acting electron is a soliton with a discrete energy spectrum, the size and geometric shape of the charge distribution depending upon the magnitude of its energy eigenvalues.

Analogously, the fundamental dynamical equations can be studied for the hydrogen atom. Relying on their solutions[23,29], we can conclude that the hydrogen atom is a system of electronic and nuclear solitons interacting with one another. Qualitatively, the picture of the energy spectrum for the stationary states may be obtained from the following simple considerations. Because of the fact that the mass of the nucleus is much greater than that of the electron, the interaction between them may be treated as a small perturbation. If we remove the electron from the atom, we shall obtain the energy spectrum of the free self-acting nucleus which is similar to that of the free self-acting electron and consists of an infinitely great number of energy levels. Since the interaction of the nucleus with the electron is small, we should expect the energy levels of the atom to be located within small vicinities of the levels for the free nucleus. This interaction has to lead to the splitting of each nuclear level into an infinitely large number of levels, forming an energy band. Thus, the Coulomb self-action of the particles in the hydrogen atom should result in appearance of the band structure of energy spectrum. One of the bands should coincide with the known Balmer's spectrum. These qualitative considerations are completely confirmed by solving the fundamental dynamical equations[29].

CONCLUSION

Based on the solutions of the fundamental equations for the free electron and for the atom, we can draw the following conclusions:

1.   The self-acting electron is a soliton which can be in different quantum states characterized by internal energy, dimensions, and geometric shape.

2.   The self-acting electron has a discrete internal energy spectrum, the size and geometric shape of the particle depending upon the value of its internal energy.

3.   The atom consists of one or several electronic solitons interacting with the nuclear soliton.

4. Discreteness of the internal energy spectrum of the nuclear soliton is responsible for the appearance of the band structure of energy spectrum for the hydrogen atom, Balmer's spectrum being one of the energy bands.

5. Discreteness of the internal energy spectrum of the electron and existence of energy bands in the atom offer great possibilities for using the quantum transitions between the internal energy levels of a particle, including the levels inside bands, for controlling intra-electron processes and producing new materials, electronic devices, and technologies.

Note that the self-acting electron cannot decay into fragments under the influence of a perturbation. The particle can merely go from one quantum state to the other, with the result that its charge distribution in space may vary considerably in size and shape. It is natural to make an attempt to carry out an experiment on electron excitation in which the electron would transfer from one internal energy level to the other. Evidently, the performance of the experiment is impossible without due development of the theory which alone is capable of making some recommendations with respect to conducting the experiment. Therefore, the immediate task of theoretical research is to obtain and investigate the solutions of the fundamental equations corresponding to both stationary and nonstationary states of the self-acting electron and to evaluate the intensity of quantum transitions between the internal energy levels. At present the theory has progressed to the point where we feel that it can lead to advances in experimental research.

REFERENCES

1. P. A. M. Dirac, *The Principles of Quantum Mechanics* (Oxford, Clarendon Press, 1958).
2. P. A. M. Dirac, *Lectures on Quantum Mechanics* (Yeshiva University, NY, 1964).
3. M. D. Crisp and E. T. Jaynes, "Radiative effects in semiclassical theory," *Phys. Rev.* 179:1253 (1969).
4. E. Schrödinger, "Quantisierung als Eigenwertproblem. Vierte Mitteilung," *Ann. der Physik* 81:109 (1926).
5. E. Schrödinger, *Collected Papers on Quantum Mechanics*, edited by L. S. Polak, (Nauka, 1976), p. 134, 199.
6. C. R. Stroud, Jr. and E. T. Jaynes, "Long-term solutions in semiclassical radiation theory," *Phys. Rev. A* 1:106 (197).
7. J. R. Ackerhalt and J. H. Eberly, "Quantum electrodynamics and radiation reaction: nonrelativistic atomic frequency shifts and lifetimes," *Phys. Rev. D* 10:3350 (1974).
8. A. O. Barut and J. Kraus, "Nonperturbative QED: The Lamb shift," *Found. of Physics* 13:189 (1983).
9. A. O. Barut and J. F. van Huele, "Quantum electrodynamics based on self-energy: Lamb shift and spontaneous emission without field quantization," *Phys. Rev. A* 32:3187 (1985).
10. A. O. Barut, "QED based on self-energy," *Physica Scripta T* 21:18 (1988).
11. A. O. Barut and J. P. Dowling, "QED based on self-energy: spontaneous emission in cavities," *Phys. Rev. A* 36:649 (1987).
12. A. O. Barut and J. P. Dowling, "Interpretation of self-field quantum electrodynamics," *Phys. Rev. A* 43:4060 (1991).
13. V. P. Oleinik, "To electrodynamics of the dielectric medium without potentials," *Quant. Electron.* 34:92 (1988).

14. V. P Oleinik, "On quantum dynamics of the self-interacting particles in electromagnetic field," *ibid.* 36:87 (1989).
15. V. P Oleinik, "On dynamics and internal energy spectrum of charged quantum particles," *ibid.* 40:75 (1991).
16. V. P Oleinik, "On internal structure of electrically charged particles due to their own Coulomb field," *ibid.* 42:68 (1992).
17. V. P Oleinik, "Quantum electrodynamics describing the internal structure of electron," *ibid.* 44:51 (1993).
18. V. P Oleinik, "To the theory of the internal structure of electron: Second quantization and energy relations," *ibid.* 45:57 (1993).
19. V. P Oleinik, "Quantum dynamics of the self-acting electron," *ibid.* 47 (1994).
20. V. P. Oleinik, "Quantum electrodynamics describing the internal structure of electron, gauge-independent and covariant theory," University of Leipzig, NTZ, *Prepr.*, No. 7 (1992), Leipzig, p. 30.
21. V. P. Oleinik, "Quantum electrodynamics describing the internal structure of electron: Energy relations and second quantization," *Prepr.*, No. 1-92, KPI, Kiev (1992), p. 40.
22. V. P. Oleinik, Yangqiang Ran, and L. P. Godenko, "Self-acting electron in external field," *Prepr.*, No. 2-93, KPI, Kiev (1993), p. 31.
23. V. P. Oleinik, *Prepr.*, No. 3-93, KPI, Kiev (1993), p. 66.
24. V. P. Oleinik, *Prepr.*, No. 4-93, KPI, Kiev (1993), p. 44.
25. G. Efimov, *Nonlocal Interactions of Quantized Fields*, (Nauka, Moscow, Russia, 1977).
26. R. Z. Sagdeev and A. A. Galeev, *Nonlinear Plasma Theory*, edited by T. O'Neil and D. Book (Benjamin, 1969).
27. G. Nicolis and I. Prigogine, *Self-Organization in Nonequilibrium Systems* (Wiley Interscience, 1977).
28. H. Haken, *Advanced Synergetics. Instability Hierarchies of Self-Organizing Systems and Devices* (Springer-Verlag, Berlin, Heidelberg, New York, Tokyo, 1983).
29. Ju. D. Arepjev, A. Ju. Buts, and V. P. Oleinik, "To the problem of internal structure of electrically charged particles. Spectra of internal energy and charge distribution for a free electron and hydrogen atom." Prepr. of the Institute of Semiconductors of the Ukraine, No. 8-91, Kiev, 1991.

COMPARISON OF TWO MODELS OF THE CLASSICAL ZITTERBEWEGUNG

AND PHOTON WAVE FUNCTION

Nuri Ünal

Akdeniz University
Physics Department
P.K.510, 07200
Topcular, Antalya, Turkey

I. INTRODUCTION

The Heisenberg equations of the Dirac electron have a helical trajectory as its natural free motion (the zitterbewegung).[1] In nineteeneighties Barut and Zanghi proposed a classical model of the electron which also has zitterbewegung oscillations.[2] Different aspects of this model have been examined.[3] In order to understand the classical analog of the zitterbewegung and its quantization we propose a simple version of this model, study the dynamical properties of this system and compare its structure with the realistic model.

In the realistic classical model (R) of the zitterbewegung the phase space consists of the external conjugate dynamical variables (x, p) and internal conjugate dynamical variables $(\bar{z}, z)$. Then, the configuration space is $M^4 \otimes C^4$. These two set of dynamical variables are related to each other by a constraint. Proca proposed a constraint for the velocity of the electron.[4] In (R) z and $\bar{z}$ are 4-component complex spinors. In this simple (S) model we choose 2-component complex spinors. Then, the configuration space is $M^4 \otimes C^2$.

In section 2 we discuss the classical dynamics of the (S) model from different points of view, in section 3 we quantize it and in section 4 compare the quantum equations of the (R) and (S) models. In section 5 we discuss the solutions of the massless and chargeless particle wave equations. Recently, there have been new interest in 'how to define a wave function for the photon'.[5] In order to answer this question we discuss the solution of the spin-1 quantum equation of the classical electron and compare this equation with the classical Maxwell equations.

II. CLASSICAL SYSTEM

The classical system has two set of conjugate dynamical variables. These are $\left(x_\mu, p_\mu\right)$ and $\left(\eta^+, -i\eta\right)$ for the external and internal dynamics respectively. Where η^+ and η are two

Electron Theory and Quantum Electrodynamics: 100 Years Later
Edited by Dowling, Plenum Press, New York, 1997

component complex spinors and the configuration space is $M^4 \otimes C^2$. The action can be written as

$$A = \int d\tau \left[-i\eta \frac{d\eta^+}{d\tau} + P_\mu \left(\frac{dx^\mu}{d\tau} - \eta^+ \sigma^\mu \eta \right) + eA_\mu \eta^+ \sigma^\mu \eta \right] \tag{1}$$

where τ is the invariant time parameter, p^μ appears as a Lagrange multiplier for the constraint between v_μ and internal dynamics and σ^μ is given in terms of (2×2) Pauli spin matrices as

$$\sigma^\mu = (1, \sigma) \tag{2}$$

We write the action in Cartan form:

$$A = \int \left[p_\mu dx^\mu - i\eta d\eta^+ - Hd\tau \right] \tag{3}$$

where H is the covariant Hamiltonian and for a minimal coupled charged particle it is

$$H = (P_\mu - eA_\mu) \eta^+ \sigma^\mu \eta \tag{4}$$

The Poisson brackets are

$$\{f,g\} = \left\{ \frac{\partial f}{\partial x^\mu} \frac{\partial g}{\partial p^\nu} - \frac{\partial g}{\partial x^\mu} \frac{\partial f}{\partial p^\nu} \right\} g_{\mu\nu} +$$

$$\left\{ \frac{\partial f}{\partial (-i\eta)_\alpha} \frac{\partial g}{\partial \eta^+_\beta} - \frac{\partial g}{\partial (-i\eta)_\alpha} \frac{\partial f}{\partial \eta^+_\beta} \right\} \delta_{\alpha\beta} \tag{5}$$

The equations of the motion are

$$\dot{\eta}^+ = -i\eta^+ \sigma^\mu \left(P_\mu - eA_\mu \right)$$
$$\dot{\eta} = i\sigma^\mu \left(P_\mu - eA_\mu \right) \eta \tag{6}$$
$$\dot{x}^\mu = \eta^+ \sigma^\mu \eta$$
$$\dot{p}^\mu = eA^{\nu,\mu} \dot{x}_\nu$$

These equations are in the Lagrangian form, Hamiltonian form or in the Poisson brackets form.

The free particle solutions are

$$p^\mu = \text{constant}$$
$$\dot{x}^0 = \eta^+ \eta = \text{constant} \tag{7}$$
$$\dot{\mathbf{x}} = v_\| \hat{\mathbf{p}} + \mathbf{v}_\perp(0) e^{i w \tau}$$

where

$$v_\|(\tau) = \mathbf{v} \cdot \hat{\mathbf{p}} = \text{constant}$$
$$\mathbf{v}_\perp(0) = \mathbf{v}(0) - v_\| \hat{\mathbf{p}}$$

and

$$\omega^2 = (2p)^2$$

In the Eq. (7) we have a helical motion. The velocity of the charge has two parts:
The first part is parellel to p and constant.
The second part is orthogonal to p and oscillates with the frequency $\pm 2p$. Thus, the centre
of the zitterbewegung has a constant velocity and the charge oscillates around this centre.
    Spin angular momentum is defined by

$$S_{\mu\nu} = \tfrac{1}{4i}\eta^*[s_\mu, s_\nu]\eta \tag{8}$$

Then

$$S_{0i} = 0$$
$$S_{ij} = +\varepsilon_{ijk}\eta^*\sigma_k\eta/2 \tag{9}$$

The total angular momentum of the particle is

$$\mathbf{J} = \mathbf{L} + \mathbf{S} = \mathbf{x} \times \mathbf{p} + \mathbf{v}/2 \tag{10}$$

and for the free particle

$$\dot{\mathbf{J}} = \dot{\mathbf{L}} + \dot{\mathbf{S}} = \mathbf{v} \times \mathbf{p} + \dot{\mathbf{v}}/2 = 0 \tag{11}$$

and $\mathbf{J}$ is a constant of the motion.

III. QUANTUM SYSTEM

    The quantum evolution of the Schrödinger states is governed by the time evolution
operator

$$U(\tau, \tau_0) = \theta(\tau - \tau_0)\exp\left\{i\int_{\tau_0}^{\tau} d\tau'\left[\eta^+\sigma^\mu\eta\left(p_\mu - eA_\mu\right)\right]\right\} \tag{12}$$

The evolution of the states is given by

$$\phi(\tau) = U(\tau - \tau_0)\phi(\tau_0) \tag{13}$$

where $\phi(\tau_0)$ is the initial state and $\phi(\tau)$ satisfies the Schrödinger equation:

$$i\frac{\partial\phi(\tau)}{\partial\tau} = H\phi(\tau) \tag{14}$$

This equation can be written explicitly in terms of the dynamical variables. It is

$$i\frac{\partial\phi(x,\eta^+;\tau)}{\partial\tau} = \left[\eta^+\eta\left(p_0 - eA_0\right) - \eta^+\sigma\eta\left(\mathbf{p} - e\mathbf{A}\right)\right]\phi(x,\eta^+;\tau) \tag{15}$$

The wave function depends on internal coordinates. In order to understand the dependence of $\phi(x,\eta^+;\tau)$ on the internal coordinates we expand the $\phi(x,\eta^+;\tau)$ in terms of η^+:

$$\phi(x,\eta^+;\tau) = \psi_0(x;\tau) + \tfrac{1}{1!}\eta_\alpha^+\psi_\alpha(x,;\tau) + \tfrac{1}{2!}\eta_\alpha^+\eta_\beta^+\psi_{\alpha\beta}(x;\tau) + \cdots \tag{16}$$

where η_α^+ and η_β^+ are commute and for this reason $\psi_{\alpha\beta}$ is symmetric. We substitute this expansion into Eq. (15). Then we obtain

$$i\frac{\partial}{\partial\tau}[\psi_0(x;\tau) + \tfrac{1}{1!}\eta_\alpha^+\psi_\alpha(x;\tau) + \tfrac{1}{2!}\eta_\alpha^+\eta_\beta^+\psi_{\alpha\beta}(x;\tau) + \cdots] =$$

$$[\eta^+\eta(p_0 - eA_0) - \eta^+\sigma\eta(p - eA)][\psi_0(x;\tau) + \tfrac{1}{1!}\eta_\alpha^+\psi_\alpha(x;\tau) + \tfrac{1}{2!}\eta_\alpha^+\eta_\beta^+\psi_{\alpha\beta}(x;\tau) + \cdots] \tag{17}$$

Each term of this expansion corresponds to eigenvalues of S^2 and S_z. Thus, they satisfy the following equations seperately:

$$i\frac{\partial\psi_0(x;\tau)}{\partial\tau} = [\eta^+\eta(p_0 - eA_0) - \eta^+\sigma\eta(\mathbf{p} - e\mathbf{A})]\psi_0(x;\tau) \tag{18}$$

$$i\eta_\alpha^+\frac{\partial}{\partial\tau}\psi_\alpha(x;\tau) = [\eta^+\eta(p_0 - eA_0) - \eta^+\sigma\eta(\mathbf{p} - e\mathbf{A})]\eta_\alpha^+\psi_\alpha(x;\tau) \tag{19}$$

$$i\frac{\partial}{\partial\tau}\eta_\alpha^+\eta_\beta^+\psi_{\alpha\beta}(x;\tau) = [\eta^+\eta(p_0 - eA_0) - \eta^+\sigma\eta(\mathbf{p} - e\mathbf{A})]\eta_\alpha^+\eta_\beta^+\psi_{\alpha\beta}(x;\tau)$$

$$\vdots$$

$$\tag{20}$$

The Eq. (18) corresponds to spin-0 case and from the solutions of the classical equations of the motion that in this case v_μ and p_μ are not independent. For this reason we represent v_μ in terms of p_μ. In order to do this we define a new velocity and momentum operators in the following way:

$$g_\mu \equiv \frac{v_\mu}{\sqrt{v^2}} \tag{21}$$

and

$$G_\mu \equiv Hg_\mu \tag{22}$$

The new dynamical variables satisfy the following relations:

$$g_\mu g^\mu = 1$$
$$G_\mu G^\mu = H^2 \tag{23}$$
$$H = G_\mu g^\mu$$

Then in terms of these new dynamical variables the Eq. (18) can be rewritten in the following form:

$$(i\partial/\partial\tau - H^{-1}G^\mu G_\mu)\psi_0(x,\tau) = 0 \tag{24}$$

In the Eqs. (19) and (20) we can represent the operator   in the Hamiltonian as

$$\frac{\partial}{\partial\eta^+} \tag{25}$$

Then we calculate the derivatives. Then the Eqs. (19) and (20) become

$$i\frac{\partial}{\partial\tau}\psi_\alpha(x,\tau) = \left[p_0 - eA_0 - \sigma\cdot(\mathbf{p} - e\mathbf{A})\right]_{\alpha\beta}\psi_\beta(x,\tau) \tag{26}$$

$$i\frac{\partial}{\partial\tau}\psi_{\alpha\beta}(x,\tau) = \left[(p_0 - eA_0)I\otimes I - (\sigma\otimes I + I\otimes\sigma)\cdot(\mathbf{p} - e\mathbf{A})\right]_{\alpha\beta,\gamma\delta}\psi_{\gamma\delta}(x,\tau) \tag{27}$$

The Eq.(26) corresponds to spin - 1/2 particles and Eq.(27) corresponds to spin-1 . Mass eigenstates are defined in the following way:

$$\left(m - m^{-1}G^\mu G_\mu\right)\psi_0(x) = 0 \tag{28}$$

$$[p_0 - eA_0 - \sigma\cdot(\mathbf{p} - e\mathbf{A})]_{\alpha\beta}\psi_\beta(x) = m\psi_\alpha(x)) \tag{29}$$

$$\left[(p_0 - eA_0)I\otimes I - (\sigma\otimes I + I\otimes\sigma)\cdot(\mathbf{p} - e\mathbf{A})\right]_{\alpha\beta,\gamma\delta}\psi_{\gamma\delta}(x) = m\psi_{\alpha\beta}(x) \tag{30}$$

where $x = x^\mu = (t,\mathbf{x})$.

IV. COMPARISON OF THE TWO MODELS

In the (R) model internal dynamics is described by 4-component spinors. The velocity is $\bar{z}\gamma^\mu z$. The internal dynamics is described by v^μ and $S^{\mu\nu}$ in the (R) model and by v^μ in the (S) model respectivel. In the (S) model v^μ and $S^{\mu\nu}$ are related to each other. The quantization of this model is also similiar to the procedure described in the previous section. Final   equations are

$$\left(m - m^{-1}G^\mu G_\mu\right)\psi_0(x) = 0 \tag{31}$$

$$[m - \gamma\cdot\hat{\pi}]_{\alpha\beta}\Psi_\beta(x) = 0 \tag{32}$$

$$[m - (\gamma\otimes 1 + 1\otimes\gamma)\hat{\pi}]_{\alpha_1\alpha_2\beta_1\beta_2}\Psi_{\beta_1\beta_2}(x) = 0 \tag{33}$$

.
.
.

$$[m - (\gamma \otimes 1 \otimes \cdots \otimes 1 + 1 \otimes 1 \otimes \cdots \otimes \gamma)\hat{\pi}]_{\alpha_1 \alpha_2 \cdots \alpha_n \beta_1 \beta_2 \cdots \beta_n} \Psi_{\beta_1 \beta_2 \cdots \beta_n}(x) = 0 \qquad (34)$$

.

.

.

Eq. (28) and (31) are Klein-Gordon equation. Eq.(32) is Dirac equation. Eq.(33) is symmetric Kemmer equation. Eq. (34) corresponds to spin-$n/2$ particle.

In Eq. (32) we write $\Psi_\alpha(x)$ in terms of two component spinors:

$$\Psi_\alpha(x) = \begin{pmatrix} \varphi(x) \\ \chi(x) \end{pmatrix} \qquad (35)$$

If

$$\varphi(x) = \chi(x)$$

and

$$m=0$$

then, Eq. (32) and (29) are the same. Thus we conclude that if the positive and negative energy solutions of the massless particle is the same then they can be represented two component equations. Neutrino and photon are the massless particles.

V. MASSLESS NEUTRINO AND PHOTON EQUATIONS

Massless neutrino is a spin-1/2 particle and satisfies the following equation:

$$[p_0 - \sigma \cdot p]_{\alpha\beta} \psi_\beta(t,x) = 0 \qquad (36)$$

In terms of the components it is

$$\left(i\frac{\partial}{\partial t} - \sigma_n p \right) \begin{pmatrix} \psi_1 \\ \psi_2 \end{pmatrix} = 0$$

where $\mathbf{n} = \mathbf{p}/p$ and $\sigma_n = \mathbf{n} \cdot \sigma$. Let us take σ_n as

$$\sigma_n = \begin{pmatrix} 1 & 0 \\ 0 & 1 \end{pmatrix} \qquad (37)$$

Then

$$\begin{pmatrix} i\dfrac{\partial}{\partial t} - p & 0 \\ 0 & i\dfrac{\partial}{\partial t} + p \end{pmatrix} \begin{pmatrix} \psi_1(\mathbf{x},t) \\ \psi_2(\mathbf{x},t) \end{pmatrix} = 0 \qquad (38)$$

284

where $p = -i\partial / \partial x_{\shortparallel}$ and $x_{\shortparallel} = x \cdot n$. We assume a solution of this equation in the following form:

$$\psi(x_{\shortparallel}, t) = \binom{a}{b} e^{-i(\omega t - px_{\shortparallel})}$$

(39)

Then the eigenvalue equation gives

$$\omega^2 = p^2$$

(40)

The eigenstates are

$$\psi^{(+)}(x_{\shortparallel}, t) = \binom{1}{0} e^{-ip(t - x_{\shortparallel})} \qquad \text{for } \omega = p$$

(41)

and

$$\psi^{(-)}(x_{\shortparallel}, t) = \binom{0}{1} e^{ip(t + x_{\shortparallel})} \qquad \text{for } \omega = -p$$

(42)

The first (second) one of these solutions is corresponding to positive (negative) helicity eigenstate.

Next we discuss the free particle solution of the spin-1 equation. In this case ψ is a 4- component spinor which can be represented as

$$\psi(x,t) = \psi_{\alpha\beta}(x,t) = \begin{pmatrix} \psi_{11}(x,t) \\ \psi_{12}(x,t) \\ \psi_{12}(x,t) \\ \psi_{22}(x,t) \end{pmatrix}$$

(43)

where $\psi_{\alpha\beta}$ is symmetric and for this reason $\psi_{21} = \psi_{12}$. The Hamiltonian is

$$H = \tfrac{1}{2}(\sigma \otimes I + I \otimes \sigma) \cdot p$$

(44)

and p is a constant of the motion. We assume p in the direction n and choose σ_n as diagonal. Then the eigenvalue condition is

$$\begin{vmatrix} \omega - p & 0 & 0 & 0 \\ 0 & \omega & 0 & 0 \\ 0 & 0 & \omega & 0 \\ 0 & 0 & 0 & \omega + p \end{vmatrix} = 0$$

(45)

285

The normalized eigenfunctions are

$$\psi^{(+)}(x_\parallel, t) = \begin{pmatrix} 1 \\ 0 \\ 0 \\ 0 \end{pmatrix} e^{ip(x_\parallel - t)} \qquad \text{for} \quad \omega = p \qquad (46)$$

$$\psi^{(-)}(x_\parallel, t) = \begin{pmatrix} 0 \\ 0 \\ 0 \\ 1 \end{pmatrix} e^{ip(x_\parallel + t)} \qquad \text{for} \quad \omega = -p \qquad (47)$$

and

$$\psi^{(0)}(x_\parallel, t) = \frac{1}{\sqrt{2}} \begin{pmatrix} 0 \\ 1 \\ 1 \\ 0 \end{pmatrix} \qquad \text{for} \quad \omega = 0 \qquad (48)$$

Finally, we compare the solutions of the spin-1 equation and Maxwell Equations. In order to compare them we try to costruct a vector function from the eigenstates of the quantum problem. In this solution there is wave vector $\mathbf{p}$ which is in the direction $\mathbf{n}$. Then there are two perpendicular directions to $\mathbf{n}$. Let us denote them by $\mathbf{e}_1$ and $\mathbf{e}_2$. We can define them two new vector for the positive and negative helicity. They are

$$\mathbf{e}_\pm = \frac{1}{\sqrt{2}}(\mathbf{e}_1 \pm i\mathbf{e}_2) \qquad (49)$$

and they satisfy the following properties:

$$\mathbf{n} \cdot \mathbf{e}_\pm = 0$$
$$\mathbf{n} \times \mathbf{e}_\pm = \pm i\mathbf{e}_\pm \qquad (50)$$

We construct a vector function in the following form:

$$\mathbf{f}(x_\parallel, t) = \mathbf{n}\psi^{(0)}(x_\parallel, t) + \mathbf{e}_+\psi^{(+)}(x_\parallel, t) + \mathbf{e}_-\psi^{(-)}(x_\parallel, t) \qquad (51)$$

where $\psi^{(+)}$ and $\psi^{(-)}$ are the positive and negative helicity solutions of the spin-1 quantum system respectively and $\psi^{(0)}$ is the constant longitudional solution. The divergence of the $\mathbf{f}(x_\parallel, t)$ is

$$\nabla \cdot \mathbf{f}(x_\parallel, t) = \frac{\partial}{\partial x_\parallel}\psi^{(0)}(x_\parallel, t) = 0 \qquad (52)$$

286

The curl of the $\mathbf{f}(x_\parallel,t)$ is

$$\nabla \times \mathbf{f}(x_\parallel,t) = \mathbf{n} \times \frac{\partial}{\partial x_\parallel}\mathbf{f}(x_\parallel,t)$$

$$= -ie_+ \frac{\partial}{\partial t}\psi^{(+)}(x_\parallel,t) - ie_- \frac{\partial}{\partial t}\psi^{(-)}(x_\parallel,t) \tag{53}$$

Combination of these two term gives

$$\nabla \times \mathbf{f}(\mathbf{x},t) = -i\frac{\partial}{\partial t}\mathbf{f}(\mathbf{x},t) \tag{54}$$

According the Eqs. (52) and (53) we have

$$\nabla \cdot \mathbf{f}(\mathbf{x},t) = 0 \tag{55.a}$$

and

$$\nabla \times \mathbf{f}(x_\parallel,t) = -i\frac{\partial}{\partial t}\mathbf{f}(x_\parallel,t) \tag{55.b}$$

In order to write the Maxwell Equations in the complex form we define

$$\mathbf{G} = \mathbf{E} - i\mathbf{B} \tag{56}$$

Then the complex form of the free space Maxwell Equations are

$$\nabla \cdot \mathbf{G}(\mathbf{x},t) = 0 \tag{57.a}$$

and

$$\nabla \times \mathbf{G}(\mathbf{x},t) = -i\frac{\partial}{\partial t}\mathbf{G}(\mathbf{x},t) \tag{57.b}$$

Thus, we have shown that chargeless and massless spin -1 particle and free space Maxwell equations have the same energy-momentum relation and they satisfy the same equations.

VI. CONCLUSION

In this work we proposed a simple model for the classical analog of the quantum zitterbewegung. We derived classical equations for the dynamical variables (x_μ, p_μ) and v^μ In this model spin and velocity are related to each other.

We quantized this model and showed that it has the same solutions with the realistic model if the particle is massless and has identical antiparticle. These conditions are satisfied by massless neutrino and photon. Thus, spin-1 quantum equations of the massless particle is equivalent to the Maxwell equations in the charge free space.

AKNOWLEDGEMENTS

This work was partially supported by TUBITAK, Scientific and Technical Research Council of Turkey.

REFERENCES

1. E.Schrödinger, *Sitzungsb. Preuss. Akad. Wiss., Phys.-Math. Kl.* **24** 418 (1930).

2. A.O.Barut and N.Zanghi, *Phys. Rev. Lett.* **52**, 2009 (1984).

3. A.O.Barut and I.H. Duru, *Phys. Rep.* **172**, 1 (1987),
   A.O.Barut and N.Pavsic, *Class. Quantum Grav.* **4**, 41 and 131 (1987),
   A.O. Barut and N.Ünal, *Phys.Rev A* **40**, 5404 (1989),
   A.O. Barut C. Önem and N.Ünal, *J. Phys. A: Math. Gen.* **23**, 1113 (1990),
   J.Vaz Jr. *These proceedings*.

4. A.Proca, *J. Phys. Radium* **15**, 5 (1954).

5. M.Scully, *These proceedings*.

Conceptual Problems and Developments of Electrodynamics

Jean Reignier

Vrije Universiteit Brussel, Theoretische Natuurkunde
Pleinlaan 2, B-1050 Brussels, Belgium.

Université Libre de Bruxelles, Département de Mathématique, CP 217
Boulevard de Triomphe, B-1050 Bruxelles, Belgique.

1. Introduction

One century ago, H.A. Lorentz (1853-1928), one of the most famous physicists of all times started a vast and ambitious scientific program based on the following combination of the laws of electromagnetism and the laws of mechanics (eq.1). This program aimed to understand all properties of matter and light, with the possible exception of gravitation. It was later on called "The theory of electrons". This refers to the title of the book that Lorentz published at the turn of the century, and that contains the main results already obtained[1]. Of course, at the very beginning of his work, Lorentz could not imagine that other types of forces exist (the discovery of radioactivity happened only in 1896), that Newton's mechanics would be soon challenged by the relativistic mechanics (partly created by Lorentz himself within this program), and that the quanta would soon appear and considerably change the further development of the program. These changes were already made at Lorentz's death, and he was then perfectly aware of the new situation. However, I guess that Lorentz would have been much surprised to learn that one century after the start, we do not yet know if the original program is feasible, i.e. whether or not the equations he used are mathematically consistent. The basic equations of the theory of electrons are:

$$div \ \vec{H} = 0 \ , \tag{a}$$

$$rot \ \vec{E} + \frac{1}{c} \ \partial_t \ \vec{H} = 0 \ , \tag{b}$$

$$div \ \vec{E} = 4\pi \rho \ , \tag{c}$$

$$rot \ \vec{H} - \frac{1}{c} \ \partial_t \ \vec{E} = \frac{4\pi}{c} \vec{j} \ , \tag{d}$$

$$\frac{d\vec{p}_\alpha}{dt} = e_\alpha \left[\vec{E}_\alpha + \frac{\vec{v}_\alpha}{c} \times \vec{H}_\alpha \right] \ , \ (\alpha = 1, 2, ...N) \ , \tag{e}$$

$$\rho = \rho(\vec{x}, t) = \sum_\alpha e_\alpha \ \delta \left(\vec{x} - \vec{x}_\alpha(t) \right) \ , \tag{f}$$

$$\vec{j} = \vec{j}(\vec{x}, t) = \sum_\alpha e_\alpha \ \vec{v}_\alpha(t) \ \delta \left(\vec{x} - \vec{x}_\alpha(t) \right) \ , \tag{g}$$

$$\tag{1}$$

(N.B. eqs. c and d imply the conservation law: $div \ \vec{j} + \partial_t \ \rho = 0$, which is obviously satisfied for the model given by eqs. f and g).

Electron Theory and Quantum Electrodynamics: 100 Years Later
Edited by Dowling, Plenum Press, New York, 1997

In these equations and in my lectures, I adopt the following conventional notations:

- $\vec{E} = \vec{E}(\vec{x}, t)$, $\vec{H} = \vec{H}(\vec{x}, t)$, $\rho = \rho(\vec{x}, t)$ and $\vec{j} = \vec{j}(\vec{x}, t)$ are respectively the electric field, the magnetic field, the electric charge density, and the electric current density at the point $\vec{x}$ at time t; c is the velocity of light in vacuum.

- The particles are named with an index α; they have an inertial mass m_α, an electric charge e_α, a position $\vec{x}_\alpha$, a velocity $\vec{v}_\alpha$ and an acceleration $\vec{a}_\alpha$.

- It can happen that we want to insist on the fact that the particle α is in the position $\vec{x}_\alpha$ at the time t; in that case we write $\vec{x}_\alpha(t)$ (id. for the velocity and the acceleration). The notation $\vec{E}_\alpha$ is a shorthand notation to represent $\vec{E}(\vec{x}_\alpha(t), t)$, i.e. the electric field at time t and point $\vec{x}$ where the particle α is at that instant of time.

These equations contain a certain number of assumptions which characterise the "classical electrodynamics" model of the world:

- The space is essentially void and this vacuum (the ether of Lorentz) has very simple electromagnetic properties: a dielectric constant ϵ and a magnetic permeability μ which remain constant under all circumstances and which therefore can be put equal to one after a convenient choice of the units (Gaussian system of units). This allows to make no difference between the electric field $\vec{E}$ and the electric displacement $\vec{D}$, and between the magnetic field $\vec{H}$ and the magnetic induction $\vec{B}$.

- As usual in physics, one separates the world into two parts: on the one hand, the system under study; and on the other hand, the rest of the world which is represented in a very simplified way by some given parameters or simple functions. In electrodynamics, the system consists into a certain number N of "electrons" and the field that they create. The rest of the world is represented by an "external part" of the electromagnetic field. This separation of the electromagnetic field into an internal (radiative) part and an external part is of course allowed by the linearity of the Maxwell equations:

$$\vec{E} = \vec{E}_{\text{rad}} + \vec{E}_{\text{ext}} , \quad \vec{H} = \vec{H}_{\text{rad}} + \vec{H}_{\text{ext}}.$$

This electrodynamical system is strongly different from an ordinary mechanical system where possible fields are a priori given, i.e. where fields are considered as belonging to the rest of the world. Of course, it is formally possible to reduce the electrodynamical system to a mechanical one but the latter tells then infinitely many degrees of freedom. This reduction is even an essential step in the process of quantization. But, if one tries to consider only the particles, i.e. to come back to a system with a finite number of degrees of freedom, with modified mechanical equations of motion which contain terms corresponding to the effect of radiation, one gets equations of motion which are incompatible with the principles of mechanics. (Cf. the problems of the runaway solutions and the acausality of these derived "mechanical" equations).

- The "electrons" are ordinary Newtonian particles, i.e. point like objects with an a priori given inertia. The only force acting on these particles is electromagnetic and it is now conventionally called the Lorentz force. In earlier works of Lorentz the "electrons" were not a priori supposed to be identical. However, after his discussion of the Zeeman effect (1896), Lorentz understood that these particles and the cathode rays electrons are alike and therefore all identical, independently of the substance under consideration. In that way, the electron became <u>the</u> fundamental constituent of matter. In the non relativistic version of the theory, the mechanical momentum $\vec{p}$ is the Newtonian momentum $m\vec{v}$, while in the relativistic version the momentum is the Planck momentum $m\vec{v}/(1 - v^2/c^2)^{1/2}$.

- The explicit writing of the electric density and current with Dirac δ-functions is modern and does not fully reproduce the richness of the Lorentz approach. It is somehow useless and pedantic because the fields created by the point-like particles are easily computed without using this notation. Furthermore, Lorentz used also in some calculations more sophisticated models of an extended electron which he eventually reduced to a point.

These equations can be used in different ways which I shall now rapidly enumerate:

- One can imagine that one gives explicitly the fields $\vec{E}$ and $\vec{H}$ and that one computes the motion of the particles corresponding to definite initial conditions. These calculations neglect completely the field created by the particles and reduce electrodynamics to a pure problem of one particle mechanics. They are currently used as a first approximation by the engineers who calculate the accelerators of particles. As a second order approximation, one generally improves the result by taking into account the averaged instantaneous influence of the set of particles on each of them, and also the phenomenological loss of energy of the particle due to radiation. These more sophisticated calculations are commonly used to-day to compute the particle accelerators and they successfully confirm the correctness of the mechanical part of electrodynamics at macroscopic scale. I shall not go further in this direction.

- One can imagine that one gives explicitly the motion of the particles and that one computes the electromagnetic field they create. This can be done exactly for one particle by means of the retarded potentials (A. Liénard 1898[2], E. Wiechert 1900[3]). Because of the linearity of the Maxwell equations the generalisation to several particles is immediate but the answer is not quite transparent for it contains as many different retarded times as the number of particles. For systems of electrons with low velocities and limited spatial extension (like an atom or a molecule) one can reduce these different retarded times to a single one by means of the multipolar perturbation technique of the Hertz-vector. We shall briefly review this technique which allows to handle successfully many problems of classical radiation theory. One checks in that way the correctness of the electromagnetic part of electrodynamics.

- One can finally try to challenge the complete problem of treating the full system of equations, i.e. to compute the motion of the particles and the radiated electromagnetic field corresponding to definite initial conditions for the particles and to some definite external field. This complete problem causes many difficulties, even for the simplest case of one single particle. One realises that the system of equations becomes a strongly non linear system. Consider for instance the mechanical equations; they contain a "force" which depends on the position and velocity of the particle at the instant of time one considers, but which depends also on the position, velocity, and acceleration of all the particles at earlier times. Such equations are not any more ordinary equations of mechanics and there is no guaranty that the problem is well posed. Also the action of a particle on itself must be carefully treated. This difficulty was already discussed by Lorentz himself who proposed a new equation of motion for the electron which is not any more a Newtonian equation (Lorentz 1892[1],[4]). The relativistic version of this equation was later on written by Abraham (1908)[5], Von Laue (1908)[6] and by Dirac (1938)[7]. This problem of the single particle in its own field was discussed at length by many authors and we shall come back on it later.

This introduction would not be complete without a word on the modifications that the discovery of the quantum structures has brought to the program of Lorentz. The discovery of the quantum of action by Planck (1900) changes our conceptual approach of the light-matter interaction. The discovery of the quanta of light (1905), followed by the idea that light might have a dual nature (1909) and later on by the idea of the photon (1917), changed considerably our representation of the electromagnetic radiation (Einstein[8]). During the same period, physicists understood that the laws of mechanics have to be replaced by those of quantum mechanics. This change happened in several steps: the Bohr-Sommerfeld quantization (1913-1915), the Heisenberg-Born-Jordan new quantum mechanics and the Schroedinger wave mechanics (1926), and the Dirac relativistic equation (1928) (for details, see Ref.(9)). Dirac also succeeded to construct a first version of a quantum electrodynamics based on a quantized representation of the electromagnetic field and on a wave mechanical description of the particles (1927)[10]. This quantum electrodynamics of Dirac is in essence a "quantum copy" of the Lorentz electrodynamics. We shall call it "The quantum theory of radiation" (QThR). This refers to the title of a book published by Heitler (1936) who masterfully exposed the results already obtained with the new theory[11]. The new theory replaces successfully classical electrodynamics in all phenomena where the quantum of action is unavoidable in the description of the particles and the field. However, it meets very serious difficulties of divergences in higher order perturbation calculations in general, and in the self field interaction problem in particular. Furthermore, the relativistic version of the theory has to face the new phenomenon of pair creation and must abandon the very old principle of permanence of the matter. This requires a complete overhaul of the theory which becomes a theory of quantized fields (Heisenberg-Pauli 1930)[12]. This new version of quantum electrodynamics does not cure the difficulties related to higher order perturbation calculations. However, a new technique called "renormalization" appears at the end of the forties (Kramers, Bethe, Tomonaga, Schwinger, Feynman, Dyson and others)[13]. It apparently solves all difficulties concerning the calculations and it allows a very detailed comparison of modern quantum electrodynamics (QED) with experiment. This very successful comparison allows to announce proudly that QED is the most precise theory that exists (in some cases, to better than one part in a hundred million).

Can we then consider that the Lorentz's program is now successfully achieved and that we have a satisfactory representation of all phenomenona concerning light and matter? The answer is not simple. On the one hand, we have indeed a very successful theory and in this sense, the answer is certainly positive. On the other hand, we must recognise that we don't yet have satisfactory answers to several fundamental questions which were posed by the original Lorentz 's program. The problem of the identity of all electrons is now solved: all electrons are alike simply because they are the quanta of one single field (the electron field) and this identity is not altered by the renormalization program. But the theory does not offer any possibility to find the value of the fundamental characteristics of the electron (i.e. its charge and its mass). Of course, one understands that a theory should contain an arbitrary mass parameter in order to fix a scale. But the *mathematical structure* of QED allows to introduce as many types of particles as pleased with arbitrary mass and charge ratios. For instance, it is clear that the old problem of the large mass of the μ meson cannot find a solution inside QED. To the question "Why does the μ-meson have such a large mass?". We can only answer: "Because the arbitrary parameter m_μ that we introduce in the field lagrangian is so large"! To the question "Is there some reason for the numerical value of the fine structure constant which determines the strength of all electromagnetic coupling?". We can only answer: "No, this is simply the common and arbitrary value we give to the lagrangian coupling constant"! These answers are not really satisfactory. But there is no hope to do better in the framework of present QED with its renormalization program. These problems could possibly be solved by constructing a non perturbative version of QED. Alternative programs already exist (cf. ref.15) ; they aim to recover the nice perturbative results obtained by QED, while preserving the fundamental nonlinearity of electrodynamics which can eventually lead to new results. However, this is not the general trend of the present development of quantum fields theory. In fact, we now understand that the world cannot be as simple as supposed in the Lorentz's program (this is certainly one of the main discoveries of this century). Many new phenomena and new particles appeared which clearly mean that QED cannot be a complete theory of matter and fields. By going on to higher and higher energies with the hope that the phenomena will finally become simpler, one has included QED into more general theories with essentially the same structure. For those who think that the electron remains <u>the</u> fundamental constituent of matter, at least up to the nuclear physics scale (say: one femtometer, hundred MeV), the more general "Electroweak theory"

(EWTh) is not a real progress with respect to QED. Incidentally, it is worth to mention that a semantic change begins to happen in the appreciation of the word "fundamental". It is now generally recognized that the word "fundamental" should not be confused with the idea of a smaller scale explanation, and that physics poses fundamental questions at each of its characteristic scales (Cf. for instance Schweber[14]). These fundamental questions should be handled within the theory corresponding to this scale and solved by methods developed for the theory at this scale. Within this conceptual scheme, one must recognize that modern QED is not a satisfactory theory.

To resume, let us now make a comparison of the answers that the different versions of electrodynamics give to the fundamental questions of the old program of Lorentz. The following list of fundamental questions and unsolved problems is somehow a personal choice and is not exhaustive. It illustrates the present state of our satisfactions and frustrations.

Question	Lorentz	QThR	QED	EWTh
Mathematical consistency	?	?	?	?
Identity of electrons	?	?	"quanta of the electron field"	
Origin of the mass	?	?	?	?
Form, charge distribution	hypothetical	irrelevant in quantum theory		
Spin	?	spin 1/2 (Dirac QM or field)		
Magnetic moment	?	$\mu = 2$ (Dirac)	"exactly computable"	
μ-meson	?	other field?		symmetry
Fine structure constant	?	?	?	irrelevant?

Another important reason of dissatisfaction is the way quantum electrodynamics treats the old problem of the duration of transitions. The original Lorentz's equations can easily accommodate for a phenomenological damping force which allows to define a damping time and a duration of transition phenomena. The restriction to a hamiltonian formalism (with infinitely many degrees of freedom) eliminates this possibility. We then meet the pitfall of the reversibility of evolution versus the irreversibility of transitions. Of course, this difficulty is not peculiar to electrodynamics: it is a general feature of the quantum dynamics and it is actively studied in relation with the theory of the measurement process. However, in view of the original goal of electrodynamics (i.e. to explain all physical phenomena up to the nuclear scale), and considering that measurement processes are ultimately of electrodynamics nature, a solution of this important question from inside electrodynamics would be particularly welcomed.

2. Lectures

In the lectures, I discussed some aspects of the early historical development which proved to be important. More precisely, I spent some time on the following topics:

1. Retarded fields created by one particle whose motion is known. Retarded radiation. Application to the accelerators. Retarded fields and retarded radiation for a system of particles slowly moving in a small domain. Application to the classical theory of radiation.

2. Self field interaction. The Lorentz model of the electron. Nature of the electron mass. The problem of the self acceleration. Acausality.

3. Hamiltonian version of the Maxwell-Lorentz equations. Quantization of the pure radi-

ation field. Quantization of the system "Radiation field and N-particles". Elementary application to the quantum theory of radiation.

4. The self energy of the electron and the Lamb-shift (early calculations).

These subjects were treated from a historical and pedagogical perspective. The scientific material concerning these lectures can be found in many text-books on electrodynamics, some of them being mentioned in the bibliography (Cf. refs.1,11,16-22). I think it is not worth to reproduce these lectures in this book. A copy of my transparencies, as well as a limited number of the old lectures notes quoted in ref.22 can be provided on demand.

Bibliography.

1. H.A. Lorentz: "The Theory of Electrons", Leipzig (1909); 2nd Ed. (1915) reprint by Dover New-York (1952).
2. A. Liénard: L'Eclairage électrique **26** (1898) 5, 53, 106.
3. E.Wiechert: Archives Néerl. (2) **5** (1900) 549.
4. H.A. Lorentz: Archives Néerl. **25** (1892) 363.
5. M. Abraham: "Theorie der Electrizität", Vol. 2, 2nd Ed.(1908) 387.
6. M. Von Laue: Verhandl. Deutsch. Phys. Ges. **10** (1908) 888.
7. P.A.M. Dirac: Proceed. Roy. Soc. **A167** (1938) 148.
8. A. Einstein: Ann. d. Phys. **17** (1905) 132 ; Phys.Zs. **10** (1909) 817 ; Phys.Zs. **18** (1917) 121.
9. J. Mehra and H. Rechenberg: "The Historical Development of Quantum Theory", 5 Vol., Springer-Verlag Berlin (1982- 1987).
10. P.A.M. Dirac: Proceed. Roy. Soc. **A114** (1927) 243.
11. W. Heitler: "The Quantum Theory of Radiation", Oxford Univ. Press (1936); 2nd Ed. (1944); 3rd Ed. (1954).
12. W. Heisenberg and W. Pauli: Zeitsc. f. Phys.**56** (1929) 1, **59** (1930) 168.
13. J. Schwinger:"Selected Papers on Electrodynamics", Dover New-York (1958).
14. S.Schweber: Phys. Today **46** (1993) 34.
15. A.O. Barut: "Brief history and recent developments in electron theory and quantum-electrodynamics", D. Hestenes and A. Weingartshofer Eds., Kluwer Acad. Publ. (1991) 105-148 ; see also 165-169.
16. E.T. Whittaker: "A History of the Theories of Aether and Electricity", Vol.1: "The Classical Theories", Vol. 2: "The Modern Theories", London (1951)-(1953), reprint by Am. Inst. of Phys. (1987).
17. A.H. Miller: "Early Quantum Electrodynamics", Cambridge Univ. Press (1994).
18. L. Rosenfeld: "Theory of Electrons", North-Holland Amsterdam (1951), reprint by Dover New-York (1965).
19. L. Landau and E. Lifchitz: "The Classical Theory of Fields", Pergamon Oxford (1971).
20. J.D. Jackson: "Classical Electrodynamics", Wiley New-York, 2nd Ed. (1976).
21. H. Arzeliés: "Rayonnement et dynamique du corpuscule chargé fortement accéléré", Gauthier-Villars Paris (1966).
22. J. Reignier: "Electrodynamique", Lectures Notes (available in French or Dutch), Vrije Universiteit Brussel (1978).

LOCALIZATION OF ELECTROMAGNETIC WAVES IN 2D RANDOM MEDIA

Arkadiusz Orłowski[1,2] and Marian Rusek[2]

[1]Arbeitsgruppe "Nichtklassische Strahlung" der Max-Planck-Gesellschaft
an der Humboldt-Universität zu Berlin, 12484 Berlin, Germany
[2]Instytut Fizyki, Polska Akademia Nauk, 02–668 Warszawa, Poland

INTRODUCTION

Recently random dielectric structures with typical length scale matching the wavelength of electromagnetic radiation in the microwave and optical part of the spectrum have attracted much attention. Propagation of electromagnetic waves in these structures resembles the properties of electrons in disordered semiconductors. Therefore many ideas concerning transport properties of light and microwaves in such media exploit the theoretical methods and concepts of solid-state physics that were developed over many decades. One of them is the concept of electron localization in noncrystalline systems such as amorphous semiconductors or disordered insulators. As shown by Anderson[1], in a sufficiently disordered *infinite* material an entire band of electronic states can be spatially localized. Thus for any energy from this band the stationary solution of the Schrödinger equation is localized for almost any realization of the random potential. Prior to the work due to Anderson, it was believed that electronic states in infinite media are either extended, by analogy with the Bloch picture for crystalline solids, or are trapped around isolated spatial regions such as surfaces and impurities[2].

The actual properties of physical systems are not observed experimentally from the properties of the stationary solutions of the Schrödinger equation which are only theoretical tools. Experiments deal rather with such quantities as a transmission coefficient, diffusion constant or a transport coefficient, e.g., the electrical conductivity. For electronic problems the natural quantity to look for is the static (dc) conductivity. Intuitively, localized states are basically staying in a finite region of space for all times, whereas extended ones are rather free to flow out of any finite region. Therefore, it is natural to expect that the material in which an entire band of electronic states is localized will be an insulator, whereas the case of extended states will correspond to conductor. In this way the phenomenon of Anderson localization may be referred to as a dramatic inhibition of the propagation of an electron when it is subject to a spatially random potential. This connection is not proven on a general basis, but it is certainly valid in reasonable physical models[4]. For many practical problems, a natural quantity to look for is the conductance of a *finite* system of the characteristic size L and its dependence with respect to L. If the waves are well localized, they will decay rapidly with L, e.g., exponentially in the case of exponential localization. If the states decay, we expect the conductance also to decay when the size of the system is growing up[4].

Let us mention, that a very common approach in investigations of the Anderson localization in the solid-state physics is to study a transport equation for the ensemble-averaged squared modulus of the wave function. Under some assumptions such a transport equation can be converted into a diffusion equation. Then the behavior of the diffusion constant D is used to recognize localization: if the diffusion constant in the scattering medium becomes zero the (strong) localization is achieved. Indeed, through an Einstein relation the dc conductivity is proportional to the diffusion constant[4]. Therefore when the fluctuations of the electronic static potential become large enough and the conductance of the system vanishes, the wavefunction ceases to diffuse and becomes localized. Thus, the Anderson transition may be viewed as a transition from particle-like behavior described by the diffusion equation to wave-like behavior described by the Schrödinger equation, which results in localization[5].

It is commonly believed, that the Anderson localization is completely based on the interference effects in multiple elastic scattering. It is obvious, however, that interference is a common property of all wave phenomena, and many generalizations of electron localization to electromagnetic waves have been proposed indeed[6-10]. So-called *weak* localization of electromagnetic waves manifesting itself as enhanced coherent backscattering is presently relatively well understood theoretically[11-13] and established experimentally[14-16] beyond any doubts. The question is whether interference effects in three-dimensional random dielectric media can reduce the diffusion constant to zero leading to *strong* localization. Despite some reasonable indications that strong localization could be possible in three-dimensional random dielectric structures (mainly some suspensions of TiO_2 spheres in air or in some low-refractive-index substances[17-21] have been considered) the convincing experimental demonstration has been given only for two dimensions[22]. In this case the strongly-scattering medium has been provided by a set of dielectric cylinders randomly placed between two parallel aluminum plates on half the sites of a square lattice.

Despite the huge amount of existing literature, there still is lack of sound theoretical models providing deeper insight into this interesting effect. To be realistic, such models should be based directly on the Maxwell equations. On the other hand, they should be simple enough to provide calculations without too many too-crude approximations. The main purpose of our paper is to construct such a model for the two-dimensional localization and to elaborate in detail its major consequences. There are two main advantages of two-dimensional localization: (a) we can use the scalar theory of light and (b) we can try to compare, at least qualitatively, the model predictions with experimental results.

PHOTONS VERSUS ELECTRONS

In the standard approach to localization of electromagnetic waves[6,23] a monochromatic wave,

$$\vec{E}(\vec{r}, t) = \text{Re}\left\{ \vec{\mathcal{E}}(\vec{r}) e^{-i\omega t} \right\}, \tag{1}$$

is called localized in a non-dissipative dielectric medium if the squared modulus of the electric field $|\vec{\mathcal{E}}(\vec{r})|^2$ is localized. This definition is based on the analogy between the Helmholtz equation,

$$\left\{ -\nabla^2 + k_0^2 \left[1 - \epsilon(\vec{r}) \right] \right\} \vec{\mathcal{E}}(\vec{r}) = k_0^2 \, \vec{\mathcal{E}}(\vec{r}), \tag{2}$$

and the time-independent Schrödinger equation. In the above formula $k_0 = 2\pi/l_0$ and l_0 denotes the wavelength in the vacuum. It is known, however, that this analogy is far from being complete. As follows from recent investigations, considerable care must be exercised in transforming results concerning localization of electrons to the case of electromagnetic waves[24-26]. Indeed, some results in condensed matter theory like the Friedel sum rule do not seem to have any analogy in the Maxwell theory[26]. To some extent the Helmholtz equation (2) may be interpreted as the eigenvalue equation for the wave function $\mathcal{E}(\vec{r})$ corresponding to the energy k_0^2. However, the presence of the energy-dependent "potential" $k_0^2[1 - \epsilon(\vec{r})]$ strongly affects dynamical properties in scattering. For example the transport velocity for multiple scattering processes can be very different from this observed for electrons[25,27]. These differences can be understood as a consequence of the fact that the counterpart of the Ward identity for electromagnetic waves contains additional "mass-enhancement" terms[25,26]. Such terms cancel out in the case of electrons[28]. Closely related is the fact that the electric field vector cannot be interpreted as a probability amplitude. The correct equivalent of the quantum-mechanical probability density is rather the energy density of the field and not the squared electric field. Therefore after some introductory remarks we would like to propose a reasonable definition of localized light waves based on the behavior of the energy density.

Since our discussion is restricted to monochromatic fields only, we shall assume that the polarization of the dielectric medium providing localization is an oscillatory function of time

$$\vec{P}(\vec{r}, t) = \mathrm{Re}\left\{\vec{\mathcal{P}}(\vec{r})e^{-i\omega t}\right\}. \tag{3}$$

Instead of solving the Helmholtz equation (2) and then checking if the resulting wave obeys the transversality condition we prefer to study integral equations[29]

$$\vec{\mathcal{E}}(\vec{r}) = \vec{\mathcal{E}}^{(0)}(\vec{r}) + \nabla \times \nabla \times \int d^3 r'\, \vec{\mathcal{P}}(\vec{r}')\, \frac{e^{ik_0|\vec{r}-\vec{r}'|}}{|\vec{r}-\vec{r}'|}, \tag{4}$$

where $\vec{\mathcal{E}}^{(0)}(\vec{r})$ denotes a solution of the Maxwell equations in vacuum and $k_0 = \omega/c$ is the wave number.

We believe that what really counts for localization is the scattering cross-section and not the geometrical shape and real size of the scatterer. Therefore we will represent the dielectric cylinders located at the points $\vec{\rho}_a$ by two-dimensional dipoles

$$\vec{\mathcal{P}}(\vec{r}) = \sum_{a=1}^{N} \vec{p}_a\, \delta^{(2)}(\vec{\rho} - \vec{\rho}_a), \tag{5}$$

with properly adjusted scattering properties. Since the polarization of our system varies only at a certain plane, we have introduced cylindric coordinates $\vec{r} = (\vec{\rho}, z)$ in the above formula.

ELASTIC SCATTERING

The crucial point is how each dipole should be coupled to the electromagnetic field. To provide a realistic and self-consistent description we must assume that the average energy is conserved in the scattering process. Therefore, if we isolate a single dipole then the time-averaged field energy flux integrated over a closed surface Σ surrounding it should vanish for an arbitrary incident wave, namely,

$$\int_\Sigma d\vec{\sigma} \cdot \vec{\mathcal{S}}(\vec{r}) = \frac{c}{4\pi}\frac{1}{2}\mathrm{Re}\int_\Sigma d\vec{\sigma} \cdot \left\{\vec{\mathcal{E}}(\vec{r}) \times \vec{\mathcal{H}}^*(\vec{r})\right\} = 0. \tag{6}$$

It is remarkable that this simple and obvious requirement gives an explicit form of the field-dipole coupling.

For the sake of simplicity let us now assume that both the free field and the medium are linearly polarized along the z-axis

$$\vec{\mathcal{E}}^{(0)}(\vec{r}) = \vec{e}_z\, \mathcal{E}^{(0)}(\vec{\rho}), \quad \vec{P}(\vec{r}) = \vec{e}_z\, P(\vec{\rho}). \tag{7}$$

It is now evident from Eq. (4) that the electric field of the wave radiated by the a-th dipole reads as

$$\vec{\mathcal{E}}_a(\vec{r}) = \vec{e}_z\, k_0^2\, p_a\, g^{(2)}(\vec{\rho} - \vec{\rho}_a), \tag{8}$$

where the Green function in two dimensions is expressed by the modified Bessel function of the second kind

$$g^{(2)}(\vec{\rho}) = \int_{-\infty}^{+\infty} dz\, \frac{e^{ik_0|\vec{r}|}}{|\vec{r}|} = 2\, K_0(-ik_0|\vec{\rho}|). \tag{9}$$

Therefore our discussion may be restricted to the scalar theory. This is impossible in three dimensions when the Green function is replaced by a tensor acting on the dipole moment and the proper description of the interaction between the field and the medium is more complicated[30].

Now we can perform the integration in Eq. (6) assuming that it is performed over a cylinder of unit height surrounding the a-th dipole. The total energy flux may be split into three terms. The first term describes the total time-averaged energy flux integrated over a closed surface for a free field and thus vanishes. The second term corresponds to the time-averaged energy radiated by the a-th dipole per unit time

$$\frac{c}{4\pi}\frac{1}{2}\mathrm{Re}\int_\Sigma d\vec{\sigma}\cdot\left\{\vec{\mathcal{E}}_a(\vec{r}) \times \vec{\mathcal{H}}_a(\vec{r})\right\} = (2\pi)^2\, k_0^3\, |p_a|^2. \tag{10}$$

To calculate the last interference term let us use the following identity fulfilled by a field $\mathcal{E}(\vec{r})$ that obeys the free Maxwell equations inside a closed surface Σ

$$\mathcal{E}(\vec{r}) = \frac{1}{4\pi}\int_\Sigma d\vec{\sigma}\cdot\left\{\frac{e^{ik_0|\vec{r}-\vec{r'}|}}{|\vec{r}-\vec{r'}|}\nabla\mathcal{E}(\vec{r'}) - \mathcal{E}(\vec{r'})\nabla\frac{e^{ik_0|\vec{r}-\vec{r'}|}}{|\vec{r}-\vec{r'}|}\right\}. \tag{11}$$

The above equation is known as the Kirchhoff integral formula[29]. After simple but lengthy calculations we finally arrive at the following conservation law which is equivalent to Eq. (6)

$$\pi k_0^2\, |p_a|^2 = \mathrm{Im}\{p_a^*\, \mathcal{E}'(\vec{\rho}_a)\}, \tag{12}$$

where the field of the wave incident on the a-th dipole,

$$\mathcal{E}'(\vec{\rho}_a) = \mathcal{E}^{(0)}(\vec{\rho}_a) + \sum_{b\neq a}\mathcal{E}_b(\vec{\rho}_a),, \tag{13}$$

is the sum of the free field and the waves radiated by other dipoles.

Assuming that the dipole moment p_a is a linear function of the electric field $\mathcal{E}'(\vec{\rho}_a)$ we get from equation (12)

$$i\pi\, k_0^2\, p_a = \frac{1}{2}(e^{i\phi_a} - 1)\mathcal{E}'(\vec{\rho}_a), \tag{14}$$

where ϕ_a is some arbitrary real number. Thus, to provide conservation of energy the dipole moment must be coupled to the electric field of the incident wave by a complex "polarizability" $(e^{i\phi_a} - 1)/2$. This fact is not specific for the considered two-dimensional case; it remains valid also in three dimensions[30]. We note, that the field of the incident wave calculated at the dipole is finite as opposed to the total field which is not defined at the dipole. Inserting (14) into (8), using (13), and introducing the following convenient notation

$$i\pi\, G_{ab} = \begin{cases} g^{(2)}(\vec{\rho}_a - \vec{\rho}_b) & \text{for } a \neq b \\ 0 & \text{for } a = b \end{cases}, \tag{15}$$

we finally arrive at the very simple set of linear algebraic equations

$$\mathcal{E}'(\vec{\rho}_a) = \mathcal{E}^{(0)}(\vec{\rho}_a) + \frac{1}{2}\sum_{b=1}^{N} G_{ab}\,(e^{i\phi_b} - 1)\,\mathcal{E}'(\vec{\rho}_b), \quad a = 1, \ldots N, \tag{16}$$

determining the field acting on each dipole $\mathcal{E}'(\vec{\rho}_a)$ for a given free field $\mathcal{E}^{(0)}(\vec{\rho}_a)$. If we solve it and calculate the dipole moments we are able to find the electromagnetic field everywhere in space using the integral equations (4).

LOCALIZED WAVES AND RESONANCES

Now we are ready to propose a definition of localized electromagnetic waves which resembles the definition of localized states in quantum mechanics and makes use of the analogy between the quantum-mechanical probability density and the energy density of the field. It seems natural to say that the monochromatic light wave is localized if the time-averaged energy density of the *total* field tends to zero far from a certain region of space

$$\mathcal{W}(\vec{r}) = \frac{1}{16\pi}\left\{ |\vec{\mathcal{E}}(\vec{r})|^2 + |\vec{\mathcal{H}}(\vec{r})|^2 \right\} \to 0 \quad \text{for} \quad |\vec{r}| \to \infty. \tag{17}$$

We used the fact that for rapidly oscillating monochromatic light waves (1) only time averages are measurable. Strictly speaking, to make the notion of the limit in Eq. (17) mathematically precise we must consider the energy density of the field outside the arbitrarily small but finite volumes surrounding the dipoles. In this way we avoid some infinities introduces by the point-scatterer approximation used. Let us emphasize that the condition (17) can be fulfilled *only* if the free field $\vec{\mathcal{E}}^{(0)}(\vec{r})$ is zero everywhere. To see this let us observe that, the localization condition (17) may be satisfied in systems (5) consisting of well separated dielectric particles only if the polarization of the medium vanishes at infinity, i.e., $\vec{\mathcal{P}}(\vec{r}) \to 0$ for $|\vec{r}| \to \infty$. Therefore, due to Eq. (8), the field of the waves radiated by the dipoles also tends to zero if $|\vec{r}| \to \infty$. Thus if the light is localized according to the definition (17) the free field $\vec{\mathcal{E}}^{(0)}(\vec{r})$ must also tend to zero in this limit. But it follows from the Kirchhoff integral formula (11) that if the free field vanishes on a closed surface it is zero everywhere inside this surface.

It is now evident that if the system of dipoles (5) provides localization of the electromagnetic wave then the system of equations (16) should have a nonzero solution for vanishing free field $\mathcal{E}^{(0)} = 0$. This means that the eigenvalues λ_j corresponding to the eigenvectors of the system (5)

$$\lambda_j\, \mathcal{E}'_j(\vec{\rho}_a) = \mathcal{E}_j^{(0)}(\vec{\rho}_a), \quad a = 1, \ldots N, \tag{18}$$

describing localized waves should be equal to zero. In a general case the eigenvalues λ_j depend on the positions of dipoles $\vec{\rho}_a$ and the phases ϕ_a describing their coupling to the field. However, assuming the same scattering properties of all dipoles, namely,

$$\phi_a = \phi, \tag{19}$$

we can express the eigenvalues as

$$\lambda_j(\phi) = 1 - \frac{e^{i\phi} - 1}{2}\lambda_j', \qquad (20)$$

where λ_j' are the eigenvalues of the G matrix which depends only on the current positions of the dipoles. Note, that in this case the eigenvectors of the system (16) are simultaneously eigenvectors of the G matrix. Thus, if $\vec{\rho}_a$ are given, then for each eigenvector of the system of equations (16) there exists a certain angle

$$\phi_j = \arg\left(1 + \frac{\lambda_j'}{2}\right) - \left(\frac{\lambda_j'}{2}\right), \qquad (21)$$

for which the modulus of the corresponding eigenvalue λ_j takes a minimal value. This means that localized waves can exist in systems (5) only when the parameter ϕ describing the scattering properties of the dipoles takes one of the *discrete* values given by Eq. (21).

Let us stress that in all experiments we can investigate only systems confined to certain finite regions of space. It is therefore reasonable to restrict our analysis to bounded media consisting of finite number N of dielectric particles. As follows from our previous considerations perfectly localized waves can exist only in two-dimensional systems (5) consisting of the *infinite* number of dipoles[31]. Nevertheless, in the case of bounded media there can exist some monochromatic resonance waves for which the field energy density inside the medium is much larger than energy density of the incident wave in the stationary regime. Note, that in the experiment on microwave localization in two-dimensional structures[22] such quasi-localized modes have been identified be direct measurement of the squared electric field.

As a simple example let us consider a system of $N = 100$ dielectric cylinders (5) distributed randomly in a square with the density n of one cylinder per wavelength squared: $n/l_0^{-2} = 1.0$. We have calculated and diagonalized numerically the G matrix (15) describing this situation. Then we have chosen a certain eigenmode of the system (16) and checked if the corresponding eigenvalue can approach zero. According to the equation (21) we have calculated the resonance value of ϕ and obtained $|\lambda_j(\phi_j)| \propto 10^{-2}$. This means that the field incident on each dipole $|\mathcal{E}_j'(\vec{r}_a)|$ can be large compared to the free field $|\mathcal{E}_j^{(0)}(\vec{r}_a)|^2$ calculated at the dipole. Thus the time-averaged field energy density inside the medium can be much larger than the energy density of the incident wave in the stationary regime[31]. To help those intuitive considerations, we have plotted at Fig. 1 the time-averaged energy density of the field corresponding to the considered eigenmode as a function of position. To avoid some nasty infinities near the dipoles, this figure shows in fact a discrete function

$$\mathcal{W}'(\vec{\rho}_a) = \frac{1}{16\pi}\left\{|\vec{\mathcal{E}}'(\vec{\rho}_a)|^2 + |\vec{\mathcal{H}}'(\vec{\rho}_a)|^2\right\}. \qquad (22)$$

Also, for simplicity, the free field has been set to zero, since, according to our assumption, a non zero field may modify the plot only by one per cent. In Fig. 2 we have a contour plot of the time-averaged energy density, corresponding to Fig. 1.

ANDERSON LOCALIZATION

Let us emphasize that, monochromatic resonance waves with energy density well localized inside the medium occur when considering one *specific* sample and than vary-

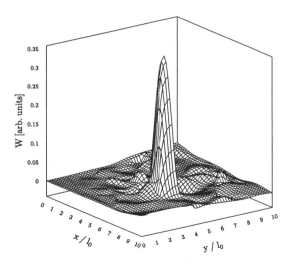

Figure 1. Time-averaged energy density of the field localized in the medium consisting of $N = 100$ randomly distributed dielectric cylinders. The density of cylinders is $n = 1$ dipole per wavelength squared.

ing the scattering properties of the dipoles ϕ. In contrast we would like now to fix first the parameter ϕ and then look at resonance waves in a corresponding *typical* medium. To do this we have diagonalized numerically the G matrix (15) for 10^4 different distributions of $N = 100$ dipoles placed randomly in a square with the uniform density $n = 1$ dipole per wavelength squared. For each value of the scattering properties of the dipoles ϕ we obtained probability distribution $P_\phi(\Lambda)$ of the minimal possible absolute value of eigenvalue calculated from *all* eigenmodes

$$\Lambda(\phi) = \min_j |\lambda_j(\phi)|. \tag{23}$$

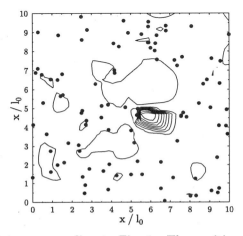

Figure 2. Contour plot corresponding to Fig. 1. The positions of the cylinders are marked by black dots.

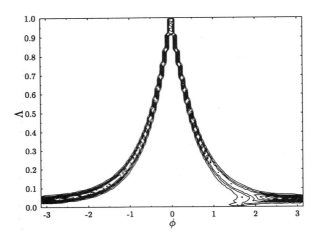

Figure 3. Contour plot of the probability distribution $P_\phi(\Lambda)$ calculated for 10^4 systems of $N = 100$ dielectric cylinders distributed randomly in a square with the density $n = 1$ dipoles per wavelength squared.

The contour plot of the resulting function of variables ϕ and Λ is presented in Fig. 3. We see that for any ϕ chosen suitably from a certain region the modulus of an eigenvalue $|\lambda_j(\phi)|$ can be very small for almost any realization of the random medium. By investigating this region of ϕ corresponding to monochromatic resonance waves with energy density well localized inside the finite medium we expect to gain some information regarding the region of ϕ corresponding to band of localized waves in an infinite medium.

Indeed, let us recall Fig. 1. In this case the energy density of the field of the monochromatic resonance wave corresponding to a certain eigenmode of the system of equations (16) was well localized inside the medium consisting of finite number of cylinders N. Therefore we may expect that adding adding additional cylinders at the boundaries of the medium will not significantly disturb the structure of the field and in the limit $N \to \infty$ this resonance wave will become localized one at some discrete value of ϕ close to the value ϕ_j calculated in the case of a finite medium. To illustrate this statement in Fig. 4 we have plotted the resonance value of the absolute value of the eigenvalue $|\lambda_j(\phi_j)|$ corresponding to the mode from Fig. 1 as a function of the number of dipoles N.

It follows from our numerical investigation that in the limit $N \to \infty$ the probability distribution presented in Fig. 3 will tend to the function

$$P_\phi(\Lambda) = \begin{cases} \delta(\Lambda) & \text{for } \phi \neq 0 \\ \delta(\Lambda - 1) & \text{for } \phi = 0 \end{cases}. \tag{24}$$

Therefore, in the considered two-dimensional case the spectrum of waves localized in the infinite medium will be the whole interval of $-\pi \leq \phi \leq \pi$ except the point $\phi = 0$. This means, that in an infinite *random* dielectric medium there indeed can exist an entire continuous *band* of spatially localized waves.

Our results are in close analogy with the scaling theory of localization[3] developed in solid state physics. According to this theory the dimension of the disordered medium providing localization is a crucial parameter. In one and two dimensions any degree of disorder will lead to localization, while in three dimensions a certain critical degree of

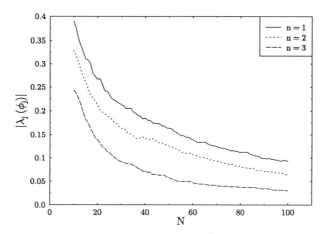

Figure 4. The resonance value of the absolute value of an eigenvalue $|\lambda_j(\phi_j)|$ as a function of the number o cylinders N for various scaled densities of the medium n.

disorder is needed before localization will set in. Note, that in the case of localization of electromagnetic waves in random suspensions of dielectric particles the "degree of disorder" may be obviously identified with the total scattering cross-section of those particles.

Last but not least let us recall the fact that according to our considerations from the previous section the countable set of $\phi = \phi_j$ corresponding to localized waves in an unbounded medium is *discrete*. On the other hand we have shown that in an dielectric medium there exist an entire *continuous* band of spatially localized waves. This naive contradiction may be removed by observing that in the case of a unbounded random medium the discrete set of $\phi = \phi_j$ corresponding to localized waves will be dense in some finite interval. Indeed, physically speaking it is difficult to separate energies or frequencies allowed to the electron or wave from energies or frequencies which may be arbitrarily near and by convention the spectrum is always a coarse-grained object[4].

SUMMARY

We have presented a novel approach to localization of electromagnetic waves in two dimensions based on the integral equations. Dielectric medium providing localization is modeled by a system of discrete dipoles. Instead of studying elastic scattering of electromagnetic waves by a point-like particle using the t-matrix formalism and several crude approximations, we presented a simple method based on the Kirchhoff theorem. It was shown that, in random media consisting of a finite number of dielectric particles, localization of light is impossible. However, for almost any random distribution of the dielectric particles there can exist certain monochromatic resonance waves with energy density well localized inside the medium. If the total number of scattering particles tends to infinity, those resonances become localized states with a continuous spectrum. The generalization of the presented concepts to the three-dimensional vector case is a subject of our paper[30]. In another forthcoming article we plan to analyze transport properties of our abstract medium and check several heuristic criteria of localization.

ACKNOWLEDGMENTS

We are very grateful to the late Asim Barut for his kind hospitality extended to one of us (AO) in Edirne. We also thank Jan Mostowski for many stimulating discussions. This paper is a modified and updated (recent references) version of the talk given by one of us (AO) in Edrine.

REFERENCES

1. P. W. Anderson, Absence of diffusion in certain random lattices, *Phys. Rev.* **109**, 1492 (1958).

2. T. V. Ramakrishnan, Electron localization, In: Ref.[32], pp. 213–304.

3. E. Abrahams, P. W. Anderson, D. C. Licciardello, and T. V. Ramakrishnan, Scaling theory of localization: Absence of quantum diffusion in two dimensions, *Phys. Rev. Lett.* **42**, 673 (1979).

4. B. Souillard, Waves and electrons in inhomogeneous media, In: Ref.[32], pp. 305–382.

5. M. Kaveh, What to expect from similarities between the Schrödinger and Maxwell equations, In: Ref.[9], pp. 21–34.

6. S. John, Electromagnetic absorption in a disordered medium near a photon mobility edge, *Phys. Rev. Lett.* **53**, 2169 (1984).

7. P. W. Anderson, The question of classical localization. A theory of white paint? *Phil. Mag. B* **52**, 505 (1985).

8. S. John, Strong localization of photons in certain disordered dielectric superlattices, *Phys. Rev. Lett.* **58**, 2486 (1987).

9. W. van Haeringen and D. Lenstra, editors, *Analogies in Optics and Micro Electronics* (Kluwer, Dordrecht, 1990).

10. C. M. Soukoulis, editor, *Photonic Band Gaps and Localization*, New York, 1993. NATO ASI Series, Plenum.

11. E. Akkermans, P. E. Wolf, and R. Maynard, Coherent backscattering of light by disordered media: Analysis of the peak line shape, *Phys. Rev. Lett.* **56**, 1471 (1986).

12. M. J. Stephen and G. Cwillich, Rayleigh scattering and weak localization: Effects of polarization, *Phys. Rev. B* **34**, 7564 (1986).

13. F. C. MacKintosh and S. John, Coherent backscattering of light in the presence of time-reversal-noninvariant and parity-nonconserving media, *Phys. Rev. B* **37**, 1884 (1988).

14. Y. Kuga and A. Ishimaru, Retroreflectance from a dense distribution of spherical particles, *J. Opt. Soc. Am. A* **1**, 831 (1984).

15. M. P. van Albada and E. Lagendijk, Observation of weak localization of light in a random medium, *Phys. Rev. Lett.* **55**, 2692 (1985).

16. P.-E. Wolf and G. Maret, Weak localization and coherent backscattering of photons in disordered media, *Phys. Rev. Lett.* **55**, 2696 (1985).

17. M. P. van Albada, A. Lagendijk, and M. B. van der Mark, Towards observation of Anderson localization of light, In: Ref.[9], pp. 85–103.

18. G. H. Watson Jr., P. A. Fleury, and S. L. McCall, Search for photon localization in the time domain, *Phys. Rev. Lett.* **58**, 945 (1987).

19. A. Z. Genack, Optical transmition in disordered media, *Phys. Rev. Lett.* **58**, 2043 (1987).

20. J. M. Drake and A. Z. Genack, Observation of nonclassical optical diffusion, *Phys. Rev. Lett.* **63**, 259 (1989).

21. A. Z. Genack and N. Garcia, Observation of photon localization in a three-dimensional disordered system, *Phys. Rev. Lett.* **66**, 2064 (1991).

22. R. Dalichaouch, J. P. Armstrong, S. Schultz, P. M. Platzman, and S. L. McCall, Microwave localization by two-dimensional random scattering, *Nature* **354**, 53 (1991).

23. S. John, Localization of light, *Physics Today* **44**, 32 (May 1991).

24. Barbara Goss Levi, Light travels more slowly through strongly scattering materials, *Physics Today* **44**, 17 (June 1991).

25. J. Kroha, C. M. Sokoulis, and P. Wölfle, Localization of classical waves in a random medium: A self-consistent theory, *Phys. Rev. B* **47**, 11093 (1993).

26. B. A. van Tiggelen and E. Kogan, Analogies between light and electrons: Density of states and Friedel's identity, *Phys. Rev. A* **49**, 708 (1994).

27. M. P. van Albada, B. A. van Tiggelen, A. Lagendijk, and A. Tip, Speed of propagation of classical waves in strongly scattering media, *Phys. Rev. Lett.* **66**, 3132 (1991).

28. G. D. Mahan, *Many-Particle Physics* (Plenum, New York, 1981).

29. M. Born and E. Wolf, *Principles of Optics* (Pergamon Press, Oxford-London, 1965).

30. M. Rusek, A. Orłowski, and J. Mostowski, Localization of light in three-dimensional random dielectric media, *Phys. Rev. E* **53**, 4122 (1996).

31. M. Rusek and A. Orłowski, Analytical approach to localization of electromagnetic waves in two-dimensional random media, *Phys. Rev. E* **51**, R2763 (1995).

32. J. Souletie, J. Vannimenus, and R. Stora, editors, *Chance and Matter* (North-Holland, Amsterdam, 1987).

THE CLASSICAL LAMB SHIFT: WHY JACKSON IS WRONG!

Jonathan P. Dowling

Weapons Sciences Directorate, AMSMI-RD-WS-ST
Research, Development, and Engineering Center
U. S. Army Missile Command
Redstone Arsenal, AL 35898-5248

INTRODUCTION

At the Edirne summer school, a discussion developed on how it seemed to be possible to use classical methods to calculate atomic spontaneous emission rates, but not the Lamb shift. This point of view has particularly been emphasized in Jackson's book, *Classical Electrodynamics*, in which he flat-out states that a classical argument based on radiation-reaction theory and the Abraham-Lorentz equation will give the correct Einstein *A* coefficient for spontaneous decay, but gives the wrong value for the Lamb shift[1]. This bit of folklore is often touted as some sort of proof that classical methods can not reproduce elements of quantum electrodynamics, even in a limiting sense. Usually, as Jackson does, the blame is placed on electromagnetic vacuum fluctuations — a manifestly nonclassical phenomenon — that is claimed to be the "true" origin of the Lamb shift. But people claim that vacuum fluctuations are also the cause of atomic spontaneous emission — how come classical radiation-reaction theory is right there? "Just good luck," is the usual reply. This seems an unsatisfactory state of affairs, for several reasons. To me, the correspondence principle presents the biggest objection to the notion that there is no classical analog to the Lamb shift. In the limit of large quantum numbers, one would expect the quantum electrodynamical Lamb shift to go over into a classical frequency pulling that an oscillating macroscopic charge experiences by interacting with its self field. An additional disturbing fact is that classical theory is quite often used to compute cavity corrections to the Lamb shift[2], and at least for excited atoms the results agree exactly with full QED calculations[3]. Why then can the classical approach get the cavity-induced Lamb shift right, but not the free-space shift? As we shall see, there is nothing wrong with the classical approach to the free-space Lamb shift; it is Jackson who is wrong!

JACKSON'S ARGUMENT

This argument is given in Jackson[1], and can even be found in the book by Barut[4]. The derivation is based on the belief that the most general

Electron Theory and Quantum Electrodynamics: 100 Years Later
Edited by Dowling, Plenum Press, New York, 1997

nonrelativistic equation of motion for a classically-charged harmonic oscillator of frequency ω_0 is given by[1-4]

$$\tau \dddot{x} - \ddot{x} - \omega_0^2 x = 0 \; ; \tag{1}$$

the Abraham-Lorentz-Dirac (ALD) equation, where

$$\tau = \frac{2}{3} \frac{e^2}{m\,c^3} \; , \tag{2}$$

is the usual radiation-reaction constant, and x is the position coordinate. Assuming a solution of the form $x = x_0\,e^{-\alpha t}$, yields a characteristic equation

$$\tau \alpha^3 + \alpha^2 + \omega_0^2 = 0 \; . \tag{3}$$

This has one real root, $-\alpha_0$, and two complex conjugate roots, α_+. The real root leads to the unphysical "runaway" solution $x = x_0\,e^{\alpha_0 t}$, which is discarded. The two remaining roots correspond to damped harmonic oscillations. In the limit of slow oscillations, $\omega_0 << 1/\tau \cong 1.67 \times 10^{23}$ Hz, the complex roots may be approximated as[1]

$$\alpha_\pm \cong \frac{\Gamma}{2} \pm i\left(\omega_0 + \Delta\omega_c\right), \tag{4}$$

where

$$\Gamma = \omega_0^2 \tau \; , \tag{5a}$$

$$\Delta\omega_c = -\frac{5}{8} \omega_0^3 \tau^2 \; . \tag{5b}$$

The most general solution in this approximation can be written

$$x(t) = x_0\,e^{-\frac{\Gamma}{2}t} \cos\left(\omega_0 + \Delta\omega_c\right), \tag{6}$$

if x_0 is the oscillator position at $t = 0$. Clearly, this indicates a damped and frequency-shifted oscillation. The classical decay constant Γ differs from the correct QED decay rate Γ_{ij} for the $|i\rangle \rightarrow |j\rangle$ transition only by the quantum mechanical oscillator strength f_{ij} that is on the order of unity or less. Hence,

$$\Gamma_{ij} = f_{ij} \Gamma \; , \tag{7}$$

and we see that the classical calculation is correct, to within at least an order of magnitude. Clearly, in addition to the classical decay constant Γ, there is apparently a classical frequency pulling $\Delta\omega_c$, Eq. (5b). However, Jackson argues that this shift is many orders of magnitude too small to be considered the classical analog of the Lamb shift $\Delta\omega_q$. The Bethe calculation of the quantum Lamb shift gives

$$\frac{\Delta\omega_q}{\omega_0} \sim \omega_0 \tau \log\left(\frac{m\,c^2}{\hbar\omega_0}\right) \sim \omega_0 \tau \tag{8}$$

if we take $\omega_0 \cong |\omega_i - \omega_j|$ in a correspondence-principle limit. However, the classical quantity from Eq. (5b) becomes

$$\frac{|\Delta\omega_c|}{\omega_0} \sim (\omega_0 \tau)^2, \tag{9}$$

where, remember, we have assumed $\omega_0 \tau << 1$. Hence, it appears that the so-called classical Lamb shift is one order of $\omega_0 \tau$ too small to be the analog of the Bethe shift. For this reason, Jackson claims that the "real" quantum shift $\Delta\omega_q$, Eq. (8), arises from a completely different mechanism. Somehow, radiation reaction and the self field are responsible for the classical shift, but "vacuum fluctuations" are responsible for the much larger quantum Lamb shift.

THE CORRECT CLASSICAL SHIFT

Several years ago I mentioned this puzzle to John Sipe that apparently classical calculations can give the correct cavity QED Lamb shift, but not the free-space one. Sipe told me that, in fact, Jackson was wrong and that Sipe had shown that a classical Lamb shift could be derived in free space that was of the same order of magnitude as the Bethe result. Unfortunately, Sipe never sent me this reference, so in Edirne, when Barut asked me how Sipe's calculation might go, I produced the following derivation and presented it on the board.

Let us work with the classical limit of Barut's self-field approach to electrodynamics. In covariant form, the self-field action density has the form,

$$W = \frac{1}{2c^3} \int\int dx^4 \, dy^4 \, j^\mu(x) \, D_{\mu\nu}(x-y) j^\nu(y), \tag{10}$$

where $j^\mu = [\rho, j]$ is the electric source current four vector, $x = x^\mu = [ct, x]$ and $y = y^\mu = [cu, y]$ are space-time four vectors, and $D_{\mu\nu}$ is the Green's function in the radiation gauge. In the full quantum Barut theory, $j^\mu(x)$ is the appropriate Dirac current. However, here, we take it to be the classical current associated with a classical point dipole oscillating at frequency ω_0. In particular

$$\rho(x, t) = 0 \tag{11a}$$

$$j(x, t) = \omega_0 \, p \, e^{i\omega_0 t} \delta(x), \tag{11b}$$

where p is the point-dipole moment. Since $j_0 = c\rho = 0$, Eq. (10) for the action becomes

$$W = \frac{1}{2c} \int\int dt \, du \int\int d^3x \, d^3y \, j(x, t) \cdot \overleftrightarrow{D} \, [x-y; t-u] \cdot j(y, t), \tag{12}$$

where j is given above in Eq. (11), and the dyadic Green's function in the radiation gauge has the form[6]

$$\overset{\leftrightarrow}{\mathbf{D}}\left[\mathbf{x}-\mathbf{y};t-u\right] = -\frac{4\pi}{(2\pi)^4}\int\frac{d\omega}{c}\int d^3k \; \frac{e^{-\omega(t-u)}e^{i\mathbf{k}\cdot(\mathbf{x}-\mathbf{y})}}{|\mathbf{k}|^2 + i\varepsilon} \times$$

$$\left[\overset{\leftrightarrow}{\delta} - \widehat{\mathbf{k}}\,\widehat{\mathbf{k}}\right]. \tag{13}$$

Hence, the action W can be written as

$$W = -\frac{1}{2}\frac{p^2\omega_0^2}{c^2}\iint dt\,du\int\frac{d\omega}{(2\pi)}\int\frac{d^3k}{(2\pi)^3}\frac{e^{i(\omega_0-\omega)t}e^{i(\omega+\omega_0)\mu}}{\omega^2/c^2-|\mathbf{k}|^2+i\varepsilon}\left[1-\left(\widehat{\mathbf{p}}\cdot\widehat{\mathbf{k}}\right)^2\right],\tag{14}$$

where we have carried out the spatial integration, and $\widehat{\mathbf{k}}$ and $\widehat{\mathbf{p}}$ are unit vectors. Carrying out the angular $\widehat{\mathbf{k}}$ integration, the ω integration and the two temporal integrals yields

$$W = -\frac{p^2\omega_0^2}{(2\pi)^2}\frac{8\pi^2}{3c^2}\int dk\;\frac{k^2\delta(\omega_0)}{\omega_0^2/c^2-k^2+i\varepsilon},\tag{15}$$

where here $k = |\mathbf{k}|$, and I have used $\delta(2\omega_0) = \delta(\omega_0)/2$. Now, according to the Barut prescription, I can extract a frequency shift $\Delta\omega$ from the action W via[5]

$$W = 2\pi\,\delta(\omega_0)\frac{\Delta\omega}{\omega_0}\,\mathscr{E},\tag{16}$$

where $\mathscr{E} = mx_0^2\,\omega_0^2/2$ is the total invariant energy of the dipole, treated as a harmonic oscillation of mass m and maximum extension x_0, and $\Delta\omega$ is the shift of frequency of oscillation. Hence, from Eq. (15), we have

$$\mathscr{E}\frac{\Delta\omega}{\omega_0} = -\frac{4\pi}{3}\frac{p^2\omega_0^2}{(2\pi)^2c^2}\int dk\left\{\frac{1}{2}\left[\frac{\omega_0/c}{\omega_0/c-k+i\varepsilon} - \frac{\omega_0/c}{\omega_0/c+k+i\varepsilon}\right]-1\right\}.\tag{17}$$

The contour integral selects out only the positive frequency solution. The infinite factor $\int dk$ contributes to mass renormalization, required even classically. The renormalized shift $\Delta\widetilde{\omega}$ then is, implicitly,

$$\mathscr{E}\frac{\Delta\widetilde{\omega}}{\omega_0} = -\frac{1}{3}\frac{p^2\omega_0^3}{c^3\pi}\int dk\left\{\mathscr{P}\left[\frac{1}{\omega_0/c-k}\right] - i\pi\,\delta(\omega_0/c-k)\right\},\tag{18}$$

where $\mathscr{P}$ stands for principle part, and an extra factor of two has been added to account for 2 degrees of polarization. Writing $\Delta\widetilde{\omega} = \Delta\Omega + i\Gamma/2$, we have $\Gamma = p^2\,\omega_0^4/(3c^3\mathscr{E})$. If we take $p = ex_0/\sqrt{2}$, and recall that $\mathscr{E} = \frac{1}{2}m\,\omega_0^2x_0^2$, then we have $\Gamma = \omega_0^2 \tau$, in agreement with the Jackson result, Eq. (5a), and hence with the quantum Einstein A coefficient.

However, the frequency shift $\Delta\Omega$ is *not* the same as Jackson's "classical" Lamb shift. Already, from Eq. (18), we can see that $\Delta\Omega \equiv \text{Re } \{\Delta\tilde{\omega}\}$ has the form of the Bethe logarithm for the Lamb shift[6]. A cut off K for the integral can be arrived at classically by noting that wavelengths $\lambda < r_0$, the classical electron radius, should not contribute much to the shift. Hence $k < 1/r_0 = mc^2/e^2$ is a reasonable cut off, and we get

$$\frac{\Delta\Omega}{\omega_0} = \frac{1}{2\pi} \, \omega_0\tau \, \ell n \left[\frac{2}{3\omega_0\tau} \right], \qquad (19)$$

where the argument of the logarithm is very large, since $\omega_0\tau << 1$. *This expression, I would claim, is then the classical Lamb shift.* It certainly does not agree with Jackson's result in Eq. (5b). We can see that the classical Lamb shift of Eq. (19) is comparable or greater to the classical linewidth of Eq. (5a), just as is the quantum Lamb shift. Hence, Jackson's classical shift, $\Delta\omega$ of Eq. (5b), is an order of $\omega_0\tau$ too small. Clearly, the shift of Eq. (19) is the correspondence-principle limit of the quantum Bethe logarithm calculation of the shift, Eq. (8). (See ref. 7.)

SUMMARY AND CONCLUSIONS

The question then becomes, what has Jackson done wrong? An expression for a classical Lamb shift similar to Eq. (19) was derived by Sipe and co-workers in a slightly different formalism.[8] These authors point out that, classically, the decay of a classical dipole arises from the interaction of the dipole with the out-of-phase portion of its own self field, and the frequency shift from the interaction with the in-phase part. However, the assumption that the ALD Eq. (1) is the correct equation of motion must be incorrect. Somehow the first-order, in-phase contribution to the classical shift, Eq. (19), is missing, and only the much smaller higher-order shift that Jackson gets, Eq. (5b), remains. It is not clear to me at this time why this term is missing from standard derivations of the ALD equation. However, these results would seem to indicate that the ALD equation, as used in Jackson and Barut and other standard texts, is not a complete classical equation of motion for a charged particle interacting with its own field.

ACKNOWLEDGMENT

I would like to dedicate this paper to the memory of Asim O. Barut, without whom the Edirne NATO Summer School would not have been possible. Asim was a great physicist, as well as my advisor, colleague, and friend. Things just won't be the same without him.

REFERENCES

1. J. D. Jackson, *Classical Electrodynamics*, Second Ed. (Wiley, New York, 1975) Sect. 17.7.
2. R. R. Chance, A. Prock, and R. Silbey, *Phys. Rev. A* 12:1448 (1975).
3. G. Barton, *Proc. R. Soc. Lond. A* 420:141 (1987).
4. A. O. Barut, *Electrodynamics and Classical Theory of Fields and Particles*, (Dover, New York, 1980) pp. 209–210.
5. A. O. Barut and J. P. Dowling, *Phys. Rev. A* 41:2284 (1990).
6. A. O. Barut and J. F. Van Hule, *Phys. Rev. A* 32:3187 (1985).

7. P. Milonni, *The Quantum Vacuum* (Academic Press, Boston, 1994), Sect. 3.5.
8. J. M. Wylie and J. E. Sipe, *Phys. Rev. A* 30:1185 (1984); *ibid.* 32:2030 (1985).

NEW MAXWELL ELECTRODYNAMICS

Edward Kapuścik

Department of Theoretical Physics,
H.Niewodniczański Institute of Nuclear Physics,
ul. Radzikowskiego 152, 31 342 Kraków, Poland
and
Institute of Physics and Informatics
Cracow Pedagogical University
ul. Podchorążych 2, Kraków, Poland.

INTRODUCTION

Maxwell electrodynamics provides a general scheme for all electromagnetic phenomena both of classical and quantum nature. In spite of its tremendous success it still suffers from some unresolved defects. Among them the most important ones are the long-standing problem of a point charge for which the standard classical theory predicts an infinite amount of energy and the need of infinite renormalization in the case of quantum electrodynamics.

According to Professor A. Barut[1] it is a duty of the present generation of theoretical physicists to resolve all the difficulties of Maxwell electrodynamics. We agree with this statement and we believe "that elementary problems should be solved before attacking sophisticated problems" and that "we cannot expect that new geometries and topologies will reveal deeper insight into the physical world if we persists in ignoring the more elementary and relevant down-to-earth problems"[2].

In our paper we present a careful analysis of the foundation of Maxwell electrodynamics with the special attention to the problems connected with distribution-valued sources and the role of the constitutive relations used to close the set of basic field equations. As a result, we arrive at a new formulation of classical electrodynamics which has many advantages over the standard ones presented in text-books. In particular, our reformulation of electrodynamics is free from troubles of the standard Maxwell theory and includes the latter as a particular case.

MAXWELL ELECTRODYNAMICS WITH DISTRIBUTION-VALUED SOURCES

Maxwell electrodynamics expresses all general laws of electromagnetism in terms of four electromagnetic fields $\mathbf{E}(\mathbf{x}, t), \mathbf{D}(\mathbf{x}, t), \mathbf{B}(\mathbf{x}, t)$ and $\mathbf{H}(\mathbf{x}, t)$. They satisfy a particular system of differential relations called Maxwell equations. In the *SI* system

Electron Theory and Quantum Electrodynamics: 100 Years Later
Edited by Dowling, Plenum Press, New York, 1997

of units these equations have the form[3]:

$$rot\ \mathbf{E}\left(\mathbf{x},t\right) = -\frac{\partial \mathbf{B}\left(\mathbf{x},t\right)}{\partial t}, \tag{1}$$

$$div\ \mathbf{B}\left(\mathbf{x},\mathbf{t}\right) = \mathbf{0}, \tag{2}$$

$$rot\ \mathbf{H}\left(\mathbf{x},t\right) = \frac{\partial \mathbf{D}\left(\mathbf{x},t\right)}{\partial t} + \mathbf{j}\left(\mathbf{x},t\right), \tag{3}$$

$$div\ \mathbf{D}\left(\mathbf{x},t\right) = \rho\left(\mathbf{x},t\right), \tag{4}$$

where $\rho\left(\mathbf{x},t\right)$ and $\mathbf{j}\left(\mathbf{x},t\right)$ are the scalar density of charge and the vector density of current, respectively. To apply the Maxwell equations to the description of any particular electromagnetic situation, we must close the system of differential relations (1 - 4) using some additional information about the electromagnetic fields. Customarily, this information is supplied by the so-called constitutive relations which describe the response of the medium to the applied electromagnetic field. The particular form of the constitutive relations contain all the relevant electromagnetic characterization of the medium. In the simplest case of the vacuum the constitutive relations have the form

$$\mathbf{D}\left(\mathbf{x},t\right) = \epsilon_0 \mathbf{E}\left(\mathbf{x},t\right), \tag{5}$$

$$\mathbf{B}\left(\mathbf{x},t\right) = \mu_0 \mathbf{H}\left(\mathbf{x},t\right), \tag{6}$$

where ϵ_0 and μ_0 are the electromagnetic constants of the vacuum. Substituting these relations into (3) and (4) we get the equations

$$rot\ \mathbf{B}\left(\mathbf{x},t\right) = \frac{1}{c^2}\frac{\partial \mathbf{E}\left(\mathbf{x},t\right)}{\partial t} + \mu_0 \mathbf{j}\left(\mathbf{x},t\right), \tag{7}$$

$$div\ \mathbf{E}\left(\mathbf{x},t\right) = \frac{1}{\epsilon_0}\rho\left(\mathbf{x},t\right), \tag{8}$$

where the relation

$$c^2 \epsilon_0 \mu_0 = 1 \tag{9}$$

has been used (c - velocity of light in vacuum). Equations (1), (2), (7) and (8) form a closed Maxwell-Lorentz system of equations for the fields $\mathbf{E}$ and $\mathbf{B}$ and are the basis of the Lorentz microscopic electrodynamics[4]. The above presented scheme works perfectly for charges and currents for which the corresponding densities ρ and $\mathbf{j}$ are represented by smooth functions of space time coordinates. Unfortunately, it crashes when both ρ and $\mathbf{j}$ are represented by generalized functions. In fact, from the Maxwell-Lorentz equations (7) and (8) it follows that for distribution-valued sources ρ and $\mathbf{j}$ the electromagnetic fields $\mathbf{E}$ and $\mathbf{B}$ also must be represented by generalized functions and the whole formalism of electrodynamics must be treated in the mathematical language of distributions. The customary discussion of the Maxwell-Lorentz electrodynamics in the framework of the theory of generalized functions [5] is physically incomplete because it involves only the linear part of electrodynamics. Apart from linear relations, electrodynamics contains a lot of important non-linear expressions, such as, for example, the Lorentz force

$$\mathbf{F}\left(t\right) = \int d^3x \left[\rho\left(\mathbf{x},t\right)\mathbf{E}\left(\mathbf{x},t\right) + \mathbf{j}\left(\mathbf{x},t\right) \times \mathbf{B}\left(\mathbf{x},t\right)\right], \tag{10}$$

the Poynting vector

$$\mathbf{S}\left(t\right) = \int d^3x\, \mathbf{E}\left(\mathbf{x}, t\right) \times \mathbf{H}\left(\mathbf{x}, t\right), \tag{11}$$

the Joule heat

$$Q\left(t\right) = \int d^3x\, \mathbf{j}\left(\mathbf{x}, t\right) \cdot \mathbf{E}\left(\mathbf{x}, t\right) \tag{12}$$

or the energy balance equation

$$\frac{d\,W\left(t\right)}{dt} = \int d^3x \left(\mathbf{E}\left(\mathbf{x}, t\right) \cdot \frac{\partial \mathbf{D}\left(\mathbf{x}, t\right)}{\partial t} + \mathbf{H}\left(\mathbf{x}, t\right) \cdot \frac{\partial \mathbf{B}\left(\mathbf{x}, t\right)}{\partial t}\right). \tag{13}$$

From the mathematical point of view all these quantities are meaningless for distribution - valued fields and sources because generalized functions cannot be multiplied[5].

The lack of a multiplication law for generalized functions is the primary source of all troubles of the conventional treatment of the problems with point charges in the framework of Maxwell electrodynamics. The fact that the fields for point charges turn out to be ordinary singular functions should not confuse us because mathematics of the field equations with distribution-valued sources uniquely requires that all these singular functions must be treated in the framework of generalized functions. Therefore, we cannot calculate the energy of the field by means of the formula

$$W = \frac{1}{2} \int d^3x \left(\epsilon_0 \mathbf{E}^2\left(\mathbf{x}, t\right) + \mu_0 \mathbf{B}^2\left(\mathbf{x}, t\right)\right) \tag{14}$$

because it is meaningless for distribution-valued fields. The whole discussion of the problem of infinite energy of the electromagnetic field of a point charge, contained in all text-books on electrodynamics, is based on an inadequate mathematics and has therefore no physical meaning because physics cannot be based on a wrong mathematics.

To see what is really going on, let us observe that the conclusion about the distribution character of the fields $\mathbf{E}$ and $\mathbf{B}$ follows not from the original Maxwell equations (1 - 4) but from the Maxwell- Lorentz equations (7 - 8). The original Maxwell equations require only that the fields $\mathbf{D}$ and $\mathbf{H}$ must be distributions for distribution-valued sources while they leave open the question of the mathematical properties of the fields $\mathbf{E}$ and $\mathbf{B}$. We may therefore use this freedom to assign meaning to all non-linear electromagnetic quantities listed above. It is easy to see that for this purpose it is sufficient to assume that the fields $\mathbf{E}$ and $\mathbf{B}$ always serve as vector-valued test functions for the vector-valued generalized functions $\mathbf{D}, \mathbf{H}, \mathbf{j}$ and the scalar generalized function ρ. This is so because all physically interesting non-linear electromagnetic quantities are always linear functionals of the fields $\mathbf{E}$ and $\mathbf{B}$. The quantities with non-linear dependence on the fields $\mathbf{E}$ and $\mathbf{B}$ always appear after using the constitutive relations. Using the notation of the theory of distributions we may rewrite the formulas (10 - 13) in the form

$$F_j\left(t\right) = <\rho_t, E_{j,t}> + \sum_{k,l=1}^{3} \epsilon_{jkl} <j_{k,t}, B_{l,t}>, \tag{15}$$

$$S_j\left(t\right) = \sum_{k,l=1}^{3} \epsilon_{jkl} <H_{k,t}, E_{l,t}>, \tag{16}$$

$$Q\left(t\right) = \sum_{k=1}^{3} <j_{k,t}, E_{k,t}>, \tag{17}$$

$$\frac{dW(t)}{dt} = \sum_{k=1}^{3} \left(< \frac{\partial D_{k,t}}{\partial t}, E_{k,t} > + < H_{k,t}, \frac{\partial B_{k,t}}{\partial t} > \right) \equiv$$

$$\equiv \sum_{k=1}^{3} \left(< H_{k,t}, \frac{\partial B_{k,t}}{\partial t} > - < D_{k,t}, \frac{\partial E_{k,t}}{\partial t} > \right)$$

(18)

where ϵ_{jkl} is the three-dimensional Levi-Civita symbol and the bracket $< f_t, \varphi_t >$ denotes the value of the generalized function $f_t(\mathbf{x}) = f(\mathbf{x}, t)$ on the test function $\varphi_t(\mathbf{x}) = \varphi(\mathbf{x}, t)$ where the time variable is treated as a parameter of both the generalized and test functions. It is now clear that all these formulas are perfectly well-defined for all distribution-valued sources provided the fields $\mathbf{E}$ and $\mathbf{B}$ will be treated as test functions of the theory. Obviously, our assumption on the role of the fields $\mathbf{E}$ and $\mathbf{B}$ is just the opposite to that usually made on these fields in electrodynamics with singular sources. The advantage of our assumption over the usual ones consists in the fact that it assign meaning to all non-linear quantities which are notoriously ill-defined in other approaches.

The difference in the mathematical properties of the electromagnetic fields, necessary for further development of the theory, is possible only in the framework of the original Maxwell approach and not in the widely used Maxwell-Lorentz microscopic electrodynamics. Since the latter arises from the former after using constitutive relations we come to the conclusion that the troubles of electrodynamics to handle distribution-valued sources are not inherent in electrodynamics itself but comes from the wrong constitutive relations. In fact, having decided that the fields $\mathbf{E}$ and $\mathbf{B}$ are the test functions for the distributions $\mathbf{D}$ and $\mathbf{H}$ (and for $\mathbf{j}$ and ρ, as well) we have lost the possibility to write relations of the type (5 - 6) because all relations like these are now meaningless. The trials to express non-trivial generalized functions in terms of test functions either fail or lead to such complicated constructions [5] that they are deprived of any physical application. Therefore, in electrodynamics with distribution-valued sources we must reject all constitutive relations which express the fields $\mathbf{D}$ and $\mathbf{H}$ in terms of the fields $\mathbf{E}$ and $\mathbf{B}$. This conclusion evidently exclude the usual way of closing Maxwell equations. In the next Section we shall show how to reformulate the original Maxwell theory so that all constitutive relations between fields may be omitted.

REFORMULATION OF MAXWELL ELECTRODYNAMICS

In the previous Section we have shown that the mathematics of Maxwell electrodynamics with distribution-valued sources is well-defined only if the fields $\mathbf{E}$ and $\mathbf{B}$ are mathematically quite different from the fields $\mathbf{D}$ and $\mathbf{H}$. Combining this with the fact that also physically the fields $\mathbf{E}$ and $\mathbf{B}$ are quite different from the fields $\mathbf{D}$ and $\mathbf{H}$, we arrive at the fundamental question :

Is it really necessary to have that asymmetry at the fundamental level of the theory ?

Trying to give an answer to this question we start with the observation that the usual Maxwell macroscopic electrodynamics is not a theory of a single medium but it is a theory of two, usually quite different, media. In fact, the fields $\mathbf{E}$ and $\mathbf{B}$ are operationally defined only in a vacuum while the fields $\mathbf{D}$ and $\mathbf{H}$ describe electromagnetism in a given medium which may be quite different from the classical vacuum understood as an empty space. In the standard approach to classical electrodynamics the classical vacuum always serves as a reference medium for all other media. The comparison of the fields in a medium with the fields in a vacuum is implemented by the constitutive relations. But the experimental verification of theses relations involves many assumptions which often cannot be really verified. Obviously, all that introduce into the theory unnecessary uncertainties and undetermined restrictions which even cannot be explicitly stated. This circumstance is always neglected in the formulation of Maxwell electrodynamics and it is assumed that this theory is applicable to all electromagnetic phenomena with an absolute accuracy !

The experience from the solid state physics unambiguously shows that it is not reasonable to describe the given medium with respect to the properties of the vacuum. On the contrary, it is much more convenient to describe each medium in its own language which incorporates the properties of the medium in the most economical way. We must also take into account the fact that our present understanding of the notion of the vacuum is quite different from the old-fashioned point of view which treated the vacuum as an empty space. Each theory should predict the properties of the vacuum and therefore at any early stage of the theory we have no right to make any assumption which may predetermine the structure of the vacuum. The real vacuum is not simpler than any other medium and therefore there is no gain to use it as a reference medium.

The above presented observation leads us to the following brave idea :

The best way of removing the aforementioned asymmetry in the foundation of electrodynamics is to resign from the fields E and B at all.

But, it is known that working with the fields **D** and **H** only we cannot extract from the theory all the information on the behavior of the medium. To proceed further, let us recall that in Maxwell theory the most general form of the constitutive relations is given by the relations

$$\mathbf{D} = \epsilon_0 \mathbf{E} + \mathbf{P}, \tag{19}$$

$$\mathbf{H} = \frac{1}{\mu_0}\mathbf{B} - \mathbf{M}, \tag{20}$$

where the two vector fields **P** $(\mathbf{x}, t)$ and **M** $(\mathbf{x}, t)$ describe the polarization and magnetization properties of the medium. In the standard approach to macroscopic electrodynamics [6] these vector fields are considered as given quantities and the relations (19 - 20) are used to eliminate the fields **D** and **H** from the theory. For the remaining fields **E** and **B** we then get the following complete system of field equations :

$$rot\ \mathbf{E} + \frac{\partial \mathbf{B}}{\partial t} = 0, \tag{21}$$

$$div\ \mathbf{B} = 0, \tag{22}$$

$$rot\ \mathbf{B} - \frac{1}{c^2}\frac{\partial \mathbf{E}}{\partial t} = \mu_0 \left(\mathbf{j} + \frac{\partial \mathbf{P}}{\partial t} + rot\ \mathbf{M} \right), \tag{23}$$

$$div\ \mathbf{E} = \frac{1}{\epsilon_0}(\rho - div\ \mathbf{P}). \tag{24}$$

It is however easy to see that in this way we arrive at a theory which uses electromagnetic fields in the vacuum but describes electromagnetism in a given medium which has nothing to do with the vacuum. The presence of the matter is taken into account solely as some corrections to the source terms. It is rather clear that such an approach is difficult to defend.

The alternative approach consists in treating the vector fields **P** $(\mathbf{x},t)$ and **M** $(\mathbf{x},t)$ not as given quantities but as fields to be determined from their own set of field equations.

To obtain these field equations, we may just use the general constitutive relations (19 - 20) not as a tool for elimination of the fields $\mathbf{D}$ and $\mathbf{H}$ but as a tool of elimination the unwanted fields $\mathbf{E}$ and $\mathbf{B}$. Proceeding in this way we arrive at a reformulation of the Maxwell electrodynamics in which the whole electromagnetism in a given medium is described in terms of two pairs of vector fields $(\mathbf{D}, \mathbf{H})$ and $(\mathbf{P}, \mathbf{M})$ both of which refer solely to the same medium without any reference to the vacuum. The fields $\mathbf{P}(\mathbf{x}, t)$ and $\mathbf{M}(\mathbf{x}, t)$ for distribution-valued sources are also generalized functions in the variable $\mathbf{x}$. Therefore the above mentioned asymmetry in the foundation of electrodynamics disappear.

Using the generalized Helmholtz theorem [7] it can be shown that the complete set of Maxwell equations for the electromagnetic fields $(\mathbf{D}, \mathbf{H})$ and $(\mathbf{P}, \mathbf{M})$ is of the form[8]:

$$div\ \mathbf{D} = \rho, \tag{25}$$

$$div\ \mathbf{H} = \rho_M, \tag{26}$$

$$rot\ \mathbf{D} + \frac{1}{c^2}\frac{\partial \mathbf{H}}{\partial t} = -\frac{1}{c^2}\mathbf{j}_\mathbf{M}, \tag{27}$$

$$rot\ \mathbf{H} - \frac{\partial \mathbf{D}}{\partial t} = \mathbf{j} \tag{28}$$

and

$$div\ \mathbf{P} = -\rho_P, \tag{29}$$

$$div\ \mathbf{M} = -\rho_M, \tag{30}$$

$$rot\ \mathbf{P} - \frac{1}{c^2}\frac{\partial \mathbf{M}}{\partial t} = -\frac{1}{c^2}\mathbf{j}_\mathbf{M}, \tag{31}$$

$$rot\ \mathbf{M} + \frac{\partial \mathbf{P}}{\partial t} = -\mathbf{j}_\mathbf{P}, \tag{32}$$

where $(\rho_P, \mathbf{j}_\mathbf{P})$ and $(\rho_M, \mathbf{j}_\mathbf{M})$ are two pairs of induced densities which depend on the properties of the medium and are related to the external densities $(\rho, \mathbf{j})$ by new type of constitutive relations.

For distribution-valued external densities all the electromagnetic quantities are also distributions. In this way we have removed the physical and mathematical asymmetries between basic electromagnetic fields. In our approach all fields are Maxwellian, i. e., the unique fields determined by Maxwell equations supplemented by the corresponding boundary or initial conditions.

TEST FUNCTIONS FOR THE ELECTROMAGNETIC FIELDS

We have arrived at the reformulation of the Maxwell electrodynamics in which all basic electromagnetic quantities are generalized functions and refer to one medium only. To make the theory complete we must find the set of test functions for the fields and establish the physical meaning of them.

It is well-known that classical electrodynamics has two interrelated aspects. The first one connects the fields with their sources and the second one describes the action of the electromagnetic field on matter. In our approach the first aspect, like in the standard approach, is contained in the sets of Maxwell equations (25 -32) and we must still discuss the second aspect.

In classical electrodynamics it is customary to consider the action of the electromagnetic field solely in the framework of classical mechanics in which a crucial role is played by the Lorentz force (10) and the Joule heat (12). This approach necessitates the introduction of the fields $\mathbf{E}$ and $\mathbf{B}$ which in our scheme may be defined only as secondary fields through the relations

$$\mathbf{E} = \epsilon_0^{-1} (\mathbf{D} - \mathbf{P}) \tag{33}$$

and

$$\mathbf{B} = \mu_0 (\mathbf{H} + \mathbf{M}). \tag{34}$$

However, we have seen that the fields $\mathbf{E}$ and $\mathbf{B}$ posses different mathematical properties than the fields $\mathbf{D}$, $\mathbf{H}$, $\mathbf{P}$ and $\mathbf{M}$ have and we arrive at an important restriction on these fields :

The differences of the distribution-valued fields D and P and the sums of the distribution-valued fields H and M must be smooth test functions.

This condition obviously introduce the physical interpretation of the test functions and provides a strong correlation between the singularities of the distribution-valued fields which up to now were quite independent. For example, for non-polarizable media we may put

$$\rho_P = \rho_M = 0 \tag{35}$$

$$\mathbf{j}_P = \mathbf{j}_M = 0 \tag{36}$$

and we obtain homogeneous field equations for the fields $\mathbf{P}$ and $\mathbf{M}$. It might seem that we may take

$$\mathbf{P} = \mathbf{M} = 0 \tag{37}$$

as solutions of these equations but in view of the restriction on the singularities of the fields this is impossible because relations (33 - 34) will not then lead to smooth fields $\mathbf{E}$ and $\mathbf{B}$. This shows that Maxwell electrodynamics possesses the following property :

Maxwell electrodynamics does not allow to make unphysical assumptions.

The impossibility of the solution (37) means that in the presence of distribution-valued sources all media, including the classical vacuum, exhibit polarization and magnetization effects. This is in sharp contradistinction with the usual assumption on the classical vacuum as an empty space. The classical vacuum may be thought as an empty space only for smooth sources. The classical vacuum gains therefore the properties of a complicated medium what is usually appreciated only on the quantum level.

Another way to describe the action of the electromagnetic field on matter is to consider wave equations for matter in which the electromagnetic interaction is implemented by the electromagnetic potentials φ and $\mathbf{A}$. It is also possible to treat these conventional potentials as primary electromagnetic quantities [9] . For our purpose we need to assume that these quantities play the role of the test functions of the theory. The relations (34 -35) may then be replaced by

$$\mathbf{D} - \mathbf{P} = \epsilon_0 \left(-\nabla \varphi - \frac{\partial \mathbf{A}}{\partial t} \right) \tag{38}$$

and

$$\mathbf{H} + \mathbf{M} = \mu_0^{-1} rot\ \mathbf{A}. \tag{39}$$

These relations are part of the constitutive relations of our theory. The remaining constitutive relations involve the relations between different sources present in the Maxwell equations. The restricted size of the present paper does not however allow us to expand all the details of the possible constitutive relations.

CONCLUSIONS

We have presented a general scheme for classical electrodynamics which is free from any troubles with distribution-valued sources encountered by the standard Maxwell theory. In particular, our approach perfectly works for point charges and gives a new insight into the long-standing problem of the energy-momentum tensor of the electromagnetic field in a medium. It also has a nice extension to the theory of gravity.

REFERENCES

1. Barut A. O., in *Frontiers of Theoretical Physics,* ed. F. Selleri, Plenum Press, N.Y., 1994.

2. Post E. J., "General Covariance in Electromagnetism", Chapter 7 of the *Delaware Seminar in the Foundation of Physics,* ed. M. Bunge, Springer-Verlag, 1967.

3. Jackson J. D., *Classical Electrodynamics,* John Wiley and Sons, N.Y., 1962.

4. Lorentz H. A., *The Theory of Electrons,* Dover, N.Y.,1952.

5. Schmeelk J., Foundations of Phys.Lett. 3 (1990) 403-423.

6. Robinson F. R., *Macroscopic Electrodynamics,* Pergamon,1973.

7. Kapuścik E., Nuovo Cimento Lett. **42** (1985) 263.

8. Kapuścik E., Comm.JINR, E2-91-272,Dubna, 1991.

9. Feynman R. P., *Lectures on Physics,* vol.II, Addison-Wesley, Reading, 1963.

GEOMETRIC THEORY OF RADIATION

Z. Ya. Turakulov[1]

[1]Institute of Nuclear Physics
Ulugbek, Tashkent 702132, Rep. of Uzbekistan, CIS
(e-mail: ULUG@pc202.suninp.tashkent.su)

Abstract

Foundations of the standard theory of radiation by moving charge are analysed. The well-known difficulty concerned with energy conservation law is considered. An exact solution of Maxwell equations obtained by the method of variables separation, which expressses the field of a charge describing hyperbolic motion, is presented. Unlike the Lienard-Wiechert potentials for this case of motion, it displays abcence of radiation and, hence, does not lead to any difficulties mentioned above. It is concluded that exact solutions of Maxwell equations constitute the only correct approach to electromagnetic field of a moving charge. The new classical theory of radiation that may be composed on the basis of exact solutions seems to be purely geometric.

INTRODUCTION

Electric current formed by a moving charge is a vector field non-zero only on the charge world line, tangent to the line and having constant norm equal to the charge value. It is thus similar to stationary current carried by a conducting wire. Therefore, the problem of field produced by a moving charge may be considered as a generalization of the well-known problem of classical magnetostatics. The generalization consists in adding the time dimension and considering time-like line of current.

The problem reduces to solving the Maxwell equations with given source. There exist two basic approaches to the problem. One of them is that of variables separation. It gives results in very few cases with the curve being of some special shape. Another one, the method of Green function integration looks to be more suitable. The classical theory of radiation used for about a century, is in fact a result of applying the retarded Green function that gives so-called retarded or Lienard-Wiechert potentials [1]. In the present work we present an exact solution of the Maxwell equations found by the method of variables separation and show that it is not the same as Lienard-Wiechert potentials for the same field. Also, we show that it is applying the retarded Green function which creats the well-known difficulties concerned with energy conservation law.

RETARDED POTENTIALS, LARMOR FORMULA AND THE PROBLEM OF ENERGY CONSERVATION

The field intensities found from the retarded Lienard-Wiechert potentials, are naturally divided to static and radiation components [1]. Consider a particle carrying a charge e, whose acceleration is $\vec{a}$ and velocity is zero at some moment of time. Due to the form of Lienard-Wiechert potentials, the radiation components of field intensities produced by the particle at this moment are

$$\vec{H}_a = \left[\vec{n} \times \vec{E}_a\right]_{ret}, \quad \vec{E}_a = \frac{e}{c}\left[\frac{\vec{n} \times [\vec{n} \times \vec{a}]}{R}\right]_{ret} \tag{1}$$

where subscripts a and ret mean "acceleration part" and "taken at retarded time" respectively, $\vec{n}$ is a unit vector towards the point of observation and R is distance from the charge. The instantaneous energy flux is given by the Poynting vector

$$\vec{S} = \frac{c}{4\pi}\vec{E}_a \times \vec{H}_a = \frac{e}{4\pi}|\vec{E}_a|^2\vec{n} \tag{2}$$

and the total instantaneous power radiated is given by the Larmor formula

$$P = \frac{2}{3}\frac{e^2}{c^3}|\vec{a}|^2. \tag{3}$$

These results remain valid also for a small particle whose field has finite total energy.

Assume that the particle moves in a smooth potential $\Phi(x)$ and its velocity becomes zero in a point x. Then, after a short period of time Δt the coordinate x and particle velocity get increments $a\Delta t^2/2$ and $a\Delta t$ respectively. Thus, its kinetic energy rises from zero to the value $ma^2\Delta t^2/2$ while its potential energy change is $(1/2)\Phi'(x)a\Delta t^2$, i.e. changes of both kinetic and potential energies have order Δt^2. During the same period of time the energy radiated given by the Larmor formula (2.3) is proportional to Δt that is much larger. As no other kinds of energy present in the system, the source of the energy radiated is not seen unless it is a portion of the particle mass.

The Larmor formula (2.3) follows directly from the expression (2.2) for the Poynting vector which is composed of field intensities (2.1) obtained exactly from the Lienard-Wiechert potentials which consitute an exact consequence of the form of retarded Green function. Since both the intensities $\vec{E}_a$ and $\vec{H}_a$ are proportional to power minus one of the distance, the Poynting vector descents exactly as power minus two of the distance and, hence, its flux through a sphere $R = const$ does not depend on R. Consequently, due to the form of Lienard-Wiechert potential, the whole of energy radiated is emitted from the particle interior. On the other hand, it is natural to believe that the phenomenon of radiation emitting is caused by the field deformation beyond the particle, as it was described in W. Thirring monograph [2].

Another difficulty with energy conservation law caused by the Larmor formula arises when considering a charged body suspended, at rest, in a static gravitational field. By principle of equaivalence, this situation is equivalent to the object being accelerated in a flat space, thus, the body emits radiation with spending neither kinetic nor potential energy. Resolution of this paradox proposed by R. Peierls reads that "amount of radiation escaping to infinity ... vanishes" [3]. However, the radiation presents while the source of its energy absents.

As no other sources of the energy emitted present, one may conclude that Lienard-Wiechert potentials are inconsistent with the presumption that mass of a charged particle may be finite, positive and constant. At the same time no contradictions between the presumption and Maxwell equations themselves are seen. Therefore, it is natural to try to obtain an exact solution by the method of variables separation and compare the result with Lienard-Wiechert potentials.

OTHER VIEWS

The field of a point charge describing hyperbolic motion was first determined by M. Born. His result presented in W. Pauli's monograph "Theory of Relativity" [4] reads that magnetic field produced by the charge vanishes at the moment of time when its velocity is zero. Since the Poynting vector is zero the radiation absents. W. Pauli concluded that "hyperbolic motion thus constitutes a special case for which there is no formation of a wave zone nor any corresponding radiation". It remains unclear, however, what features of this special case of motion eliminate the vector product $\left[\vec{n} \times \vec{E}_a\right]_{ret}$ in each point of the space.

More recently A. Sommerfeld wrote in his monograph "Elektrodynamik" that it is power minus two of interval which should be taken as the Green function, and called Lienard-Wiechert potentials "Lienard-Wiechert approximation" ("Lienard-Wiechertsche Näherung") [5]. Consider the standard procedure of deriving the Green function for a massless scalar field on the basis of Fourier transform [1]:

$$G(\vec{x}, t; \vec{x}', t') = \frac{1}{(2\pi)^4} \int d^3k \int d\omega \frac{e^{i(\vec{k}\vec{r} - \omega t)}}{k^2 - \frac{1}{c^2}\omega^2}$$

To evaluate this integral one introduces complex variable ω and choses a path of integration on the complex ω-plane. It should be pointed out that this result acually depends on a certain choice of the path. In fact it is unnecessary to take into account the singularity because the integrand is not singular. Indeed, accounting that the Green function depends on the only variable $c^2\tau^2 - R^2$ allows one to chose a reference frame so that the value of R for a point under consideration is zero, and introduce the mass-shell coordinates for the momentum space referred to the frame:

$$m^2 = \omega^2 - k^2, \quad \xi = arctanh(k/\omega).$$

Then,

$$\omega t - \vec{k} \cdot \vec{R} = m\zeta \cosh \xi$$

and the integral takes the following form:

$$G(\tau, R) = \frac{1}{4\pi^3} \int dm \ m^3 \sinh \xi d\xi \sin \theta d\theta d\varphi \frac{e^{im\zeta \cosh \xi}}{m^2}$$

in which absence of any singularities is apparent. Consequently, there exist the only way for evaluating it (without additions like $i\varepsilon$) and the only result of the evaluation. The result coincides with the causal Green function, because substitution $y = m\zeta$ extracts the ζ-dependence:

$$G(\tau, R) = \frac{\pi^2}{\zeta^2} \int dy \ y \int d\xi \sinh^2 \xi e^{\imath y \cosh \xi}.$$

Now, integrating predetermines solely a constant factor at ζ^{-2} and one can select any path of integration in the complex plane. In other words, the retarded Green function and, hence, retarded potentials are created by inserting a non-existing singulartity into the integrand. Note that retarded Green function is used solely in the classical theory of radiation.

Applying these two Green functions give obviously different results. Indeed, since the retarded Green function is manifestly zero inside the light cone it gives the field depending on the current vector direction solely in one point of the charge world line, i. e., pure retardance, whereas the function like ζ^{-2} is non-zero there and gives the field depending on the directions at all the preceding points. Apparently, only one of these two dependences is correct. Another difference between these two Green functions is purely practical: the retarded Green function is much easier to be integrated, because it is nothing but δ-function.

EXACT SOLUTION

One exact solution of the Maxwell equations for the field of a charge describing hyperbolic motion, was found by the variables separation method in our recent work [6, 7]. The field has the following form:

$$A_t = \frac{z}{\zeta^2} \frac{\zeta^2 + \rho^2 + a^2}{\sqrt{[(\zeta - a)^2 + \rho^2][(\zeta + a)^2 + \rho^2]}}$$

$$A_z = -\frac{t}{\zeta^2} \frac{\zeta^2 + \rho^2 + a^2}{\sqrt{[(\zeta - a)^2 + \rho^2][(\zeta + a)^2 + \rho^2]}},$$

where $\zeta = \sqrt{z^2 - t^2}$. This solution is valid only for $z \geq t$. Although it is not the same as M. Born's field, both solutions are similar, differ from the Lienard-Wiechert potentials and give zero magnetic field at the moment of zero velocity. They do not lead to any difficulties with energy conservation law and related troubles, because they show that no radiation is emitted under uniformly accelerated motion. As seen from our result, the only way to have correct expressions for electromagnetic field of a moving charge is to solve Maxwell equations as they stand. As for the classical theory of radiation, the correct version may be composed only on exact solutions obtained properly.

Exact solutions of Maxwell equations for the field of uniformly accelerated charge are obtained by using purely geometric approaches. As the field does not contain any radiation part it is natural to expect the radiation to take place at moments of changes of the acceleration. Thus, the first problem arising now is to find out the radiation field under small changes of the particle acceleration. This is also a purely geometric problem. Therefore, we conclude that the phenomenon of radiation as well as the new classical theory of radiation based on exact solutions of Maxwell equations we intend to propose, are purely geometric. Foundations of the new theory are presented in our recent work [8].

ACKNOWLEDGMENTS

The author is grateful to D. Bambusi, A. O. Barut, P. Fortini, L. Galgani, A. Laufer, D. Noja, A. Orlowski, J. Reignier, E. Remiddi, I. Sassarini and J. Vaz for helpful discussions.

References

[1] J.D. Jackson, *Classical Electrodynamics* (Wiley, New York, 1962) 658.

[2] W. Thirring. *Classical Field Theory* (Springer, New York 1979) 3.

[3] R. Peierls. *Surprising in Theoretical Physics* (Princeton University Press, New Jersey, 1979) 166.

[4] W. Pauli, *Theory of Relativity* (Pergamon Press, New York, 1958) 93.

[5] A. Sommerfeld *Elektrodynamik* (Geesst & Portig, Leipzig, 1949) 256.

[6] Z. Y. Turakulov, *Turkish J. of Phys.* 18 (1994) 479

[7] Z. Y. Turakulov, *Geometry and Physics* 14 (1994) 305

[8] Z. Y. Turakulov, *What Description of Radiation Follows from M.Born Solution of 1909?* JINR preprint, to be published.

IS THERE ANY RELATIONSHIP BETWEEN
MAXWELL AND DIRAC EQUATIONS ?

Jayme Vaz, Jr.[1] and Waldyr A. Rodrigues, Jr.[2]

[1]DFESCM - Instituto de Física "Gleb Wataghin"
Universidade Estadual de Campinas
CP 6165, 13081-970, Campinas, S.P., Brazil
[2]Departamento de Matemática Aplicada - IMECC
Universidade Estadual de Campinas
CP 6065, 13081-970, Campinas, S.P., Brazil

INTRODUCTION

The electron is an intriguing object. In spite of all our misconceptions, along the last 100 years an extraordinary technological development emerged based on the concept of electron, which makes it difficult to doubt about its reality. On the other hand, electromagnetism and relativistic quantum mechanics are two of our most successful theories, but neither quantum mechanics nor electromagnetism help us to picture the electron, and to understand its role as a basic block of electric charge and matter.

In this paper we try to contribute to this fundamental problem by focusing our attention into the basic equations of those theories, namely Maxwell equations and Dirac equation. It is natural to expect some relationship between Maxwell and Dirac equations if there is some relationship between electromagnetism and quantum mechanics. An interesting and recent review of these problems was given by Keller[1]. Our objective is to show that Maxwell equations can be put in a form which is identical to Dirac equation!

First of all, before addresing our specific problem, we shall introduce the mathematical background used in this paper (sec.2). In particular, we shall use in this paper the Clifford algebra formalism and the concept of Dirac-Hestenes spinor. One can ask if there is a necessity of doing this, if our calculations cannot be performed with usual mathematical tools, etc. In this way we observe that Clifford algebras seems to be a natural framework for physics, and in particular for our problem; calculations are much more easy to be done using Clifford algebras than with the traditional spinor and tensor calculus – an example of how powerful and natural they are is that the use of the Clifford algebra of spacetime enable us to write the Maxwell's eight scalar equations into a single equation.

In sec.3 we shall look for a spinorial representation of Maxwell equations. There are in the literature several different spinorial representations of Maxwell equations. These spinorial representations, as a rule, use as components of the spinor field the components of the electromagnetic field. The deficiency of this representations is obvious: the

Electron Theory and Quantum Electrodynamics: 100 Years Later
Edited by Dowling, Plenum Press, New York, 1997

action of the Lorentz group on the spinor field does not give the correct transformation laws for the electric and magnetic fields. A detailed study of these kind of spinorial representation of Maxwell equations can be found in [2]. The spinorial representation of Maxwell equations we shall use is based on the concept of Dirac-Hestenes spinor, and does not suffer the deficiency indicated above.

Once we find a spinorial representation of Maxwell equations we study it in details (sec.4), and we show that it can be reduced to a form which is identical to Dirac equation; then we study certain conditions under which this equation can be really interpreted as Dirac equation.

MATHEMATICAL PRELIMINARIES: CLIFFORD ALGEBRAS AND DIRAC-HESTENES SPINORS

(A) Exterior, Grassmann and Clifford Algebras

Let V be a vector space of dimension n endowed with an interior product g : $V \times V \to \mathbb{R}$. Let $\wedge$ be the exterior (or wedge or Grassmann) product, that is, an associative, bilinear and skew-symmetric product of vectors: $(a \wedge b) \wedge c = a \wedge (b \wedge c)$, $(a + \alpha b) \wedge c = a \wedge c + \alpha b \wedge c$, $a \wedge b = -b \wedge a$, $(\forall a, b, c \in V)$. If $\{e_1, \cdots, e_n\}$ is a basis for V, then $\{e_1 \wedge e_2, \cdots, e_1 \wedge e_n; e_2 \wedge e_3, \cdots, e_{n-1} \wedge e_n\}$ is a basis for the vector space $\bigwedge^2(V)$ whose elements are called bivectors (2-vectors). In this way $\wedge : V \times V \to \bigwedge^2(V)$. We can naturally extend the definition of the exterior product for vectors and bivectors, giving trivectors (3-vectors), and so on. We denote by $\bigwedge^k(V)$ $(0 \le k \le n)$ the vector space of k-vectors, which is of dimension $\binom{n}{k}$ (we adopt the convention $\bigwedge^0(V) = \mathbb{R}$ and $\bigwedge^1(V) = V$). We have $\wedge : \bigwedge^k(V) \times \bigwedge^l(V) \to \bigwedge^{k+l}(V)$, and if $A_k \in \bigwedge^k(V)$ and $A_l \in \bigwedge^l(V)$ then $A_k \wedge B_l = (-1)^{kl} B_l \wedge A_k$. A k-vector is said to be simple if it is the external product of k 1-vectors. Consider the direct sum $\oplus_{k=0}^n \bigwedge^k(V) = \bigwedge(V)$. An element of $\bigwedge(V)$ is called a multivector, and $(\bigwedge(V), \wedge)$ is called exterior algebra. Let $\langle \ \rangle_k$ denote the projector $\langle \ \rangle_k : \bigwedge(V) \to \bigwedge^k(V)$. If $\langle A \rangle_k = A_k$ the multivector A is said to be homogeneous, and k is its grade.

We can extend g to $\bigwedge(V)$. First, define the extension of g to $\bigwedge^k(V)$ as

$$g_k(a_1 \wedge \cdots \wedge a_k, b_1 \wedge \cdots \wedge b_k) = \det\{g(a_i, b_j)\} \tag{1}$$

for simple k-vectors, and then extend to all $\bigwedge^k(V)$ by linearity (by g_0 we mean ordinary multiplication of scalars). We denote by G the extension of g to $\bigwedge(V)$ given by $G = \sum_{k=0}^n g_k$, with $G(A_k, B_l) = 0$ when $k \ne l$. $(\bigwedge(V), \wedge, G)$ is called Grassmann algebra. The structure of Grassmann algebra is obviously richer than the one of exterior algebra. In order to see this, let us first introduce two involutions. The first, called reversion, and denoted by a tilde, is defined by $\tilde{A}_k = (-1)^{k(k-1)/2} A_k$; the name reversion is due to the fact that it reverses the order of the exterior product of the vectors in a simple k-vector, i.e., $(a_1 \wedge \cdots \wedge a_k)\tilde{} = a_k \wedge \cdots \wedge a_1 = (-1)^{k(k-1)/2}(a_1 \wedge \cdots \wedge a_k)$. The second one, called graded involution, and denoted by a hat, is defined by $\hat{A}_k = (-1)^k A_k$. Now, let us define the contraction. The left contraction $\lrcorner$ is defined by [3]

$$G(A \lrcorner B, C) = G(B, \tilde{A} \wedge C), \quad \forall C \in \bigwedge(V), \tag{2}$$

and the right contraction $\llcorner$ is defined by

$$G(A \llcorner B, C) = G(A, C \wedge \tilde{B}), \quad \forall C \in \bigwedge(V). \tag{3}$$

Left and right contractions are related by

$$A_r \lrcorner B_s = (-1)^{r(s-r)} B_s \llcorner A_r, \quad (s > r), \tag{4}$$

328

and satisfies the following properties:

$$a \lrcorner b = g(a,b), \tag{5}$$

$$a \lrcorner (B \wedge C) = (a \lrcorner B) \wedge C + \hat{B} \wedge (a \lrcorner C), \tag{6}$$

$$a \lrcorner (b \lrcorner C) = (a \wedge b) \lrcorner C, \tag{7}$$

where $a, b \in V$, $B, C \in \bigwedge(V)$. In what follows, whenever there is no danger of confusion, we denote the contraction by a dot : $a \cdot B = a \lrcorner B$ (the dot must not be confused with internal product in spite that for vectors we have eq.(5)).

Let us introduce the Clifford (or geometrical) product $\vee$; given $a \in V$ and $B \in \bigwedge(V)$ we define $a \vee B$ by

$$a \vee B = a \lrcorner B + a \wedge B, \tag{8}$$

and extend this definition to all $\bigwedge(V)$ by associativity. $(\bigwedge(V), \wedge, G, \vee)$ is called a Clifford algebra (denoted by $Cl(V,g)$). In order to simplify the notation, we denote the Clifford product simply by justaposition, i.e.:

$$aB = a \lrcorner B + a \wedge B. \tag{9}$$

Let $\{e_1, \cdots, e_n\}$ be an orthonormal basis for V. In this case we have

$$e_i^2 = e_i e_i = e_i \cdot e_i + e_i \wedge e_i = e_i \cdot e_i = g(e_i, e_i), \tag{10}$$

$$e_{ij} = e_i e_j = e_i \cdot e_j + e_i \wedge e_j = e_i \wedge e_j \quad (i \neq j). \tag{11}$$

Since exterior, Grassmann and Clifford algebras are isomorphic as vector spaces, a general element of $Cl(V,g)$ is of the form

$$A = a_0 + a^i e_i + a^{ij} e_{ij} + \cdots + a^{1 \cdots n} e_{1 \cdots n}, \tag{12}$$

where, in particular, $e_{1 \cdots n} = e_1 \cdots e_n = e_1 \wedge \cdots \wedge e_n = \tau$ is the volume element of V. We observe that Clifford algebra is Z_2-graded algebra. In fact, let Cl_+ (Cl_-) denote the set of elements of Cl with even (odd) grade; we have $Cl_+ Cl_+ \subset Cl_+$, $Cl_+ Cl_- \subset Cl_-$, $Cl_- Cl_+ \subset Cl_-$, $Cl_- Cl_- \subset Cl_+$. This fact shows that Cl_+ is a sub-algebra of Cl, called the even sub-algebra.

The great advantage of Clifford algebra over Grassmann algebra follows from the fact that the Clifford product contains more information than Grassmann product. Moreover, under the Clifford product it is possible to "divide" multivectors. In fact, define the norm $|A|$ of a multivector A as

$$|A|^2 = \langle \tilde{A} A \rangle_0. \tag{13}$$

If $\tilde{A}A = |A|^2 \neq 0$ we define $A^{-1} = \tilde{A} |A|^{-2}$; in fact $A^{-1}A = AA^{-1} = |A|^{-2} \tilde{A}A = 1$. Another advantage is that Clifford algebras are isomorphic to matrix algebras over $\mathbb{R}$, $\mathbb{C}$ or $\mathbb{H}$ (we shall see examples below)[4].

Finally, we say that an algebraic element e is an idempotent if $e^2 = e$; it is called primitive if it cannot be written as the sum of two mutually annihilating idempotents (that is: $e \neq e' + e''$ with $(e')^2 = e'$, $(e'')^2 = e''$, $e'e'' = e''e' = 0$). The sub-algebra I_E of an algebra A is called a left ideal if given $i \in I_E$ we have $xi \in I_E$, $\forall x \in A$ (similarly for right ideals). An ideal is said to be minimal if it contains only trivial sub-ideals. Now, one can prove that the minimal left ideals of a Clifford algebra $Cl(V,g)$ are of the form $Cl(V,g)e$, where e is a primitive idempotent[5]. This is an important result to be used later.

(B) Spacetime Algebra

Let $V = \mathbb{R}^{1,3}$ be Minkowski vector space, and choose a basis $\{\gamma_\mu\}$ ($\mu = 0, 1, 2, 3$) such that $g(\gamma_\mu, \gamma_\nu) = \eta_{\mu\nu} = \text{diag}(1, -1, -1, -1)$. The spacetime algebra (STA)[6] is the Clifford algebra of $(\mathbb{R}^{1,3}, \eta)$, denoted by $\mathbb{R}_{1,3}$. Observe that $\gamma_0^2 = -\gamma_1^2 = -\gamma_2^2 = -\gamma_3^2 = 1$. A general element of $\mathbb{R}_{1,3}$ is of the form

$$A = a + a^\mu \gamma_\mu + a^{\mu\nu} \gamma_{\mu\nu} + a^{\mu\nu\sigma} \gamma_{\mu\nu\sigma} + a^{0123} \gamma_5, \tag{14}$$

where $\gamma_{\mu\nu} = \gamma_\mu \wedge \gamma_\nu$ ($\mu \neq \nu$) and $\gamma_5 = \gamma_0 \gamma_1 \gamma_2 \gamma_3 = \gamma_0 \wedge \gamma_1 \wedge \gamma_2 \wedge \gamma_3$ is the volume element. Note that $\gamma_5^2 = -1$ and $\gamma_5 \gamma_\mu = -\gamma_\mu \gamma_5$.

The STA is particularly useful in formulating the Lorentz rotations. Let $\mathbb{R}_{1,3}^+$ denote the even sub-algebra of $\mathbb{R}_{1,3}$, and let $\mathcal{N}$ be the norm map, i.e., $\mathcal{N}(L) = |L|^2 = \langle \tilde{L} L \rangle_0$. The double covering of the restricted Lorentz group $\text{SO}_+(1,3)$ is $\text{Spin}_+(1,3)$ defined as

$$\text{Spin}_+(1,3) = \{R \in \mathbb{R}_{1,3}^+ \mid \mathcal{N}(R) = 1\}. \tag{15}$$

An arbitrary Lorentz rotation is therefore given by $a \mapsto RaR^{-1} = Ra\tilde{R}$, with $R \in \text{Spin}_+(1,3)$. One can also prove that any $R \in \text{Spin}_+(1,3)$ can be written in the form $R = \pm e^B$ with $B \in \wedge^2(\mathbb{R}^{1,3})$, and the choice of the sign can always be the positive one except when $R = -e^B$ with $B^2 = 0$. When B is a timelike bivector ($B^2 > 0$) R describes a boost, while when B is a spacelike bivector ($B^2 < 0$) R describes a spatial rotation.

We said that Clifford algebras are isomorphic to matrix algebras. In the case of $\mathbb{R}_{1,3}$ it is isomorphic to $\mathcal{M}(2, \mathbb{H})$, the algebra of 2×2 matrices over the quaternions. This isomorphism defines representations of STA. One representation is:

$$\gamma_0 \leftrightarrow \begin{pmatrix} 1 & 0 \\ 0 & -1 \end{pmatrix}, \gamma_1 \leftrightarrow \begin{pmatrix} 0 & i \\ i & 0 \end{pmatrix}, \gamma_2 \leftrightarrow \begin{pmatrix} 0 & j \\ j & 0 \end{pmatrix}, \gamma_3 \leftrightarrow \begin{pmatrix} 0 & k \\ k & 0 \end{pmatrix}, \tag{16}$$

where $i^2 = j^2 = k^2 = -1$ and $ij = k$, $jk = i$, $ki = j$.

Finally, consider the idempotent $e = \frac{1}{2}(1 + \gamma_0)$. $I_{1,3} = \mathbb{R}_{1,3}e$ is a minimal left ideal. One can easily prove that

$$I_{1,3} = \mathbb{R}_{1,3}e = \mathbb{R}_{1,3}^+ e, \tag{17}$$

which is an important result to be used later.

(C) Dirac Algebra

Consider the vector space $\mathbb{R}^{4,1}$ and a basis $\{E_a\}$ ($a = 0, 1, 2, 3, 4$) such that $E_1^2 = E_2^2 = E_3^2 = E_4^2 = -E_0^2 = 1$. Let $\mathbb{R}_{4,1}$ be its Clifford algebra and $i = E_0 E_1 E_2 E_3 E_4$ be the volume element. Note that $i^2 = -1$, but differently from STA now we have $E_a i = i E_a$ ($\forall a$), that is, $\mathbb{R} \oplus i\mathbb{R}$ is the center of $\mathbb{R}_{4,1}$. The volume element i plays therefore the role of imaginary unity. Let us define

$$\gamma_\mu = E_\mu E_4 \quad (\mu = 0, 1, 2, 3). \tag{18}$$

One can easily see from the above map and with i playing the role of imaginary unity that $\mathbb{R}_{4,1}$ is isomorphic to the complexified STA:

$$\mathbb{R}_{4,1} \simeq \mathbb{C} \otimes \mathbb{R}_{1,3}. \tag{19}$$

Moreover, the even subalgebra of $\mathbb{R}_{4,1}$ is isomorphic to $\mathbb{R}_{1,3}$:

$$\mathbb{R}_{4,1}^+ \simeq \mathbb{R}_{1,3}. \tag{20}$$

The algebra $\mathbb{R}_{4,1}$, or equivalently, the complexified STA, is called Dirac algebra. In fact, they are isomorphic to $\mathcal{M}(4,\mathbb{C})$ – the algebra of 4×4 matrices over the complexes. One representation (the standard one) of γ_μ in eq.(18) is:

$$\gamma_0 \leftrightarrow \begin{pmatrix} 1 & 0 & 0 & 0 \\ 0 & 1 & 0 & 0 \\ 0 & 0 & -1 & 0 \\ 0 & 0 & 0 & -1 \end{pmatrix}, \quad \gamma_1 \leftrightarrow \begin{pmatrix} 0 & 0 & 0 & -1 \\ 0 & 0 & -1 & 0 \\ 0 & 1 & 0 & 0 \\ 1 & 0 & 0 & 0 \end{pmatrix},$$

$$\gamma_2 \leftrightarrow \begin{pmatrix} 0 & 0 & 0 & i \\ 0 & 0 & -i & 0 \\ 0 & -i & 0 & 0 \\ i & 0 & 0 & 0 \end{pmatrix}, \quad \gamma_3 \leftrightarrow \begin{pmatrix} 0 & 0 & -1 & 0 \\ 0 & 0 & 0 & 1 \\ 1 & 0 & 0 & 0 \\ 0 & -1 & 0 & 0 \end{pmatrix}. \tag{21}$$

Consider the idempotent $f = \frac{1}{2}(1+\gamma_0)\frac{1}{2}(1+i\gamma_{12}) = e\frac{1}{2}(1+i\gamma_{12})$, where $e = \frac{1}{2}(1+\gamma_0)$ is a primitive idempotent of $\mathbb{R}_{1,3}$. Then $I_{4,1} = \mathbb{R}_{4,1}f$ is a minimal left ideal, and one can show that

$$I_{4,1} = \mathbb{R}_{4,1}f \simeq \mathbb{R}_{4,1}^+f. \tag{22}$$

(D) Dirac-Hestenes Spinors

Let us consider a Dirac spinor $|\Psi\rangle \in \mathbb{C}^4$. There is an obvious isomorphism between $\mathbb{C}^4$ and minimal left ideals of $\mathcal{M}(4,\mathbb{C})$, given by

$$\mathbb{C}^4 \ni |\Psi\rangle = \begin{pmatrix} \psi_1 \\ \psi_2 \\ \psi_3 \\ \psi_4 \end{pmatrix} \leftrightarrow \begin{pmatrix} \psi_1 & 0 & 0 & 0 \\ \psi_2 & 0 & 0 & 0 \\ \psi_3 & 0 & 0 & 0 \\ \psi_4 & 0 & 0 & 0 \end{pmatrix} = \Psi \in \text{ideal of } \mathcal{M}(4,\mathbb{C}). \tag{23}$$

One can, of course, work with Ψ instead of $|\Psi\rangle$, and since $\mathcal{M}(4,\mathbb{C})$ is a representation of Dirac algebra $\mathbb{R}_{4,1} \simeq \mathbb{C} \otimes \mathbb{R}_{1,3}$, one can work with the corresponding ideal of Dirac algebra. Note that

$$\begin{pmatrix} \psi_1 & 0 & 0 & 0 \\ \psi_2 & 0 & 0 & 0 \\ \psi_3 & 0 & 0 & 0 \\ \psi_4 & 0 & 0 & 0 \end{pmatrix} = \underbrace{\begin{pmatrix} \psi_1 & \cdot & \cdot & \cdot \\ \psi_2 & \cdot & \cdot & \cdot \\ \psi_3 & \cdot & \cdot & \cdot \\ \psi_4 & \cdot & \cdot & \cdot \end{pmatrix}}_{\in \mathcal{M}(4,C)} \underbrace{\begin{pmatrix} 1 & 0 & 0 & 0 \\ 0 & 0 & 0 & 0 \\ 0 & 0 & 0 & 0 \\ 0 & 0 & 0 & 0 \end{pmatrix}}_{f}, \tag{24}$$

where f is a matrix representation of the idempotent $f = \frac{1}{2}(1+\gamma_0)\frac{1}{2}(1+i\gamma_{12})$. One can work therefore with the ideal $I_{4,1} = \mathbb{R}_{4,1}f$ instead of $\mathbb{C}^4$. But the isomorphisms discussed in the preceeding subsections tell us that

$$I_{4,1} = \mathbb{R}_{4,1}f = (\mathbb{C} \otimes \mathbb{R}_{1,3})f \simeq \mathbb{R}_{4,1}^+f \simeq \mathbb{R}_{1,3}f = (\mathbb{R}_{1,3}e)\frac{1}{2}(1+i\gamma_{12}). \tag{25}$$

Note that in the last equality we have a minimal left ideal $\mathbb{R}_{1,3}e$ of STA. These equalities show that all informations we obtain from an element of the ideal $I_{4,1}$ can be obtained from an element of the ideal $I_{1,3} = \mathbb{R}_{1,3}e$. Moreover, we have that

$$if = \gamma_{21}f. \tag{26}$$

These results mean that we can work with the ideal $\mathbb{R}_{1,3}e$ once we identify γ_{21} as playing in STA the role of the imaginary unity i.

Now, we saw that

$$\mathbb{R}_{1,3}e = \mathbb{R}_{1,3}^+ e. \tag{27}$$

What the idempotent makes is "kill" redundant degrees of freedom. Since $\dim \mathbb{R}_{1,3}^+ = 8$ we can work with $\mathbb{R}_{1,3}^+$ instead of $\mathbb{R}_{1,3}e$ (this is not the case for $\mathbb{R}_{1,3}$ or $\mathbb{C} \otimes \mathbb{R}_{1,3}$ since $\dim \mathbb{R}_{1,3} = 16$ and $\dim(\mathbb{C} \otimes \mathbb{R}_{1,3}) = 32$). We established therefore the isomorphism

$$\mathbb{C}^4 \simeq \mathbb{R}_{1,3}^+. \tag{28}$$

The Dirac spinor $|\Psi\rangle$ is related to $\psi \in \mathbb{R}_{1,3}^+$ by

$$|\Psi\rangle = \psi \frac{1}{2}(1 + \gamma_0)\frac{1}{2}(1 + i\gamma_{12}). \tag{29}$$

Such ψ will be called Dirac-Hestenes spinor. Its (standard) matrix representation is:

$$\psi = \begin{pmatrix} \psi_1 & -\psi_2^* & \psi_3 & \psi_4^* \\ \psi_2 & \psi_1^* & \psi_4 & -\psi_3^* \\ \psi_3 & \psi_4^* & \psi_1 & -\psi_2^* \\ \psi_4 & -\psi_3^* & \psi_2 & \psi_1^* \end{pmatrix}. \tag{30}$$

Suppose now that ψ is non-singular, i.e., $\psi\tilde{\psi} \neq 0$. Since $\psi \in \mathbb{R}_{1,3}^+$ we have

$$\psi\tilde{\psi} = \sigma + \gamma_5\omega, \tag{31}$$

where σ and ω are scalars. Define quantities $\rho = \sqrt{\sigma^2 + \omega^2}$ and $\tan\beta = \omega/\sigma$. Then ψ can be written as

$$\psi = \sqrt{\rho}e^{\gamma_5\beta/2}R, \tag{32}$$

where $R \in \text{Spin}_+(1,3)$, $\rho \in \mathbb{R}_+$ and $0 \leq \beta < 2\pi$. This is the canonical decomposition of Dirac-Hestenes spinors in terms of a Lorentz rotation R, a dilation ρ and a duality transformation by an angle β (called Yvon-Takabayasi angle).

Finally, let us write Dirac equation in terms of ψ. Since $|\Psi\rangle = \psi f$, $\gamma_{21}f = if$ and $\gamma_0 f = f$, the (free) Dirac equation $i\partial |\Psi\rangle = (mc/\hbar) |\Psi\rangle$ is written in STA as

$$\partial\psi\gamma_{21} = \frac{mc}{\hbar}\psi\gamma_0, \tag{33}$$

which we call Dirac-Hestenes equation.

SPINORIAL REPRESENTATION OF MAXWELL EQUATIONS

The spinorial representation of Maxwell equations we shall give in this section is based on the following theorem:

Theorem: Any electromagnetic field $F \in \bigwedge^2(\mathbb{R}^{1,3})$ can be written in the form

$$F = \psi\gamma_{21}\tilde{\psi}, \tag{34}$$

where ψ is a Dirac-Hestenes spinor field.

The proof of this theorem can be divided in three steps: (i) $F^2 \neq 0$; (ii) $F^2 = 0$, $F \neq 0$; (ii) $F^2 = 0$, $F = 0$. In the first case the proof is based on a theorem by Rainich[7], and reconsidered by Misner and Wheeler[8].

Theorem (Rainich-Misner-Wheeler): Let an extremal field be an electromagnetic field for which the electric [magnetic] field vanishes and the magnetic [electric] field is parallel to a given spatial direction. Then, at any point of spacetime, any non-null ($F^2 \neq 0$) electromagnetic field F can be transformed in an extremal field by means of a Lorentz transformation and a duality transformation.

An easy proof of the theorem of Rainich-Misner-Wheeler can be found in [9]. Now consider a Lorentz transformation $F \mapsto F' = LF\tilde{L}$, and a duality transformation $F' \mapsto F'' = e^{\gamma_5 \alpha} F'$. According to the theorem of RMW the electromagnetic field F'' is extremal; let us suppose it of magnetic type along the $\vec{k}$ direction, that is: $F'' = \rho \gamma_{21}$, where ρ is the extremal field intensity. We have therefore that $\rho \gamma_{21} = e^{\gamma_5 \alpha} LF\tilde{L}$. Let us define $R = \tilde{L}$ and $\beta = -\alpha$; then we have that

$$F = \psi \gamma_{21} \tilde{\psi}, \qquad (35)$$

where

$$\psi = \sqrt{\rho} e^{\gamma_5 \beta/2} R, \qquad (36)$$

which we recognize as the canonical decomposition of Dirac-Hestenes spinor.

In order to prove our theorem for case (ii) we observe that since $F^2 = 0$ we have $\vec{E} \cdot \vec{H} = 0$ and $\vec{E}^2 = \vec{H}^2$; we can make therefore a spatial rotation $\mathcal{R}$ such that $E_1' = H_1' = 0$ and $H_3' = \pm E_2' = \eta_1$ and $H_2' = \pm E_3' = \eta_2$; then for $F' = (1/2)(F')^{\mu\nu}\gamma_{\mu\nu}$ we have $F' = (\eta_1 + \gamma_5 \eta_2)(1/2)(1 \pm \gamma_{01})\gamma_{21}$. If we take $R = \tilde{\mathcal{R}}$ we have for $F = RF'\tilde{R}$ that $F = (\eta_1 + \gamma_5 \eta_2)R(1/2)(1 \pm \gamma_{01})\gamma_{21}\tilde{R}$. Remember that $(1/2)(1 \pm \gamma_{01})$ is an idempotent; defining $\eta_1 = \eta \cos\varphi$ and $\eta_2 = \eta \sin\varphi$ it follows that

$$F = \psi_M \gamma_{21} \tilde{\psi}_M, \qquad (37)$$

where

$$\psi_M = \sqrt{\eta} e^{\gamma_5 \varphi/2} R \frac{1}{2}(1 \pm \gamma_{01}) = \psi \frac{1}{2}(1 \pm \gamma_{01}), \qquad (38)$$

which proves our assertion in this case. ψ_M is a particular type of Dirac-Hestenes spinor known as Majorana spinor[3].

Now for case (iii) ($F = 0$) $\psi \gamma_{21} \tilde{\psi} = -\psi \gamma_{21} \tilde{\psi} = \gamma_5 \psi \gamma_{21} \tilde{\psi} \gamma_5 = \gamma_5 \psi \gamma_{21} \gamma_{21} \gamma_{12} \tilde{\psi} \gamma_5$ is satisfied for $\psi = \pm \gamma_5 \psi \gamma_{21}$. It follows therefore that

$$F = \psi_W \gamma_{21} \tilde{\psi}_W = 0, \qquad (39)$$

where

$$\psi_W = \frac{1}{2}(\psi \pm \gamma_5 \psi \gamma_{21}). \qquad (40)$$

This particular kind of Dirac-Hestenes spinor is called a Weyl spinor[3].

We have now proved our theorem. It remains, of course, the question of the constants in eq.(34) since the units of F are charge $\times$ (length)$^{-2}$ and the units of ψ are (length)$^{-3/2}$. A combination of constants we need can be

$$F = \frac{e\hbar}{2mc} \psi \gamma_{21} \tilde{\psi}, \qquad (41)$$

which gives correct units for F and ψ. In this expression the symbols have their usual meaning, that is: e is the elementary electric charge, $\hbar$ is the Planck constant, m is the electron mass and c the velocity of light in vacuum. In what follows we will work, to simplify the notation, with eq.(34) instead of eq.(41).

The idea now is to use $F = \psi \gamma_{21} \tilde{\psi}$ in Maxwell equations and obtain from it an equivalent equation for ψ. Maxwell equations using STA are written as an unique equation, namely

$$\partial F = \mathcal{J}, \qquad (42)$$

333

where $\partial = \gamma^\mu \partial_\mu$ is the Dirac operator and $\mathcal{J}$ is the electromagnetic current (an electric current J_e plus a magnetic monopolar current $\gamma_5 J_m$ in the general case). If we use eq.(34) in Maxwell equation (42) we obtain

$$\partial(\psi\gamma_{21}\tilde{\psi}) = \gamma^\mu \partial_\mu(\psi\gamma_{21}\tilde{\psi}) = \gamma^\mu(\partial_\mu\psi\gamma_{21}\tilde{\psi} + \psi\gamma_{21}\partial_\mu\tilde{\psi}) = \mathcal{J}. \qquad (43)$$

But $\psi\gamma_{21}\partial\tilde{\psi} = -(\partial_\mu\psi\gamma_{21}\tilde{\psi})\tilde{}$, and since reversion does not change the sign of scalars and of pseudo-scalars (4-vectors), we have that

$$2\gamma^\mu\langle\partial_\mu\psi\gamma_{21}\tilde{\psi}\rangle_2 = \mathcal{J}. \qquad (44)$$

There is a more convenient way of rewriting the above equation. Note that

$$\gamma^\mu\langle\partial_\mu\psi\gamma_{21}\tilde{\psi}\rangle_2 = \partial\psi\gamma_{21}\tilde{\psi} - \gamma^\mu\langle\partial_\mu\psi\gamma_{21}\tilde{\psi}\rangle_0 - \gamma^\mu\langle\partial_\mu\psi\gamma_{21}\tilde{\psi}\rangle_4, \qquad (45)$$

and if we define the vectors

$$j = \gamma^\mu\langle\partial_\mu\psi\gamma_{21}\tilde{\psi}\rangle_0, \qquad (46)$$

$$g = \gamma^\mu\langle\partial_\mu\psi\gamma_5\gamma_{21}\tilde{\psi}\rangle_0, \qquad (47)$$

we can rewrite eq.(44) as

$$\partial\psi\gamma_{21}\tilde{\psi} = \left[\frac{1}{2}\mathcal{J} + (j + \gamma_5 g)\right]. \qquad (48)$$

If correct units have been used, that is, if we used eq.(41), then instead of $(1/2)\mathcal{J}$ we would obtained $(mc/e\hbar)\mathcal{J}$. Eq.(48) is the spinorial representation of Maxwell equations we were looking for. In the case where ψ is non-singular (which corresponds to non-null electromagnetic fields) we have

$$\partial\psi\gamma_{21} = \frac{e^{\gamma_5\beta}}{\rho}\left[\frac{1}{2}\mathcal{J} + (j + \gamma_5 g)\right]\psi. \qquad (49)$$

Eq.(49) has been proved [9] to be equivalent to the spinorial representation of Maxwell equations obtained originally by Campolattaro[10, 11] in terms of the usual covariant Dirac spinor.

IS THERE ANY RELATIONSHIP BETWEEN MAXWELL AND DIRAC EQUATIONS ?

The spinorial eq.(49) that represents Maxwell ones, as written in that form, does not appear to have any relationship with Dirac equation (33). However, we shall make some modifications on it in such a way to put it in a form that suggests a very interesting and intriguing relationship between them, and consequently between electromagnetism and quantum mechanics.

Since ψ is supposed to be non-singular (F non-null) we can use the canonical decomposition (32) of ψ and write

$$\partial_\mu\psi = \frac{1}{2}\left(\partial_\mu \ln \rho + \gamma_5\partial_\mu\beta + \Omega_\mu\right)\psi, \qquad (50)$$

where we defined

$$\Omega_\mu = 2(\partial_\mu R)\tilde{R}. \qquad (51)$$

Using this expression for $\partial_\mu\psi$ into the definitions of the vectors j and g (eqs.(46,47)) we obtain that

$$j = \gamma^\mu(\Omega_\mu \cdot S)\rho\cos\beta + \gamma_\mu[\Omega_\mu \cdot (\gamma_5 S)]\rho\sin\beta, \qquad (52)$$

$$g = \gamma^\mu [(\Omega_\mu \cdot (\gamma_5 S)] \rho \cos\beta - \gamma_\mu (\Omega_\mu \cdot S) \rho \sin\beta, \tag{53}$$

where we defined the bivector S by

$$S = \frac{1}{2} \psi \gamma_{21} \tilde\psi. \tag{54}$$

A more convenient expression can be written. Let v be given by $\rho v = J = \psi \gamma_0 \tilde\psi$, and $v_\mu = v \cdot \gamma_\mu$. Define the bivector $\Omega = v^\mu \Omega_\mu$ and the scalars Λ and K by

$$\Lambda = \Omega \cdot S, \tag{55}$$

$$K = \Omega \cdot (\gamma_5 S). \tag{56}$$

Using these definitions we have that

$$\Omega_\mu \cdot S = \Lambda v_\mu, \tag{57}$$

$$\Omega_\mu \cdot (\gamma_5 S) = K v_\mu, \tag{58}$$

and for the vectors j and g:

$$j = \Lambda v \rho \cos\beta + K v \rho \sin\beta = \lambda \rho v, \tag{59}$$

$$g = K v \rho \cos\beta - \Lambda v \rho \sin\beta = \kappa \rho v, \tag{60}$$

where we defined

$$\lambda = \Lambda \cos\beta + K \sin\beta, \tag{61}$$

$$\kappa = K \cos\beta - \Lambda \sin\beta. \tag{62}$$

The spinorial representation of Maxwell equations are written now as

$$\partial \psi \gamma_{21} = \frac{e^{\gamma_5 \beta}}{2\rho} J + \lambda \psi \gamma_0 + \gamma_5 \kappa \psi \gamma_0. \tag{63}$$

If $\mathcal{J} = 0$ (free case) we have that

$$\partial \psi \gamma_{21} = \lambda \psi \gamma_0 + \gamma_5 \kappa \psi \gamma_0, \tag{64}$$

which is very similar to Dirac equation (33).

In order to go a step further into the relationship between those equations, we remember that the electromagnetic field has six degrees of freedom, while a Dirac-Hestenes spinor field has eight degrees of freedom; we are free therefore to impose two constrains on ψ if it is to represent an electromagnetic field[11]. We choose these two "gauge conditions" as

$$\partial \cdot j = 0 \quad \text{and} \quad \partial \cdot g = 0. \tag{65}$$

Using eqs.(59,60) these two constrains become

$$\partial \cdot j = \rho \dot\lambda + \lambda \partial \cdot J = 0, \tag{66}$$

$$\partial \cdot g = \rho \dot\kappa + \kappa \partial \cdot J = 0, \tag{67}$$

where $J = \rho v$ and $\dot\lambda = (v \cdot \partial)\lambda$, $\dot\kappa = (v \cdot \partial)\kappa$. These conditions imply that

$$\kappa \dot\lambda = \lambda \dot\kappa, \tag{68}$$

which gives $(\lambda \neq 0)$:

$$\frac{\kappa}{\lambda} = \text{const} = -\tan\beta_0, \tag{69}$$

or from eqs.(61,62):

$$\frac{K}{\Lambda} = \tan(\beta - \beta_0). \tag{70}$$

Now we observe that β is the angle of the duality rotation from F to $F' = e^{\gamma_5 \beta} F$. If we perform another duality rotation by β_0 we have $F \mapsto e^{\gamma_5(\beta+\beta_0)} F$, and for the Yvon-Takabayasi angle $\beta \mapsto \beta + \beta_0$. If we work therefore with an electromagnetic field duality rotated by an additional angle β_0, the above relationship becomes

$$\frac{K}{\Lambda} = \tan\beta. \tag{71}$$

This is, of course, just a way to say that we can choose the constant β_0 in eq.(69) to be zero. Now, this expression gives

$$\lambda = \Lambda \cos\beta + \Lambda \tan\beta \sin\beta = \frac{\Lambda}{\cos\beta}, \tag{72}$$

$$\kappa = \Lambda \tan\beta \cos\beta - \Lambda \sin\beta = 0, \tag{73}$$

and the spinorial representation (64) of the free Maxwell equations becomes

$$\partial\psi\gamma_{21} = \lambda\psi\gamma_0. \tag{74}$$

Note that λ is such that

$$\rho\dot{\lambda} = -\lambda\partial \cdot J. \tag{75}$$

The current $J = \psi\gamma_0\tilde{\psi}$ is not conserved unless λ is constant. If we suppose also that

$$\partial \cdot J = 0 \tag{76}$$

we must have

$$\lambda = \text{const.} \tag{77}$$

Now, throughout these calculations we have assumed $\hbar = c = 1$. We observe that in eq.(74) λ has the units of $(\text{length})^{-1}$, and if we introduce the constants $\hbar$ and c we have to introduce another constant with unit of mass. If we denote this constant by m such that

$$\lambda = \frac{mc}{\hbar}, \tag{78}$$

then eq.(74) assumes a form which is identical to Dirac equation:

$$\partial\psi\gamma_{21} = \frac{mc}{\hbar}\psi\gamma_0. \tag{79}$$

It is true that we didn't proved that eq.(79) is really Dirac equation since the constant m has to be identified in this case with the electron's mass. However, we shall make some remarks concerning this identification which are very interesting and intriguing. First, if in analogy to eq.(78) we write $\Lambda = Mc/\hbar$, then eq.(72) reads

$$m = \frac{M}{\cos\beta}, \tag{80}$$

or

$$M = m\cos\beta = \frac{m}{\sqrt{1 + \sigma^2/\omega^2}}, \tag{81}$$

where σ and ω are the invariants of Dirac theory and given by eq.(31). If m is constant, the above expression defines a variable mass M.

On the other hand, de Broglie introduced in his interpretation of quantum mechanics a variable mass M related to the constant one m by de Broglie-Vigier formula[12] $M = m\sqrt{1 + \sigma^2/\omega^2}$, which is very similar the our one. However, the difference in these formulas are unimportant for the free case where we have for the plane wave solutions that[13] $\cos\beta = \pm 1$. Another interesting fact comes from eq.(55). If we write $\psi = \psi_0\exp(\gamma_{21}\omega t)$, where ψ_0 is a constant spinor (which is the case again for plane waves), then eq.(55) gives $\Lambda = \omega$, or, introducing the constants $\hbar$ and c:

$$Mc^2 = \hbar\omega. \tag{82}$$

The variable mass M now appears to be related to energy of some "internal" vibration, and eq.(82) is just another formula of de Broglie[14], who suggested that mass is related to the frequency of an internal clock supposed associated to a particle.

Acknowledgments: We are grateful for discussions to G. Cabrera, S. Ragusa, E. Recami, Q. A. G. Souza, W. Seixas, and at NATO-ASI to D. Bambusi, A. Barut, E. Hynds, A. Laufer, A. Orlowski, J. Ralph, J. Reignier, F. Stumpf, D. Taylor, Z. Turakulov, N. Unal, T. Waite and A. Weis. We are also grateful to CNPq for finnancial support.

References

[1] Keller, J., Adv. Appl. Clifford Algebras **3**, 147 (1993).

[2] Rodrigues, Jr., W. A. and Oliveira, E. C., Int. J. Theor. Phys. **29**, 397 (1990).

[3] Lounesto, P., Found. Phys. **23**, 1203 (1993).

[4] Figueiredo, V. L., Oliveira, E. C. and Rodrigues, Jr., W. A., Int. J. Theor. Phys. **29**, 371 (1990).

[5] Porteous, I., *Topological Geometry*, 2nd. ed., Cambridge University Press (1981).

[6] Hestenes, D., *Spacetime Algebra*, Gordon & Breach (1966).

[7] Rainich, G. Y., Trans. Am. Math. Soc. **27**, 106 (1925).

[8] Misner, C. W. and Wheeler, J. A., Ann. Phys. **2**, 525 (1957).

[9] Vaz, Jr., J. and Rodrigues, Jr., W. A., Int. J. Theor. Phys. **32**, 945 (1993).

[10] Campolattaro, A. A., Int. J. Theor. Phys. **19**, 99 (1980); **19**, 127 (1980).

[11] Campolattaro, A. A. Int. J. Theor. Phys. **29**, 141 (1990).

[12] de Broglie, L., *Ondes Électromagnétiques et Photons*, Gauthier-Villars (1967).

[13] Hestenes, D., Found. Phys. **20**, 1213 (1990).

[14] de Broglie, L. Found. Phys. **1**, 5 (1970).

[15] Rodrigues, Jr., W. A., Vaz, Jr., J. and Recami, E., Found. Phys. **23**, 469 (1993).

CONTRIBUTORS LIST

Agarwal, Girish S.
Physical Research Laboratory,
Navrangpura
Ahmedabad 380 009
India

Arsenovic, Dušan
P. O. Box 57 i 68
11001 Belgrade
Yugoslavia

Bambusi, D.
Dipartimento di Matematica dell'Università
Via Saldini 50
20133 Milano
Italy

Barut[†], Asim O.
Department of Physics
University of Colorado
Boulder, Colorado 80309
USA

Benson, O.
Max-Planck-Institut für Quantenoptik
Ludwig-Prandtl-Straße 10 , Postfach 1513
D-85748 Garching
Germany

Božić, Mirjana
Institut za Fiziku
P. O. Box 57 i 68
11001 Belgrade
Yugoslavia
Tel: [381] (11) 10 71 07
Fax: [381] (11) 10 81 98
Email: epopom@etf.bg.ac.yu

Carati, A.
Dipartimento di Matematica dell'Università
Via Saldini 50
20133 Milano
Italy

Chizhov, Alexei
Joint Institute for Nuclear Research
Bogolubov Laboratory of Theoretical Physics
141980 Dubna
Russia
Fax: [7] (9621) 65084
Email: chizhov@thsun1.jinr.dubna.su

Clark, T. D.
Mathematical and Physical Sciences
University of Sussex
Falmer BN1 9QH
United Kingdom

Diggins, J.
Mathematical and Physical Sciences
University of Sussex
Falmer BN1 9QH
United Kingdom

Dowling, Jonathan P.
U.S. Army Missile Command, AMSMI-RD-WS-ST
Weapons Sciences Directorate, Bldg. 7804, Rm 241B
Research, Development, and Engineering Center
Redstone Arsenal, Alabama 35898-5428
USA
Tel: [001] (205) 842-9734
Fax: [001] (205) 955-7216
Email: jdowling@ssdd.redstone.army.mil
Homepage: http://hwilwww.rdec.redstone.army.mil/MICOM/wsd/ST/ST.html

Duru, I. H.
Trakya University
Department of Mathematics
P. K. 126
41470 Edirne
Turkey
Fax: [90] (284) 212-0934
Email: duru@yunus.mam.tubitak.gov.tr

Galgini, L.
Dipartimento di Matematica dell'Università
Via Saldini 50
20133 Milano
Italy
Email: galgani@vaxmi.mi.infn.it

Hänsch, T. W.
Max-Planck-Institut für Quantenoptik
Hans-Kopfermann-Straße 1
D-85748 Garching
Germany

Hofmann, C. R.
Institut für Theoretische Physik
Technische Universität
D-01062 Dresden, Germany

Huang, Hu
Department of Physics
Texas A&M University
College Station, TX 77843-4242
USA

Huber, A.
Max-Planck-Institut für Quantenoptik
Hans-Kopfermann-Straße 1
D-85748 Garching
Germany

Jaekel, Marc-Thierry
Laboratoire de Physique Théorique de l'Ecole Normale Supérieure (CNRS)
24 rue Lhomond
F75231 Paris Cedex 05
France
Tel: [33] (1) 47 07 71
Fax: [33] (1) 43 36 76 66
Email: jaekel@physique.ens.fr

Kapuścik, Edward
Cracow Pedagogical University
Institute of Physics and Informatics
ul. Podchorazych 2
30-084 Kraków
Poland
Tel: [48](12) 37 02 22
Fax: [48] (12) 37 54 41
Email: sfkapusc@cyf-kr.edu.pl

König, W.
Max-Planck-Institut für Quantenoptik
Hans-Kopfermann-Straße 1
D-85748 Garching
Germany

Leibfried, D.
Max-Planck-Institut für Quantenoptik
Hans-Kopfermann-Straße 1
D-85748 Garching, Germany

Man'ko, Olga
Lebedev Physical Institute
Leninsky Prospekt 53
Moscow 117924
Russia
Fax: [7] (95) 938-2251
Email: manko@sci.fian.msk.su

Marić, Zvonko
Institut za Fiziku
P. O. Box 57 i 68
11001 Belgrade
Yugoslavia

Meyer, Georg M.
Department of Physics
Texas A&M University
College Station, Texas 77843-4242
USA
Tel: [001] (409) 845-7017
Fax: [001] (409) 845-2590
Email: meyerg@phys@phys.tamu.edu

Murzakhmetov, Bolat
Joint Institute for Nuclear Research
Bogolubov Laboratory of Theoretical Physics
141980 Dubna
Russia

Noja, D.
Dipartimento di Matematica dell'Università
Via Saldini 50
20133 Milano
Italy

Oleinik, V. P.
Department of General and Theoretical Physics
Kiev Polytechnic Institute
Prospect Pobedy 37
Kiev 252056
Ukraine
Tel: [7] (44) 267-5808
Fax: [7] (44) 274-5932

Orlowski, Arkadiusz
Arbeitsgruppe "Nichtklassische Strahlung" der Max-Planck-Gesellschaft
an der Humboldt-Universität zu Berlin
12484 Berlin
Germany
Email: Orlowski@huhepl.harvard,edu

Pachucki, K.
Max-Planck-Institut für Quantenoptik
Hans-Kopfermann-Straße 1
D-85748 Garching
Germany
Email: Krzysztof.Pachucki@fuw.edu.pl

Plunien, Günter
Institut für Theoretische Physik
Technische Universität
D-01062 Dresden
Germany

Prance, H.
Mathematical and Physical Sciences
University of Sussex
Falmer BN1 9QH
United Kingdom

Prance, R. J.
Mathematical and Physical Sciences
University of Sussex
Falmer BN1 9QH
United Kingdom

Raithel, Georg
National Institute of Standards & Technology
Gaithersburg Maryland 20899-0001
USA
Tel: [001] (301) 975-4116
Fax: [001] (301) 975-3038
Email: graithel@enh.nist.gov

Ralph, Jason F.
Scuola Normale Superiore
Piazza dei Cavalieri 7
56126 Pisa
Italy
Tel: [39] (50) 56 18 11
Fax: [39] (50) 56 18 11
Email: ralph@aix35ns.sns.it

Rathe, Ulrich W.
Department of Physics
Texas A&M University
College Station, Texas 77843
USA

Recami, Erasmo
Facoltà di Ingegneria
5 viale Marconi
24044 Dalmine (BG)
Italy
Tel: [39] (35) 277 313 / 308
Fax: [39] (35) 562 779
Email: recami@mi.infn.it

Reignier, Jean
Vrije Universiteit Brussel, Theoretische Natuurkunde
Pleinlaan 2
B-1050 Brussels
Belgium
Tel [32] (2) 629 32 41
Fax: [32] (2) 629 22 76
Email: jreignie@tena1.vub.ac.be

Reynaud, Serge
Laboratoire Kastler-Brossel (UPMC-ENS-CNRS), case 74
4 place Jussieu
F75252 Paris Cedex 05
France

Rodrigues Jr., Waldyr A.
Department of Applied Mathematics - IMECC
State University at Campinas (UNICAMP)
CP 6065
13081-970 Campinas, SP
Brazil

Rusek, Marian
Instytut Fizyki, Polska Akademia Nauk, 02-668
Warszawa
Poland

Salesi, Giovanni
Dipart. di Fisica
Università Statale di Catania
57 Corsitalia, Catania
Italy

Sassarini, J.
Dipartimento di Matematica dell'Università
Via Saldini 50
20133 Milano
Italy

Schneider, S. M.
Institut für Theoretische Physik
J. W. Goethe-Universität
D-60065 Frankfurt am Main
Germany

Scully, Marlan O.
Department of Physics
Texas A&M University
College Station, TX 77843-4242
USA

Soff, Gerhard
Institut für Theoretische Physik
Technische Universität
D-01062 Dresden
Germany
Tel: [49] (351) 463-3842
Fax: [49] (351) 463-7299
Email: soff@physik.tu-dresden.de

Spiller, T. P.
Mathematical and Physical Sciences
University of Sussex
Falmer BN1 9QH
United Kingdom

Turakulov, Z. Ya.
Institute of Nuclear Physics
Ulugbek
Tashkent 702132
Republic of Uzbekistan
CIS
Email: ulug@pc202.suninp.tashkent.su

Ünal, Nuri
Akdeniz University
Physics Department
P.K. 510
07200 Antalya
Turkey
Tel: [90] (242) 323-2360
Fax: [90] (242) 323-2363

Vairo, Antonio
I.N.F.N. - Sez di Bologna and Dip. di Fisica
Università di Bologna, Via Irnerio 46
I-40126 Bologna
Italy
Fax: [39] (51) 24 42 01
Email: vairo@alte01.bo.infn.it

Vaz Jr., Jayme
Department of Applied Mathematics - IMECC
State University at Campinas (UNICAMP)
CP 6065,
13081-970 Campinas, SP
Brazil
Tel: [55] (192) 39 79 15
Fax: [55] (192) 39 58 08
Email: vaz@ime.unicamp.br

Waite, Tom
13778 Shablow Ave.
Sylmar, California 91342
USA
Tel: [001] (818) 364-1125

Walther, Herbert
Max-Planck-Institut für Quantenoptik
Ludwig-Prandtl-Straße 10, Postfach 1513
D-85748 Garching, Germany

Weis, Antoine
Max-Planck-Institut für Quantenoptik
H. Kopfermannstr. 1
85748 Garching
Germany
Tel: [49] (89) 3 29 05 0
Fax: [49] (89) 3 29 05 200
Email: trw@:pp-garching.mpg.de

Weitz, M.
Max-Planck-Institut für Quantenoptik
Hans-Kopfermann-Straße 1
D-85748 Garching
Germany

Whiteman, R.
Mathematical and Physical Sciences
University of Sussex
Falmer BN1 9QH
United Kingdom

Widom, A.
Physics Department
Northeastern University
Boston, Massachusetts 02115
USA

Yelin, Susanne F.
Department of Physics
Texas A&M University
College Station, Texas 77843
USA
Tel: [001] (409) 845-7034
Fax: [001] (409) 845-2590
Email: yelin@phys.tamu.edu

Zeni, José R.
Depto. Ciencas Naturai
FUNREI
Sao Jao Del Rei
MG 36300
Brazil

INDEX